Probabilistic Methods

for Bioinformatics

with an Introduction to Bayesian Networks

Probabilistic Methods
for Bioinformatics

with an Introduction to Bayesian Networks

Richard E. Neapolitan

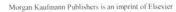
AMSTERDAM • BOSTON • HEIDELBERG • LONDON
NEW YORK • OXFORD • PARIS • SAN DIEGO
SAN FRANCISCO • SINGAPORE • SYDNEY • TOKYO

ELSEVIER

Morgan Kaufmann Publishers is an imprint of Elsevier

MORGAN
KAUFMANN

Morgan Kaufmann Publishers is an imprint of Elsevier.
30 Corporate Drive, Suite 400
Burlington, MA 01803

This book is printed on acid-free paper.

Library of Congress Cataloging-in-Publication Data
Application submitted.

ISBN: 978-0-12-370476-4

For information on all Morgan Kaufmann publications,
visit our Web site at *www.bookselsevier.com*

Printed and bound by CPI Group (UK) Ltd, Croydon, CR0 4YY
Transferred to Digital Print 2011

Contents

Preface

For the past 10 to 15 years, I have been investigating molecular genetics, and the application of Bayesian networks in this area. I decided to summarize this material into an interdisciplinary course on *probabilistic bioinformatics* at Northeastern Illinois University (NEIU). When searching for a textbook for it, I learned that there is not really anything appropriate. The closest thing was the edited volume *Probabilistic Modelling in Bioinformatics and Medical Informatics*, by Dirk Husmeier, Richard Dybowski, and Stephen Roberts. However, that book is not sufficiently accessible to be useful for a college-level course, at least at a mainstream university such as NEIU. So I decided to write this book, which serves not only as a text for an upper-level undergraduate or graduate course in probabilistic bioinformatics, but also should be a useful reference for a practitioner in the field.

This is not a book on biology, but rather one on the application of probability and statistics, in particular Bayesian networks, to genetics; so, there is no effort to discuss chemical structure or behavior. For example, a DNA sequence is simply treated as a string composed of the characters A, G, C, and T. I do, however, present some material on high-level features of biological processes so as to provide a context for the applications.

As in most books I write, I have taken the following measures to make the content accessible. First, ordinarily I do not introduce a concept or prove a theorem concerning the most general case. Rather, I discuss a specific, concrete case and then mention that the result can be generalized. Second, I provide ample figures and examples. A picture may be worth a 1000 words, but an example is worth 10,000 words.

The book has three parts. Part I provides background material on probability, statistics, and genetics, which the student or reader may already know. Many readers should be able to skip this part. Part II discusses Bayesian networks, inference in Bayesian networks, and learning Bayesian networks from data. Part III covers applications to bioinformatics. Specifically, Chapter 9 concerns non-molecular evolutionary genetics, Chapter 10 presents molecular evolutionary genetics, Chapter 11 shows an application of Bayesian networks to molecular phylogenetics, Chapter 12 introduces analyzing gene expression data, and Chapter 13 concerns gene linkage analysis.

For the purpose of classification, it seemed best to group the material into these three parts. However, I do not cover the material in this order in the course. If the material were covered in the same order as here, one would not get to bioinformatics until the course was half-over. By that time, rather than being motivated by the interesting applications of probability and statistics to genetics, the student might be motivated to drop the course. So I cover the material in an order that gets to applications to genetics as quickly as possible. That is, I do not discuss any particular background material until that material is necessary to the understanding of a chapter on bioinformatics. The order is as follows:

> Chapter 1, 2 all
> Chapter 3 Sections 3.1, 3.3
> Chapter 4 all
> Chapter 9 Sections 9.1 − 9.2
> Chapter 3, Section 3.2
> Chapter 9, Sections 9.3 − 9.5
> Chapter 10, all
> Chapters 5, 6, 7, 8 all
> Chapters 11, 12, 13 all

Reader too may prefer to read the book in this order.

> Rich Neapolitan
> RE-Neapolitan@neiu.edu

About the Author

Rich Neapolitan is Professor and Chair of Computer Science at Northeastern Illinois University. His research interests include probability and statistics (in particular, Bayesian networks and probabilistic modeling), expert systems, cognitive science (in particular, human learning and processing of causal knowledge), and applications of Bayesian networks and probabilistic modeling to fields such as medicine, biology, and finance.

Rich is a prolific author and has been publishing in the most prestigious journals in the broad area of reasoning under uncertainty since 1987. He has previously written four books including the seminal 1990 Bayesian network text *Probabilistic Reasoning in Expert Systems*. More recently, he wrote *Learning Bayesian Networks* (2004); the textbook *Foundations of Algorithms* (1996, 1998, 2003), which has been translated to three languages and is one of the most widely-used algorithms texts worldwide; and *Probabilistic Methods for Financial and Marketing Informatics* (2007). Rich's approach to textbook writing is innovative; his method engages students with the logical flow of the material and his books have the reputation for making difficult concepts easy to understand while still remaining thought-provoking.

Part I

Background

Chapter 1

Probabilistic Informatics

Informatics programs in the United States go back to at least the 1980s, when Stanford University offered a Ph.D. in medical informatics. Since that time, a number of informatics programs in other disciplines have emerged at universities throughout the United States. These programs go by various names, including bioinformatics, medical informatics, chemical informatics, music informatics, marketing informatics, and so on. What do these programs have in common? To answer that question we must articulate what we mean by the term *informatics*. Because other disciplines are usually referenced when we discuss informatics, some define *informatics* as the application of information technology in the context of another field. However, such a definition does not really tell us the focus of informatics itself. Here, we first explain what we mean by the term *informatics*; then we discuss why we have chosen to concentrate on the probabilistic approach; finally, we provide an outline of the material that will be covered in the rest of the book.

1.1   What Is Informatics?

In much of Western Europe, informatics has come to mean the rough translation of the English *computer science*, which is the discipline that studies computable processes. Certainly, there is overlap between computer science and informatics programs, but they are not the same. Informatics programs ordinarily investigate subjects such as biology and medicine, whereas computer science programs do not. So the European definition does not suffice for the way the word is currently used in the United States.

To gain insight into the meaning of informatics, let us consider the suffix *-ics*, which means the science, art, or study of some entity. For example, *linguistics* is the study of the nature of language, *economics* is the study of the production and distribution of goods, and *photonics* is the study of electromagnetic energy that has as its basic unit the photon. Given this, informatics should be the study of information. Indeed, WordNet 2.1 defines it as "the science concerned with gathering, manipulating, storing, retrieving, and classifying recorded information." To proceed from this definition, we need to define the word *information*. Most dictionary definitions do not help as far as giving us anything concrete; that is, they define information either as knowledge or as a collection of data, which means we are left with the situation of determining the meaning of knowledge and data. To arrive at a concrete definition of informatics, let's define data, information, and knowledge first.

By **datum** we mean a character string that can be recognized as a unit. For example, the nucleotide G in the nucleotide sequence GATC is a datum, the field *cancer* in a record in a medical database is a datum, and the field *Gone with the Wind* in a movie database is a datum. Note that a single character, a word, or a group of words can be a datum, depending on the particular application. **Data**, then, are more than one datum. By **information** we mean the meaning given to data. For example, in a medical database the data *Joe Smith* and *cancer* in the same record mean that Joe Smith has cancer. By **knowledge** we mean dicta that enable us to infer new information from existing information. For example, suppose we have the following item of knowledge (dictum):[1]

IF the stem of the plant is woody

  AND the position is upright

  AND there is one main trunk

  THEN the plant is a tree.

---

[1] Such an item of knowledge would be part of a rule-based expert system.

Suppose further that you are looking at a plant in your backyard and you observe that its stem is woody, its position is upright, and it has one main trunk. Then, using the preceding knowledge item, you can deduce the new information that the plant in your backyard is a tree.

Finally, we define **informatics** as the discipline that applies the methodologies of science and engineering to information. It concerns organizing data into information, learning knowledge from information, learning new information from existing information and knowledge, and making decisions based on the knowledge and information learned. We use engineering to develop the algorithms that learn knowledge from information and that learn information from information and knowledge. We use science to test the accuracy of these algorithms.

Next, we show several examples that illustrate how informatics pertains to other disciplines.

Example 1.1 *(**medical informatics**) Suppose we have a large data file of patient records as follows:*

Patient	Smoking History	Bronchitis	Lung Cancer	Fatigue	Positive Chest X-Ray
1	Yes	Yes	Yes	No	Yes
2	No	No	No	No	No
3	No	No	Yes	Yes	No
⋮	⋮	⋮	⋮	⋮	⋮
10,000	Yes	No	No	No	No

From the information in this data file we can use the methodologies of informatics to obtain knowledge, such as "25% of people with smoking history have bronchitis" and "60% of people with lung cancer have positive chest X-rays." Then from this knowledge and the information that "Joe Smith has a smoking history and a positive chest X-ray," we can use the methodologies of informatics to obtain the new information that "there is a 5% chance Joe Smith also has lung cancer."

Example 1.2 *(**bioinformatics**) Suppose we have long homologous DNA sequences from a human, a chimpanzee, a gorilla, an orangutan, and a rhesus monkey. From this information we can use the methodologies of informatics to obtain the new information that it is most probable that the human and the chimpanzee are the most closely related of the five species.*

Example 1.3 *(**marketing informatics**) Suppose we have a large data file of movie ratings as follows:*

Person	Aviator	Shall We Dance	Dirty Dancing	Vanity Fair
1	1	5	5	4
2	5	1	1	2
3	4	1	2	1
4	2	5	4	5
⋮	⋮	⋮	⋮	⋮
10,000	1	4	5	5

This means, for example, that Person 1 rated *Aviator* the lowest (1) and *Shall We Dance* the highest (5). From the information in this data file, we can develop a knowledge system that will enable us to estimate how an individual will rate a particular movie. For example, suppose Kathy Black rates *Aviator* as 1, *Shall We Dance* as 5, and *Dirty Dancing* as 5. The system could estimate how Kathy will rate *Vanity Fair*. Just by eyeballing the data in the five records shown, we see that Kathy's ratings on the first three movies are similar to those of Persons 1, 4, and 10,000. Since they all rated *Vanity Fair* high, based on these five records, we would suspect Kathy would rate it high. An informatics algorithm can formalize a way to make these predictions. This task of predicting the utility of an item to a particular user based on the utilities assigned by other users is called **collaborative filtering**.

1.2 Bioinformatics

This book concentrates on **bioinformatics**, which applies the methods of informatics to solving problems in biology using biological data sets. The problems investigated are usually at the molecular level, including sequence alignment, genome assembly, models of evolution and phylogenetic trees, analyzing gene expression data, and gene linkage analysis. Sometimes the terms *bioinformatics* and *computational biology* are used interchangeably. However, according to our definition, bioinformatics can be considered a subdiscipline of computational biology. Indeed, Wikipedia defines computational biology as follows:

> **Computational biology** is an interdisciplinary field that applies the techniques of computer science, applied mathematics, and statistics to address problems inspired by biology. Major fields in biology that use computational techniques include:

- Bioinformatics, which applies algorithms and statistical techniques to biological datasets that typically consist of large numbers of DNA, RNA, or protein sequences. Examples of specific techniques include sequence alignment, which is used for both sequence database searching and for comparison of homologous sequences; gene finding; and prediction of gene expression. (The term computational biology is sometimes used as a synonym for bioinformatics.)

- Computational biomodeling, a field within biocybernetics concerned with building computational models of biological systems.

- Computational genomics, a field within genomics which studies the genomes of cells and organisms by high-throughput genome sequencing that requires extensive post-processing known as genome assembly, and which uses DNA microarray technologies to perform statistical analyses on the genes expressed in individual cell types. Mathematical foundations have also been developed for sequencing.

- Molecular modeling, a field dealing with theoretical methods and computational techniques to model or mimic the behavior of molecules, ranging from descriptions of a molecule of few atoms, to small chemical systems, to large biological molecules and material assemblies.

- Systems biology, which aims to model large-scale biological interaction networks (also known as the interactome).

- Protein structure prediction and structural genomics, which attempt to systematically produce accurate structural models for three-dimensional protein structures that have not been solved experimentally.

- Computational biochemistry and biophysics, which make extensive use of structural modeling and simulation methods such as molecular dynamics and Monte Carlo method-inspired Boltzmann sampling methods in an attempt to elucidate the kinetics and thermodynamics of protein functions.

- Wikipedia

This definition of bioinformatics as a subdiscipline within computational biology is consistent with our definition. Notice that there are many other subdisciplines of computational biology including, for example, molecular modeling. A difference between bioinformatics and a discipline, such as molecular modeling, is that bioinformatics concerns learning from large data sets, whereas disciplines such as molecular modeling do not.

1.3   Probabilistic Informatics

As shown in Examples 1.1, 1.2, and 1.3, the knowledge we use to process information often does not consist of IF-THEN rules, such as the one concerning plants we discussed earlier. Rather, we only know relationships such as "smoking makes lung cancer more likely." Similarly, our conclusions are uncertain. For example, we feel it is most likely that the closest living relative of the human is the chimpanzee, but we are not certain of this. So ordinarily we must reason under uncertainty when handling information and knowledge. In the 1960s and 1970s a number of new formalisms for handling uncertainty were developed, including certainty factors, the Dempster-Shafer Theory of Evidence, fuzzy logic, and fuzzy set theory. Probability theory has a long history of representing uncertainty in a formal, axiomatic way. Neapolitan [1990] contrasts the various approaches and argues for the use of probability theory.[2] We will not present that argument here. Rather, we accept probability theory as being the way to handle uncertainty and explain why we choose to describe informatics algorithms that use the model-based probabilistic approach.

A **heuristic algorithm** uses a commonsense rule to solve a problem. Ordinarily, heuristic algorithms have no theoretical basis and therefore do not enable us to prove results based on assumptions concerning a system. Examples of heuristic algorithms are the distance matrix methods for learning phylogenetic trees, discussed in Section 11.2.

An **abstract model** is a theoretical construct that represents a physical process with a set of variables and a set of quantitative relationships (axioms) among them. We use models so we can reason within an idealized framework and thereby make predictions and determinations about a system. We can mathematically prove these predictions and determinations are "correct," but they are correct only to the extent that the model accurately represents the system. A **model-based algorithm** therefore makes predictions and determinations within the framework of some model. Algorithms that make predictions and determinations within the framework of probability theory are model-based algorithms. We can prove results concerning these algorithms based on the axioms of probability theory, which are discussed in Chapter 2. We concentrate on such algorithms in this book. In particular, we present algorithms that use Bayesian networks to reason within the framework of probability theory. Bayesian networks have become one of the most important statistical tools used in bioinformatics.

1.4   Outline of This Book

Part I provides the background necessary to read the rest of the book. Chapter 2 reviews probability theory, Chapter 3 discusses statistics, and Chapter 4 reviews the basics of genetics. Part II concerns Bayesian networks. In Chapter 5, we introduce Bayesian networks and inference in Bayesian networks. Chapter 6 covers further properties of Bayesian networks, Chapter 7 concerns learning the parameters in a Bayesian network from data, and finally, Chapter 8 addresses learning the structure of a Bayesian network from data.

Applications to bioinformatics appear in Part III. Chapter 9 concerns non-molecular evolutionary genetics, which investigates evolution at the allele (gene) level. Chapter 10 then discusses molecular evolutionary genetics, which concerns how nucleotide sequences (the components of genes) change over time. Chapter 11 concerns molecular phylogenetics, which uses molecular structure to construct phylogenetic trees. (A phylogenetic tree is a tree representing the history of how species have evolved over time.) Chapter 12 covers analyzing gene expression data, which investigates how the expression level of one gene has a causal effect on the expression level of another gene. Finally, Chapter 13

---

[2]Fuzzy set theory and fuzzy logic model a different class of problems than probability theory and therefore complement probability theory rather than compete with it (see [Zadeh, 1995] or [Neapolitan, 1992]).

discusses genetic linkage analysis, which investigates how the locus of genes on the same chromosome are physically connected and thus tend to stay together during meiosis, which means that the genes are genetically linked. Genetic linkage analysis is used to develop a genetic-linkage map, which shows the order of genes on a chromosome and the relative distances between those genes.

Chapter 2

Probability Basics

This chapter reviews the probability that you need to read the remainder of this book. Section 2.1 presents the basics of probability theory; Section 2.2 reviews random variables; Section 2.3 briefly discusses the meaning of probability; and Section 2.4 shows how random variables are used in practice.

2.1 Probability Basics

After defining probability spaces, we discuss conditional probability, independence and conditional independence, and Bayes' Theorem.

2.1.1   Probability Spaces

You may recall using probability in situations such as drawing the top card from a deck of playing cards, tossing a coin, or drawing a ball from an urn. We call the process of drawing the top card or tossing a coin an **experiment**. Probability theory has to do with experiments that have a set of distinct **outcomes**. The set of all outcomes is called the **sample space** or **population**. Mathematicians ordinarily say sample space, while social scientists ordinarily say population. We will say sample space. In this simple review, we assume that the sample space is finite. Any subset of a sample space is called an **event**. A subset containing exactly one element is called an **elementary event**.

Example 2.1 *Suppose we have the experiment of drawing the top card from an ordinary deck of cards. Then the set*

$$\mathsf{E} = \{jack\ of\ hearts,\ jack\ of\ clubs,\ jack\ of\ spades,\ jack\ of\ diamonds\}$$

is an event, and the set

$$\mathsf{F} = \{jack\ of\ hearts\}$$

is an elementary event.

The meaning of an event is that one of the elements of the subset is the outcome of the experiment. In the preceding example, the meaning of the event E is that the card drawn is one of the four jacks, and the meaning of the elementary event F is that the card is the jack of hearts.

We articulate our certainty that an event contains the outcome of the experiment with a real number between 0 and 1. This number is called the **probability** of the event. When the sample space is finite, a probability of 0 means we are certain the event does not contain the outcome, whereas a probability of 1 means we are certain it does. Values in between represent varying degrees of belief. The following definition formally defines probability for a finite sample space.

Definition 2.1 *Suppose we have a sample space Ω containing n distinct elements; that is,*

$$\Omega = \{e_1, e_2, \ldots, e_n\}.$$

A function that assigns a real number $P(\mathsf{E})$ to each event $\mathsf{E} \subseteq \Omega$ is called a **probability function** *on the set of subsets of Ω if it satisfies the following conditions:*

1. *$0 \leq P(e_i) \leq 1 \quad for\ 1 \leq i \leq n.$*

2. *$P(e_1) + P(e_2) + \ldots + P(e_n) = 1.$*

3. *For each event that is not an elementary event, $P(\mathsf{E})$ is the sum of the probabilities of the elementary events whose outcomes are in E. For example, if*

$$\mathsf{E} = \{e_3, e_6, e_8\}$$

then

$$P(\mathsf{E}) = P(e_3) + P(e_6) + P(e_8).$$

*The pair (Ω, P) is called a **probability space**.*

Because probability is defined as a function whose domain is a set of sets, we should write $P(\{e_i\})$ instead of $P(e_i)$ when denoting the probability of an elementary event. However, for the sake of simplicity, we do not do this. In the same way, we write $P(e_3, e_6, e_8)$ instead of $P(\{e_3, e_6, e_8\})$.

The most straightforward way to assign probabilities is to use the **Principle of Indifference**, which says that outcomes are to be considered equiprobable if we have no reason to expect one over the other. According to this principle, when there are n elementary events, each has probability equal to $1/n$.

Example 2.2 *Let the experiment be tossing a coin. Then the sample space is*

$$\Omega = \{heads, tails\},$$

and, according to the Principle of Indifference, we assign

$$P(heads) = P(tails) = .5.$$

We stress that there is nothing in the definition of a probability space that says we must assign the value of .5 to the probabilities of heads and tails. We could assign $P(heads) = .7$ and $P(tails) = .3$. However, if we have no reason to expect one outcome over the other, we give them the same probability.

Example 2.3 *Let the experiment be drawing the top card from a deck of 52 cards. Then Ω contains the faces of the 52 cards, and, according to the Principle of Indifference, we assign $P(e) = 1/52$ for each $e \in \Omega$. For example,*

$$P(jack\ of\ hearts) = \frac{1}{52}.$$

The event

$$E = \{jack\ of\ hearts,\ jack\ of\ clubs,\ jack\ of\ spades,\ jack\ of\ diamonds\}$$

means that the card drawn is a jack. Its probability is

$$
\begin{aligned}
P(E) &= P(jack\ of\ hearts) + P(jack\ of\ clubs) + \\
&\quad P(jack\ of\ spades) + P(jack\ of\ diamonds) \\
&= \frac{1}{52} + \frac{1}{52} + \frac{1}{52} + \frac{1}{52} = \frac{1}{13}.
\end{aligned}
$$

We have Theorem 2.1 concerning probability spaces. Its proof is left as an exercise.

Theorem 2.1 *Let* (Ω, P) *be a probability space. Then*

1. $P(\Omega) = 1$.

2. $0 \leq P(\mathsf{E}) \leq 1$ *for every* $\mathsf{E} \subseteq \Omega$.

3. *For every two subsets* E *and* F *of* Ω *such that* $\mathsf{E} \cap \mathsf{F} = \varnothing$,

$$P(\mathsf{E} \cup \mathsf{F}) = P(\mathsf{E}) + P(\mathsf{F}),$$

where $\varnothing$ *denotes the empty set.*

Example 2.4 *Suppose we draw the top card from a deck of cards. Denote by* Queen *the set containing the 4 queens and by* King *the set containing the 4 kings. Then*

$$P(\mathsf{Queen} \cup \mathsf{King}) = P(\mathsf{Queen}) + P(\mathsf{King}) = \frac{1}{13} + \frac{1}{13} = \frac{2}{13}$$

because Queen $\cap$ King $= \varnothing$. *Next denote by* Spade *the set containing the 13 spades. The sets* Queen *and* Spade *are not disjoint, so their probabilities are not additive. However, it is not hard to prove that, in general,*

$$P(\mathsf{E} \cup \mathsf{F}) = P(\mathsf{E}) + P(\mathsf{F}) - P(\mathsf{E} \cap \mathsf{F}).$$

So

$$
\begin{aligned}
P(\mathsf{Queen} \cup \mathsf{Spade}) &= P(\mathsf{Queen}) + P(\mathsf{Spade}) - P(\mathsf{Queen} \cap \mathsf{Spade}) \\
&= \frac{1}{13} + \frac{1}{4} - \frac{1}{52} = \frac{4}{13}.
\end{aligned}
$$

2.1.2 Conditional Probability and Independence

We start with a definition.

Definition 2.2 *Let* E *and* F *be events such that* $P(\mathsf{F}) \neq 0$. *Then the* ***conditional probability*** *of* E *given* F, *denoted* $P(\mathsf{E}|\mathsf{F})$, *is given by*

$$P(\mathsf{E}|\mathsf{F}) = \frac{P(\mathsf{E} \cap \mathsf{F})}{P(\mathsf{F})}.$$

We can gain intuition for this definition by considering probabilities that are assigned using the Principle of Indifference. In this case, $P(\mathsf{E}|\mathsf{F})$, as defined previously, is the ratio of the number of items in $\mathsf{E} \cap \mathsf{F}$ to the number of items in F. We show this as follows: Let n be the number of items in the sample space, n_F be the number of items in F, and n_EF be the number of items in $\mathsf{E} \cap \mathsf{F}$. Then

$$\frac{P(\mathsf{E} \cap \mathsf{F})}{P(\mathsf{F})} = \frac{n_\mathsf{EF}/n}{n_\mathsf{F}/n} = \frac{n_\mathsf{EF}}{n_\mathsf{F}},$$

which is the ratio of the number of items in $\mathsf{E} \cap \mathsf{F}$ to the number of items in F. As far as the meaning is concerned, $P(\mathsf{E}|\mathsf{F})$ is our belief that event E contains the outcome (i.e. E occurs) when we already know that event F contains the outcome (i.e. F occurred).

Example 2.5 *Again, consider drawing the top card from a deck of cards. Let* Jack *be the set of the 4 jacks,* RedRoyalCard *be the set of the 6 red royal cards,[1] and* Club *be the set of the 13 clubs. Then*

$$P(\mathsf{Jack}) = \frac{4}{52} = \frac{1}{13}$$

$$P(\mathsf{Jack}|\mathsf{RedRoyalCard}) = \frac{P(\mathsf{Jack} \cap \mathsf{RedRoyalCard})}{P(\mathsf{RedRoyalCard})} = \frac{2/52}{6/52} = \frac{1}{3}$$

$$P(\mathsf{Jack}|\mathsf{Club}) = \frac{P(\mathsf{Jack} \cap \mathsf{Club})}{P(\mathsf{Club})} = \frac{1/52}{13/52} = \frac{1}{13}.$$

Notice in the previous example that $P(\mathsf{Jack}|\mathsf{Club}) = P(\mathsf{Jack})$. This means that finding out the card is a club does not change the likelihood that it is a jack. We say that the two events are independent in this case, which is formalized in the following definition.

Definition 2.3 *Two events* E *and* F *are* ***independent*** *if one of the following holds:*

1. $P(\mathsf{E}|\mathsf{F}) = P(\mathsf{E})$ *and* $P(\mathsf{E}) \neq 0,\ P(\mathsf{F}) \neq 0.$

2. $P(\mathsf{E}) = 0$ *or* $P(\mathsf{F}) = 0.$

Notice that the definition states that the two events are independent even though it is in terms of the conditional probability of E given F. The reason is that independence is symmetric. That is, if $P(\mathsf{E}) \neq 0$ and $P(\mathsf{F}) \neq 0$, then $P(\mathsf{E}|\mathsf{F}) = P(\mathsf{E})$ if and only if $P(\mathsf{F}|\mathsf{E}) = P(\mathsf{F})$. It is straightforward to prove that E and F are independent if and only if $P(\mathsf{E} \cap \mathsf{F}) = P(\mathsf{E})P(\mathsf{F})$.

If you've previously studied probability, you should have already been introduced to the concept of independence. However, a generalization of independence, called **conditional independence**, is not covered in many introductory texts. This concept is important to the applications discussed in this book. We discuss it next.

Definition 2.4 *Two events* E *and* F *are* ***conditionally independent*** *given* G *if* $P(\mathsf{G}) \neq 0$ *and one of the following holds:*

1. $P(\mathsf{E}|\mathsf{F} \cap \mathsf{G}) = P(\mathsf{E}|\mathsf{G})$ *and* $P(\mathsf{E}|\mathsf{G}) \neq 0,\ P(\mathsf{F}|\mathsf{G}) \neq 0.$

2. $P(\mathsf{E}|\mathsf{G}) = 0$ *or* $P(\mathsf{F}|\mathsf{G}) = 0.$

Notice that this definition is identical to the definition of independence except that everything is conditional on G. The definition entails that E and F are independent once we know that the outcome is in G. The next example illustrates this.

---

[1] A royal card is a jack, queen, or king.

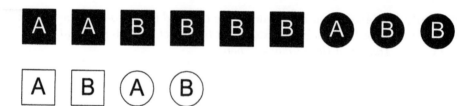

Figure 2.1: Using the Principle of Indifference, we assign a probability of 1/13 to each object.

Example 2.6 *Let Ω be the set of all objects in Figure 2.1. Using the Principle of Indifference, we assign a probability of 1/13 to each object. Let* Black *be the set of all black objects,* White *be the set of all white objects,* Square *be the set of all square objects, and* A *be the set of all objects containing an A. We then have that*

$$P(\mathsf{A}) \;=\; \frac{5}{13}$$
$$P(\mathsf{A}|\mathsf{Square}) \;=\; \frac{3}{8}.$$

So A *and* Square *are not independent. However,*

$$P(\mathsf{A}|\mathsf{Black}) \;=\; \frac{3}{9} = \frac{1}{3}$$
$$P(\mathsf{A}|\mathsf{Square} \cap \mathsf{Black}) \;=\; \frac{2}{6} = \frac{1}{3}.$$

We see that A *and* Square *are conditionally independent given* Black*. Furthermore,*

$$P(\mathsf{A}|\mathsf{White}) \;=\; \frac{2}{4} = \frac{1}{2}$$
$$P(\mathsf{A}|\mathsf{Square} \cap \mathsf{White}) \;=\; \frac{1}{2}.$$

So A *and* Square *are also conditionally independent given* White*.*

Next, we discuss an important rule involving conditional probabilities. Suppose we have n events $\mathsf{E}_1, \mathsf{E}_2, \ldots, \mathsf{E}_n$ such that

$$\mathsf{E}_i \cap \mathsf{E}_j = \varnothing \qquad \text{for } i \neq j$$

and

$$\mathsf{E}_1 \cup \mathsf{E}_2 \cup \ldots \cup \mathsf{E}_n = \Omega.$$

Such events are called **mutually exclusive and exhaustive**. Then the **Law of Total Probability** says that for any other event F,

$$P(\mathsf{F}) = P(\mathsf{F} \cap \mathsf{E}_1) + P(\mathsf{F} \cap \mathsf{E}_2) + \cdots + P(\mathsf{F} \cap \mathsf{E}_n). \qquad (2.1)$$

You are asked to prove this rule in the exercises. If $P(\mathsf{E}_i) \neq 0$, then

$$P(\mathsf{F} \cap \mathsf{E}_i) = P(\mathsf{F}|\mathsf{E}_i)P(\mathsf{E}_i).$$

Therefore, if $P(\mathsf{E}_i) \neq 0$ for all i, the law is often applied in the following form:

$$P(\mathsf{F}) = P(\mathsf{F}|\mathsf{E}_1)P(\mathsf{E}_1) + P(\mathsf{F}|\mathsf{E}_2)P(\mathsf{E}_2) + \cdots + P(\mathsf{F}|\mathsf{E}_n)P(\mathsf{E}_n). \tag{2.2}$$

Example 2.7 *Suppose we have the objects discussed in Example 2.6. Then, according to the Law of Total Probability,*

$$\begin{aligned} P(\mathsf{A}) &= P(\mathsf{A}|\mathsf{Black})P(\mathsf{Black}) + P(\mathsf{A}|\mathsf{White})P(\mathsf{White}) \\ &= \left(\frac{1}{3}\right)\left(\frac{9}{13}\right) + \left(\frac{1}{2}\right)\left(\frac{4}{13}\right) = \frac{5}{13}. \end{aligned}$$

2.1.3 Bayes' Theorem

We can compute conditional probabilities of events of interest from known probabilities using the following theorem.

Theorem 2.2 *(**Bayes**) Given two events* E *and* F *such that* $P(\mathsf{E}) \neq 0$ *and* $P(\mathsf{F}) \neq 0$, *we have*

$$P(\mathsf{E}|\mathsf{F}) = \frac{P(\mathsf{F}|\mathsf{E})P(\mathsf{E})}{P(\mathsf{F})}. \tag{2.3}$$

Furthermore, given n mutually exclusive and exhaustive events $\mathsf{E}_1, \mathsf{E}_2, \ldots, \mathsf{E}_n$ *such that* $P(\mathsf{E}_i) \neq 0$ *for all i, we have for* $1 \leq i \leq n$,

$$P(\mathsf{E}_i|\mathsf{F}) = \frac{P(\mathsf{F}|\mathsf{E}_i)P(\mathsf{E}_i)}{P(\mathsf{F}|\mathsf{E}_1)P(\mathsf{E}_1) + P(\mathsf{F}|\mathsf{E}_2)P(\mathsf{E}_2) + \cdots P(\mathsf{F}|\mathsf{E}_n)P(\mathsf{E}_n)}. \tag{2.4}$$

Proof. *To obtain Equality 2.3, we first use the definition of conditional probability as follows:*

$$P(\mathsf{E}|\mathsf{F}) = \frac{P(\mathsf{E} \cap \mathsf{F})}{P(\mathsf{F})} \qquad and \qquad P(\mathsf{F}|\mathsf{E}) = \frac{P(\mathsf{F} \cap \mathsf{E})}{P(\mathsf{E})}.$$

Next we multiply each of these equalities by the denominator on its right side to show that

$$P(\mathsf{E}|\mathsf{F})P(\mathsf{F}) = P(\mathsf{F}|\mathsf{E})P(\mathsf{E})$$

because they both equal $P(\mathsf{E} \cap \mathsf{F})$. Finally, we divide this last equality by $P(\mathsf{F})$ to obtain our result.

To obtain Equality 2.4, we place the expression for F, obtained using the Law of Total Probability (Equality 2.2), in the denominator of Equality 2.3. ∎

Both of the formulas in the preceding theorem are called **Bayes' Theorem** because the original version was developed by Thomas Bayes, published in 1763. The first enables us to compute $P(\mathsf{E}|\mathsf{F})$ if we know $P(\mathsf{F}|\mathsf{E})$, $P(\mathsf{E})$, and $P(\mathsf{F})$; the second enables us to compute $P(\mathsf{E}_i|\mathsf{F})$ if we know $P(\mathsf{F}|\mathsf{E}_j)$ and $P(\mathsf{E}_j)$ for $1 \leq j \leq n$. The next example illustrates the use of Bayes' Theorem.

Example 2.8 *Let Ω be the set of all objects in Figure 2.1, and assign each object a probability of 1/13. Let* A *be the set of all objects containing an* A, B *be the set of all objects containing a* B, *and* Black *be the set of all black objects. Then, according to Bayes' Theorem,*

$$
\begin{aligned}
P(\mathsf{Black}|\mathsf{A}) &= \frac{P(\mathsf{A}|\mathsf{Black})P(\mathsf{Black})}{P(\mathsf{A}|\mathsf{Black})P(\mathsf{Black}) + P(\mathsf{A}|\mathsf{White})P(\mathsf{White})} \\
&= \frac{\left(\frac{1}{3}\right)\left(\frac{9}{13}\right)}{\left(\frac{1}{3}\right)\left(\frac{9}{13}\right) + \left(\frac{1}{2}\right)\left(\frac{4}{13}\right)} = \frac{3}{5},
\end{aligned}
$$

which is the same value we get by computing $P(\mathsf{Black}|\mathsf{A})$ *directly.*

In the previous example we can just as easily compute $P(\mathsf{Black}|\mathsf{A})$ directly. We will see a useful application of Bayes' Theorem in Section 2.4.

2.2   Random Variables

In this section we present the formal definition and mathematical properties of a random variable. In Section 2.4 we show how they are developed in practice.

2.2.1   Probability Distributions of Random Variables

Definition 2.5 *Given a probability space* (Ω, P), *a **random variable** X is a function whose domain is* Ω.

The range of X is called the **space** of X.

Example 2.9 *Let Ω contain all outcomes of a throw of a pair of six-sided dice, and let P assign 1/36 to each outcome. Then Ω is the following set of ordered pairs:*

$$\Omega = \{(1,1),(1,2),(1,3),(1,4),(1,5),(1,6),(2,1),(2,2),\ldots(6,5),(6,6)\}.$$

Let the random variable X assign the sum of each ordered pair to that pair, and let the random variable Y assign odd to each pair of odd numbers and even to a pair if at least one number in that pair is an even number. The following table shows some of the values of X and Y.

e	$X(e)$	$Y(e)$
$(1,1)$	2	odd
$(1,2)$	3	even
$\ldots$	$\ldots$	$\ldots$
$(2,1)$	3	even
$\ldots$	$\ldots$	$\ldots$
$(6,6)$	12	even

The space of X is $\{2,3,4,5,6,7,8,9,10,11,12\}$, and that of Y is $\{odd, even\}$.

For a random variable X, we use $X = x$ to denote the subset containing all elements $e \in \Omega$ that X maps to the value of x. That is,

$$X = x \quad \text{represents the event} \quad \{e \text{ such that } X(e) = x\}.$$

Note the difference between X and x. Small x denotes any element in the space of X, whereas X is a function.

Example 2.10 Let Ω, P, and X be as in Example 2.9. Then

$$X = 3 \quad \text{represents the event} \quad \{(1,2),(2,1)\} \text{ and}$$

$$P(X = 3) = \frac{1}{18}.$$

Notice that

$$\sum_{x \in space(X)} P(X = x) = 1.$$

Example 2.11 Let Ω, P, and Y be as in Example 2.9. Then

$$\sum_{y \in space(Y)} P(Y = y) \quad = \quad P(Y = odd) + P(Y = even)$$

$$= \quad \frac{9}{36} + \frac{27}{36} = 1.$$

We call the values of $P(X = x)$ for all values x of X the **probability distribution** of the random variable X. When we are referring to the probability distribution of X, we write $P(X)$.

We often use x alone to represent the event $X = x$, and so we write $P(x)$ instead of $P(X = x)$ when we are referring to the probability that X has value x.

Example 2.12 Let Ω, P, and X be as in Example 2.9. Then if $x = 3$,

$$P(x) = P(X = x) = \frac{1}{18}.$$

If we want to refer to all values of, for example, the random variables X, we sometimes write $P(X)$ instead of $P(X = x)$ or $P(x)$.

Example 2.13 Let Ω, P, and X be as in Example 2.9. Then for all values of X

$$P(X) > 1.$$

Given two random variables X and Y, defined on the same sample space Ω, we use $X = x, Y = y$ to denote the subset containing all elements $e \in \Omega$ that are mapped both by X to x and by Y to y. That is,

$$X = x, Y = y \quad \text{represents the event}$$

$$\{e \text{ such that } X(e) = x\} \cap \{e \text{ such that } Y(e) = y\}.$$

Example 2.14 *Let Ω, P, X, and Y be as in Example 2.9. Then*

$$X = 4, Y = odd \quad represents \ the \ event \quad \{(1,3),(3,1)\},$$

and so

$$P(X = 4, Y = odd) = 1/18.$$

We call $P(X = x, Y = y)$ the **joint probability distribution** of X and Y. If $\mathsf{A} = \{X, Y\}$, we also call this the joint probability distribution of A. Furthermore, we often just say *joint distribution* or *probability distribution*.

For brevity, we often use x, y to represent the event $X = x, Y = y$, and so we write $P(x, y)$ instead of $P(X = x, Y = y)$. This concept extends to three or more random variables. For example, $P(X = x, Y = y, Z = z)$ is the joint probability distribution function of the random variables X, Y, and Z, and we often write $P(x, y, z)$.

Example 2.15 *Let Ω, P, X, and Y be as in Example 2.9. Then, if $x = 4$ and $y = odd$,*

$$P(x, y) = P(X = x, Y = y) = 1/18.$$

Similar to the case of a single random variable, if we want to refer to all values of, for example, the random variables X and Y, we sometimes write $P(X, Y)$ instead of $P(X = x, Y = y)$ or $P(x, y)$.

Example 2.16 *Let Ω, P, X, and Y be as in Example 2.9. It is left as an exercise to show that for all values of x and y we have*

$$P(X = x, Y = y) < 1/2.$$

For example, as shown in Example 2.14

$$P(X = 4, Y = odd) = 1/18 < 1/2.$$

We can restate this fact as follows: for all values of X and Y we have that

$$P(X, Y) < 1/2.$$

If, for example, we let $\mathsf{A} = \{X, Y\}$ and $\mathsf{a} = \{x, y\}$, we use

$$\mathsf{A} = \mathsf{a} \quad to \ represent \quad X = x, Y = y,$$

and we often write $P(\mathsf{a})$ instead of $P(\mathsf{A} = \mathsf{a})$.

Example 2.17 *Let Ω, P, X, and Y be as in Example 2.9. If $\mathsf{A} = \{X, Y\}$, $\mathsf{a} = \{x, y\}$, $x = 4$, and $y = odd$, then*

$$P(\mathsf{A} = \mathsf{a}) = P(X = x, Y = y) = 1/18.$$

Recall the Law of Total Probability (Equalities 2.1 and 2.2). For two random variables X and Y, these equalities are as follows:

$$P(X = x) = \sum_y P(X = x, Y = y). \tag{2.5}$$

$$P(X = x) = \sum_y P(X = x | Y = y)P(Y = y). \tag{2.6}$$

It is left as an exercise to show this.

Example 2.18 *Let Ω, P, X, and Y be as in Example 2.9. Then, owing to Equality 2.5,*

$$
\begin{aligned}
P(X = 4) &= \sum_y P(X = 4, Y = y) \\
&= P(X = 4, Y = odd) + P(X = 4, Y = even) = \frac{1}{18} + \frac{1}{36} = \frac{1}{12}.
\end{aligned}
$$

Example 2.19 *Again, let Ω, P, X, and Y be as in Example 2.9. Then, due to Equality 2.6,*

$$
\begin{aligned}
P(X = 4) &= \sum_y P(X = x | Y = y)P(Y = y) \\
&= P(X = 4 | Y = odd)P(Y = odd) + \\
&\quad\ P(X = 4 | Y = even)P(Y = even) \\
&= \frac{2}{9} \times \frac{9}{36} + \frac{1}{27} \times \frac{27}{36} = \frac{1}{12}.
\end{aligned}
$$

In Equality 2.5 the probability distribution $P(X = x)$ is called the **marginal probability distribution** of X relative to the joint distribution $P(X = x, Y = y)$ because it is obtained using a process similar to adding across a row or column in a table of numbers. This concept also extends in a straightforward way to three or more random variables. For example, if we have a joint distribution $P(X = x, Y = y, Z = z)$ of X, Y, and Z, the marginal distribution $P(X = x, Y = y)$ of X and Y is obtained by summing over all values of Z. If $A = \{X, Y\}$, we also call this the **marginal probability distribution** of A.

The next example reviews the concepts covered so far concerning random variables.

Example 2.20 *Let Ω be a set of 12 individuals, and let P assign $1/12$ to each individual. Suppose the sexes, heights, and wages of the individuals are as follows:*

Case	Sex	Height (inches)	Wage ($)
1	female	64	30,000
2	female	64	30,000
3	female	64	40,000
4	female	64	40,000
5	female	68	30,000
6	female	68	40,000
7	male	64	40,000
8	male	64	50,000
9	male	68	40,000
10	male	68	50,000
11	male	70	40,000
12	male	70	50,000

Let the random variables S, H, and W, respectively, assign the sex, height, and wage of an individual to that individual. Then the probability distributions of the three random variables are as follows (recall that, for example, $P(s)$ represents $P(S = s)$).

s	$P(s)$
female	1/2
male	1/2

h	$P(h)$
64	1/2
68	1/3
70	1/6

w	$P(w)$
30,000	1/4
40,000	1/2
50,000	1/4

The joint distribution of S and H is as follows:

s	h	$P(s,h)$
female	64	1/3
female	68	1/6
female	70	0
male	64	1/6
male	68	1/6
male	70	1/6

The following table also shows the joint distribution of S and H and illustrates that the individual distributions can be obtained by summing the joint distribution over all values of the other variable.

s \ h	64	68	70	Distribution of S
female	1/3	1/6	0	1/2
male	1/6	1/6	1/6	1/2
Distribution of H	1/2	1/3	1/6	

The table that follows shows the first few values in the joint distribution of S, H, and W. There are 18 values in all, many of which are 0.

s	h	w	$P(s, h, w)$
female	64	30,000	1/6
female	64	40,000	1/6
female	64	50,000	0
female	68	30,000	1/12
...	...	...	...

We close with the **chain rule** for random variables, which says that given n random variables $X_1, X_2, \ldots, X_n$, defined on the same sample space Ω,

$$P(x_1, x_2, \ldots, x_n) = P(x_n | x_{n-1}, x_{n-2}, \ldots, x_1) \cdots \times P(x_2 | x_1) \times P(x_1)$$

whenever $P(x_1, x_2, \ldots, x_n) \neq 0$. It is straightforward to prove this rule using the rule for conditional probability.

Example 2.21 *Suppose we have the random variables in Example 2.20. Then, according to the chain rule for all values s, h, and w of S, H, and W,*

$$P(s, h, w) = P(w | h, s) P(h | s) P(s).$$

There are eight combinations of values of the three random variables. The table that follows shows that the equality holds for two of the combination.

s	h	w	$P(s, h, w)$	$P(w\|h, s) P(h\|s) P(s)$
female	64	30,000	$\frac{1}{6}$	$\left(\frac{1}{2}\right)\left(\frac{2}{3}\right)\left(\frac{1}{2}\right) = \frac{1}{6}$
female	64	40,000	$\frac{1}{12}$	$\left(\frac{1}{2}\right)\left(\frac{1}{3}\right)\left(\frac{1}{2}\right) = \frac{1}{12}$

It is left as an exercise to show that the equality holds for the other six combinations.

2.2.2 Independence of Random Variables

The notion of independence extends naturally to random variables.

Definition 2.6 *Suppose we have a probability space (Ω, P) and two random variables X and Y defined on Ω. Then X and Y are **independent** if, for all values x of X and y of Y, the events $X = x$ and $Y = y$ are independent. When this is the case, we write*

$$I_P(X, Y),$$

where I_P stands for independent in P.

Example 2.22 *Let Ω be the set of all cards in an ordinary deck, and let P assign 1/52 to each card. Define random variables as follows:*

Variable	Value	Outcomes Mapped to This Value
R	r_1	All royal cards
	r_2	All nonroyal cards
S	s_1	All spades
	s_2	All nonspades

Then the random variables R and S are independent. That is,

$$I_P(R, S).$$

To show this, we need show for all values of r and s that

$$P(r|s) = P(r).$$

The following table shows that this is the case.

s	r	$P(r)$	$P(r|s)$
s_1	r_1	$\frac{12}{52} = \frac{3}{13}$	$\frac{3}{13}$
s_1	r_2	$\frac{40}{52} = \frac{10}{13}$	$\frac{10}{13}$
s_2	r_1	$\frac{12}{52} = \frac{3}{13}$	$\frac{9}{39} = \frac{3}{13}$
s_2	r_2	$\frac{40}{52} = \frac{10}{13}$	$\frac{30}{39} = \frac{10}{13}$

The concept of conditional independence also extends naturally to random variables.

Definition 2.7 *Suppose we have a probability space (Ω, P), and three random variables X, Y, and Z defined on Ω. Then X and Y are **conditionally independent** given Z if for all values x of X, y of Y, and z of Z, whenever $P(z) \neq 0$, the events $X = x$ and $Y = y$ are conditionally independent given the event $Z = z$. When this is the case, we write*

$$I_P(X, Y|Z).$$

Example 2.23 *Let Ω be the set of all objects in Figure 2.1, and let P assign $1/13$ to each object. Define random variables S (for shape), L (for letter), and C (for color) as follows:*

Variable	Value	Outcomes Mapped to This Value
L	l_1	All objects containing an A
	l_2	All objects containing a B
S	s_1	All square objects
	s_2	All circular objects
C	c_1	All black objects
	c_2	All white objects

Then L and S are conditionally independent given C. That is,

$$I_P(L, S|C).$$

To show this, we need to show for all values of l, s, and c that

$$P(l|s, c) = P(l|c).$$

There is a total of eight combinations of the three variables. The table that follows shows that the equality holds for two of the combinations:

c	s	l	$P(l\|s,c)$	$P(l\|c)$
c_1	s_1	l_1	$\frac{2}{6}=\frac{1}{3}$	$\frac{3}{9}=\frac{1}{3}$
c_1	s_1	l_2	$\frac{4}{6}=\frac{2}{3}$	$\frac{6}{9}=\frac{2}{3}$

It is left as an exercise to show that it holds for the other combinations.

Independence and conditional independence can also be defined for sets of random variables.

Definition 2.8 *Suppose we have a probability space (Ω, P) and two sets* A *and* B *containing random variables defined on Ω. Let* a *and* b *be sets of values of the random variables in* A *and* B, *respectively. The sets* A *and* B *are said to be* **independent** *if, for all values of the variables in the sets* a *and* b, *the events* A = a *and* B = b *are independent. When this is the case, we write*

$$I_P(\mathsf{A}, \mathsf{B}),$$

where I_P stands for independent in P.

Example 2.24 *Let Ω be the set of all cards in an ordinary deck, and let P assign $1/52$ to each card. Define random variables as follows:*

Variable	Value	Outcomes Mapped to This Value
R	r_1	All royal cards
	r_2	All nonroyal cards
T	t_1	All tens and jacks
	t_2	All cards that are neither tens nor jacks
S	s_1	All spades
	s_2	All nonspades

Then the sets $\{R, T\}$ and $\{S\}$ are independent. That is,

$$I_P(\{R, T\}, \{S\}). \tag{2.7}$$

To show this, we need to show for all values of r, t, and s that

$$P(r, t|s) = P(r, t).$$

There are eight combinations of values of the three random variables. The table that follows shows that the equality holds for two of the combinations.

s	r	t	$P(r,t\|s)$	$P(r,t)$
s_1	r_1	t_1	$\frac{1}{13}$	$\frac{4}{52}=\frac{1}{13}$
s_1	r_1	t_2	$\frac{2}{13}$	$\frac{8}{52}=\frac{2}{13}$

It is left as an exercise to show that it holds for the other combinations.

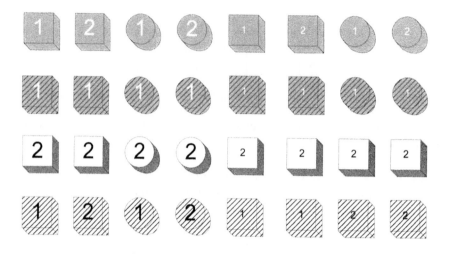

Figure 2.2: Objects with five properties.

When a set contains a single variable, we do not ordinarily show the braces. For example, we write Independency 2.7 as

$$I_P(\{R, T\}, S).$$

Definition 2.9 *Suppose we have a probability space* (Ω, P) *and three sets* A, B, *and* C *containing random variables defined on* Ω. *Let* a, b, *and* c *be sets of values of the random variables in* A, B, *and* C, *respectively. Then the sets* A *and* B *are said to be* **conditionally independent** *given the set* C *if, for all values of the variables in the sets* a, b, *and* c, *whenever* $P(c) \neq 0$, *the events* A = a *and* B = b *are conditionally independent given the event* C = c. *When this is the case, we write*

$$I_P(\mathsf{A}, \mathsf{B}|\mathsf{C}).$$

Example 2.25 *Suppose we use the Principle of Indifference to assign probabilities to the objects in Figure 2.2, and we define random variables as follows:*

Variable	Value	Outcomes Mapped to This Value
V	v_1	All objects containing a 1
	v_2	All objects containing a 2
L	l_1	All objects covered with lines
	l_2	All objects not covered with lines
C	c_1	All gray objects
	c_2	All white objects
S	s_1	All square objects
	s_2	All circular objects
F	f_1	All objects containing a number in a large font
	f_2	All objects containing a number in a small font

It is left as an exercise to show for all values of v, l, c, s, and f that

$$P(v, l|s, f, c) = P(v, l|c).$$

So we have

$$I_P(\{V, L\}, \{S, F\}|C).$$

2.3   The Meaning of Probability

When one does not have the opportunity to study probability theory in depth, one is often left with the impression that all probabilities are computed using ratios. Next, we discuss the meaning of probability in more depth and show that this is not how probabilities are ordinarily determined.

2.3.1   Relative Frequency Approach to Probability

A classic textbook example of probability concerns tossing a coin. Because the coin is symmetrical, we use the Principle of Indifference to assign

$$P(\text{Heads}) = P(\text{Tails}) = .5.$$

Suppose that instead we toss a thumbtack. It can also land one of two ways. That is, it could land on its flat end, which we will call "heads," or it could land with the edge of the flat end and the point touching the ground, which we will call "tails." Because the thumbtack is not symmetrical, we have no reason to apply the Principle of Indifference and assign probabilities of .5 to both outcomes. How then should we assign the probabilities? In the case of the coin, when we assign $P(heads) = .5$, we are implicitly assuming that if we tossed the coin a large number of times it would land heads about half the time. That is, if we tossed the coin 1000 times, we would expect it to land heads about 500 times. This notion of repeatedly performing the experiment gives us a method for computing (or at least estimating) the probability. That is, if we repeat an experiment many times, we are fairly certain that the probability of an outcome is about equal to the fraction of times the outcome occurs. For example, a student tossed a thumbtack 10,000 times and it landed heads 3761 times. So

$$P(\text{Heads}) \approx \frac{3761}{10,000} = .3761.$$

Indeed, in 1919 Richard von Mises used the limit of this fraction as the definition of probability. That is, if n is the number of tosses and S_n is the number of times the thumbtack lands heads, then

$$P(\text{Heads}) \equiv \lim_{n \to \infty} \frac{S_n}{n}.$$

This definition assumes that a limit actually is approached. That is, it assumes that the ratio does not fluctuate. For example, there is no reason a priori to assume that the ratio is not .5 after 100 tosses, .1 after 1000 tosses, .5 after

10,000 tosses, .1 after 100,000 tosses, and so on. Only experiments in the real world can substantiate that a limit is approached. In 1946 J. E. Kerrich conducted many such experiments using games of chance in which the Principle of Indifference seemed to apply (e.g., drawing a card from a deck). His results indicated that the relative frequency does appear to approach a limit and that this limit is the value obtained using the Principle of Indifference.

This approach to probability is called the **relative frequency approach to probability**, and probabilities obtained using this approach are called **relative frequencies**. A **frequentist** is someone who feels this is the only way we can obtain probabilities. Note that, according to this approach, we can never know a probability for certain. For example, if we tossed a coin 10,000 times and it landed heads 4991 times, we would estimate

$$P(\text{Heads}) \approx \frac{4991}{10,000} = .4991.$$

On the other hand, if we used the Principle of Indifference, we would assign $P(\text{Heads}) = .5$. In the case of the coin, the probability might not actually be .5 because the coin might not be perfectly symmetrical. For example, Kerrich [1946] found that the six came up the most in the toss of a die and that one came up the least. This makes sense because, at that time, the spots on the die were hollowed out of the die, so the die was lightest on the side with a six. On the other hand, in experiments involving cards or urns, it seems we can be certain of probabilities obtained using the Principle of Indifference.

Example 2.26 *Suppose we toss an asymmetrical six-sided die, and in 1000 tosses we observe that each of the six sides comes up the following number of times.*

Side	Number of Times
1	250
2	150
3	200
4	70
5	280
6	50

So we estimate $P(1) \approx .25$, $P(2) \approx .15$, $P(3) \approx .2$, $P(4) \approx .07$, $P(5) \approx .28$, and $P(6) \approx .05$.

Repeatedly performing an experiment (so as to estimate a relative frequency) is called **sampling**, and the set of outcomes is called a **random sample** (or simply a **sample**). The set from which we sample is called a **population**.

Example 2.27 *Suppose our population is all males in the United States between the ages of 31 and 85, and we are interested in the probability of such* *males having high blood pressure. Then, if we sample 10,000 males, this set of males is our sample. Furthermore, if 3210 have high blood pressure, we estimate*

$$P(\textit{High Blood Pressure}) \approx \frac{3210}{10{,}000} = .321.$$

Technically, we should not call the set of all current males in this age group the population. Rather, the theory says that there is a propensity for a male in this group to have high blood pressure and that this propensity is the probability. This propensity might not be equal to the fraction of current males in the group who have high blood pressure. In theory, we would have to have an infinite number of males to determine the probability exactly. The current set of males in this age group is called a **finite population**. The fraction of them with high blood pressure is the probability of obtaining a male with high blood pressure when we sample him from the set of all males in the age group. This latter probability is simply the ratio of males with high blood pressure.

When doing statistical inference, we sometimes want to estimate the ratio in a finite population from a sample of the population, and at other times we want to estimate a propensity from a finite sequence of observations. For example, TV raters ordinarily want to estimate the actual fraction of people in a nation watching a show from a sample of those people. On the other hand, medical scientists want to estimate the propensity with which males tend to have high blood pressure from a finite sequence of males. One can create an infinite sequence from a finite population by returning a sampled item back to the population before sampling the next item. This is called **sampling with replacement**. In practice, it is rarely done, but ordinarily the finite population is so large that statisticians make the simplifying assumption that it is done. That is, they do not replace the item but still assume that the ratio is unchanged for the next item sampled.

In sampling, the observed relative frequency is called the **maximum likelihood estimate (MLE)** of the probability (limit of the relative frequency) because it is the estimate of the probability that makes the observed sequence most probable when we assume that the trials (repetitions of the experiment) are probabilistically independent. See Section 3.1.3 for further discussion of this estimate.

Another facet of von Mises' relative frequency approach is that a random process is generating the sequence of outcomes. According to von Mises' theory, a **random process** is defined as a repeatable experiment for which the infinite sequence of outcomes is assumed to be a random sequence. Intuitively, a **random sequence** is one that shows no regularity or pattern. For example, the finite binary sequence 1011101100 appears random, whereas the sequence 1010101010 does not, because it has the pattern 10 repeated five times. There is evidence that experiments such as coin tossing and dice throwing are indeed random processes. In 1971 Iversen et al. ran many experiments with dice indicating the sequence of outcomes is random. It is believed that unbiased sampling also yields a random sequence and is therefore a random process. See [van Lambalgen, 1987] for a formal treatment of random sequences.

2.3.2 Subjective Approach to Probability

If we tossed a thumbtack 10,000 times and it landed heads 6000 times, we would estimate $P(heads)$ to be .6. Exactly what does this number approximate? Is there some probability, accurate to an arbitrary number of digits, of the thumbtack landing heads? It seems not. Indeed, as we toss the thumbtack, its shape will slowly be altered, changing any propensity for landing heads. As another example, is there really an exact propensity for a male in a certain age group to have high blood pressure? Again, it seems not. So it seems that, outside of games of chance involving cards and urns, the relative frequency notion of probability is only an idealization. Regardless, we obtain useful insights concerning our beliefs from this notion. For example, after the thumbtack lands heads 6000 times out of 10,000 tosses, we believe it has about a .6 chance of landing heads on the next toss, and we bet accordingly. That is, we would consider it fair to win $0.40 if the thumbtack landed heads and to lose $1 - $0.40 = $0.60 if the thumbtack landed tails. Since the bet is considered fair, the opposite position, namely, to lose $0.40 if the thumbtack landed heads and to win $0.60 if it landed tails, would also be considered fair. Hence, we would take either side of the bet. This notion of a probability as a value that determines a fair bet is called a **subjective approach to probability**, and probabilities assigned within this frame are called **subjective probabilities** or **beliefs**. A **subjectivist** is someone who feels we can assign probabilities within this framework. More concretely, in this approach the **subjective probability** of an uncertain event is the fraction p of units of money we would agree it is fair to give (lose) if the event does not occur in exchange for the promise to receive (win) $1 - p$ units if it does occur.

Example 2.28 *Suppose we estimate that $P(\text{Heads}) = .6$. This means that we would agree it is fair to give $0.60 if heads does not occur for the promise to receive $0.40 if it does occur. Notice that if we repeated the experiment 100 times and heads did occur 60% of the time (as we expected), we would win $60($0.40) = 24 and lose $40($0.60) = 24. That is, we would break even.*

Unlike the relative frequency approach to probability, the subjective approach allows us to compute probabilities of events that are not repeatable. A classic example concerns betting at the racetrack. To decide how to bet, we must first determine how likely we feel it is that each horse will win. A particular race has never run before and never will be run again, so we cannot look at previous occurrences of the race to obtain our belief. Rather, we obtain this belief from a careful analysis of the horses' overall previous performance, of track conditions, of jockeys, and so on. Clearly, not everyone will arrive at the same probabilities based on their analyses. This is why these probabilities are called *subjective*. They are particular to individuals. In general, they do not have objective values in nature on which we all must agree. Of course, if we did do an experiment such as tossing a thumbtack 10,000 times and it landed heads 6000 times, most would agree that the probability of heads is about .6. Indeed, de Finetti [1937] showed that if we make certain reasonable assumptions about your beliefs, this would have to be your probability.

Before pursuing this matter further, we discuss a concept related to probability, namely, odds. Mathematically, if $P(E)$ is the probability of event E, then the **odds** $O(E)$ are defined by

$$O(E) = \frac{P(E)}{1 - P(E)}.$$

As far as betting, $O(E)$ is the amount of money we would consider it fair to lose if E did not occur in return for gaining \$1 if E did occur.

Example 2.29 *Let E be the event that the horse Oldnag wins the Kentucky Derby. If we feel $P(E) = .2$, then*

$$O(E) = \frac{P(E)}{1 - P(E)} = \frac{.2}{1 - .2} = .25.$$

This means we would consider it fair to lose \$0.25 if the horse did not win in return for gaining \$1 if it did win.

If we state the fair bet in terms of probability (as discussed previously), we would consider it fair to lose \$0.20 if the horse did not win in return for gaining \$0.80 if it did win. Notice that with both methods the ratio of the amount won to the amount lost is 4, so they are consistent in the way they determine betting behavior.

At the racetrack, the betting odds shown are the odds against the event. That is, they are the odds of the event not occurring. If $P(E) = .2$ and $\neg E$ denotes that E does not occur, then

$$O(\neg E) = \frac{P(\neg E)}{1 - P(\neg E)} = \frac{.8}{1 - .8} = 4,$$

and the odds shown at the racetrack are 4 to 1 against E. If you bet on E, you will lose \$1 if E does not occur and win \$4 if E does occur. Note that these are the track odds based on the betting behavior of all participants. If you believe $P(E) = .5$, for you the odds against E are 1 to 1 (even money), and you should jump at the chance to get 4 to 1.

Some individuals are uncomfortable at being forced to consider wagering to assess a subjective probability. There are other methods for ascertaining these probabilities. One of the most popular is the following, which was suggested by Lindley in 1985. This method says an individual should liken the uncertain outcome to a game of chance by considering an urn containing white and black balls. The individual should determine for what fraction of white balls the individual would be indifferent between receiving a small prize if the uncertain event E happened (or turned out to be true) and receiving the same small prize if a white ball was drawn from the urn. That fraction is the individual's probability of the outcome. Such a probability can be constructed using binary cuts. If, for example, you were indifferent when the fraction was .8, for you

$P(E) = .8$. If someone else were indifferent when the fraction was .6, for that individual $P(E) = .6$. Again, neither individual is right or wrong.

It would be a mistake to assume that subjective probabilities are only important in gambling situations. Actually, they are important in all the applications discussed in this book. In the next section we illustrate interesting uses of subjective probabilities.

See [Neapolitan, 1990] for more on the two approaches to probability presented here.

2.4   Random Variables in Applications

Although it is mathematically elegant to first specify a sample space and then define random variables on the space, in practice this is not what we ordinarily do. In practice some single entity or set of entities has features, the states of which we want to determine but that we cannot determine for certain. So we settle for determining how likely it is that a particular feature is in a particular state. An example of a single entity is a jurisdiction in which we are considering introducing an economically beneficial chemical that might be carcinogenic. We would want to determine the relative risk of the chemical versus its benefits. An example of a set of entities is a set of patients with similar diseases and symptoms. In this case, we would want to diagnose diseases based on symptoms. As mentioned in Section 2.3.1, this set of entities is called a *population*, and technically it is usually not the set of all currently existing entities, but rather is, in theory, an infinite set of entities.

In these applications, a random variable represents some feature of the entity being modeled, and we are uncertain as to the value of this feature. In the case of a single entity, we are uncertain as to the value of the feature for that entity, whereas in the case of a set of entities, we are uncertain as to the value of the feature for some members of the set. To help resolve this uncertainty, we develop probabilistic relationships among the variables. When there is a set of entities, we assume the entities in the set all have the same probabilistic relationships concerning the variables used in the model. When this is not the case, our analysis is not applicable. In the case of the scenario concerning introducing a chemical, features may include the amount of human exposure and the carcinogenic potential. If these are our features of interest, we identify the random variables *HumanExposure* and *CarcinogenicPotential*. (For simplicity, our illustrations include only a few variables. An actual application ordinarily includes many more than this.) In the case of a set of patients, features of interest might include whether or not diseases such as lung cancer are present, whether or not manifestations of diseases such as a chest X-ray are present, and whether or not causes of diseases such as smoking are present. Given these features, we would identify the random variables *ChestXray*, *LungCancer*, and *SmokingHistory*, respectively.

After identifying the random variables, we distinguish a set of mutually exclusive and exhaustive values for each of them. The possible values of a random variable are the different states that the feature can take. For example, the state of *LungCancer* could be *present* or *absent*, the state of *ChestXray* could be *positive* or *negative*, and the state of *SmokingHistory* could be *yes* or *no*, where *yes* might mean the patient has smoked one or more packs of cigarettes every day during the past 10 years.

After distinguishing the possible values of the random variables (i.e., their spaces), we judge the probabilities of the random variables having their values. However, in general, we do not directly determine values in a joint probability distribution of the random variables. Rather, we ascertain probabilities concerning relationships among random variables that are accessible to us. We can then reason with these variables using Bayes' Theorem to obtain probabilities of events of interest. The next example illustrates this idea.

Example 2.30 *Suppose Sam plans to marry, and to obtain a marriage licence in the state in which he resides, one must take the blood test enzyme-linked immunosorbent assay (ELISA), which tests for the presence of human immunodeficiency virus (HIV). Sam takes the test and it comes back positive for HIV. How likely is it that Sam is infected with HIV? Without knowing the accuracy of the test, Sam really has no way of knowing how probable it is that he is infected with HIV.*

The data we ordinarily have on such tests are the true positive rate (sensitivity) and the true negative rate (specificity). The true positive rate is the number of people who both have the infection and test positive divided by the total number of people who have the infection. For example, to obtain this number for ELISA, 10,000 people who were known to be infected with HIV were identified. This was done using the Western Blot, which is the gold standard test for HIV. These people were then tested with ELISA, and 9990 tested positive. Therefore, the true positive rate is .999. The true negative rate is the number of people who both do not have the infection and test negative divided by the total number of people who do not have the infection. To obtain this number for ELISA 10,000 nuns who denied risk factors for HIV infection were tested. Of these, 9980 tested negative using the ELISA test. Furthermore, the 20 positive-testing nuns tested negative using the Western Blot test. So, the true negative rate is .998, which means that the false positive rate is .002. We therefore formulate the following random variables and subjective probabilities:

$$P(ELISA = positive | HIV = present) = .999 \qquad (2.8)$$

$$P(ELISA = positive | HIV = absent) = .002. \qquad (2.9)$$

*You might wonder why we called these **subjective probabilities** when we obtained them from data. Recall that the frequentist approach says that we can never know the actual relative frequencies (objective probabilities); we can only estimate them from data. However, within the subjective approach, we can make our beliefs (subjective probabilities) equal to the fractions obtained from the data.*

It might seem that Sam almost certainly is infected with HIV, since the test is so accurate. However, notice that neither the probability in Equality 2.8 nor the one in Equality 2.9 is the probability of Sam being infected with HIV. Since we know that Sam tested positive on ELISA, that probability is

$$P(HIV = present | ELISA = positive).$$

We can compute this probability using Bayes' Theorem if we know $P(HIV = present)$. Recall that Sam took the blood test simply because the state required it. He did not take it because he thought for any reason he was infected with HIV. So, the only other information we have about Sam is that he is a male in the state in which he resides. Therefore if 1 in 100,000 men in Sam's state is infected with HIV, we assign the following subjective probability:

$$P(HIV = present) = .00001.$$

We now employ Bayes' Theorem to compute

$P(present | positive)$

$$= \frac{P(positive|present)P(present)}{P(positive|present)P(present) + P(positive|absent)P(absent)}$$

$$= \frac{(.999)(.00001)}{(.999)(.00001) + (.002)(.99999)}$$

$$= .00497.$$

Surprisingly, we are fairly confident that Sam is not infected with HIV.

A probability such as $P(HIV = present)$ is called a **prior probability** because, in a particular model, it is the probability of some event prior to updating the probability of that event, within the framework of that model, using new information. Do not mistakenly think it means a probability prior to any information. A probability such as $P(HIV = present | ELISA = positive)$ is called a **posterior probability** because it is the probability of an event after its prior probability has been updated, within the framework of some model, based on new information. In the previous example the reason the posterior probability is small, even though the test is fairly accurate, is that the prior probability is extremely low. The next example shows how dramatically a different prior probability can change things.

Example 2.31 Suppose Mary and her husband have been trying to have a baby and she suspects she is pregnant. She takes a pregnancy test that has a true positive rate of .99 and a false positive rate of .02. Suppose further that 20% of all women who take this pregnancy test are indeed pregnant. Using Bayes' Theorem we then have

$P(present | positive)$

$$= \frac{P(positive|present)P(present)}{P(positive|present)P(present) + P(positive|absent)P(absent)}$$

$$= \frac{(.99)(.2)}{(.99)(.2) + (.02)(.8)}$$

$$= .92523.$$

Even though Mary's test was far less accurate than Sam's test, she probably is pregnant, whereas he probably is not infected with HIV. This is due to the prior information. There was a significant prior probability (.2) that Mary was pregnant, because only women who suspect they are pregnant on other grounds take pregnancy tests. Sam, on the other hand, took his test simply because he wanted to get married. We had no previous information indicating he could be infected with HIV.

In the previous two examples we obtained our beliefs (subjective probabilities) directly from the observed fractions in the data. Although this is often done, it is not necessary. In general, we obtain our beliefs from our information about the past, which means that these beliefs are a composite of all our experience rather than merely observed relative frequencies. We will see examples of this throughout the book. The following example illustrates such a case.

Example 2.32 *Suppose you feel there is a .4 probability the NASDAQ will go up at least 1% today. This is based on your knowledge that, after trading closed yesterday, excellent earnings were reported by several big companies in the technology sector, and that U.S. crude oil supplies unexpectedly increased. Furthermore, if the NASDAQ does go up at least 1% today, you feel there is a .1 probability that your favorite stock NTPA will go up at least 10% today. If the NASDAQ does not go up at least 1% today, you feel there is only a .02 probability NTPA will go up at least 10% today. You have these beliefs because you know from the past that NTPA's performance is linked to overall performance in the technology sector. You checked NTPA after the close of trading, and you noticed it went up over 10%. What is the probability that the NASDAQ went up at least 1%? Using Bayes' Theorem we have*

$$P(NASDAQ = up\ 1\%|NTPA = up\ 10\%)$$

$$= \frac{P(up\ 10\%|up\ 1\%)P(up\ 1\%)}{P(up\ 10\%|up\ 1\%)P(up\ 1\%) + P(up\ 10\%|not\ up\ 1\%)P(not\ up\ 1\%)}$$

$$= \frac{(.1)(.4)}{(.1)(.4) + (.02)(.6)} = .769.$$

In the previous three examples we used Bayes' Theorem to compute posterior subjective probabilities from known subjective probabilities. In a rigorous sense, we can only do this within the subjective framework. That is, since strict frequentists say we can never know probabilities for certain, they cannot use Bayes' Theorem. They can only do analyses such as the computation of a confidence interval for the value of an unknown probability based on the data. These techniques are discussed in any classic statistics text such as [Hogg and Craig, 1972]. Since subjectivists are the ones who use Bayes' Theorem, they are often called **Bayesians**.

EXERCISES

Section 2.1

Exercise 2.1 Let the experiment be drawing the top card from a deck of 52 cards. Let Heart be the event a heart is drawn, and RoyalCard be the event a royal card is drawn.

1. Compute $P(\mathsf{Heart})$.

2. Compute $P(\mathsf{RoyalCard})$.

3. Compute $P(\mathsf{Heart} \cup \mathsf{RoyalCard})$.

Exercise 2.2 Prove Theorem 2.1.

Exercise 2.3 Example 2.5 showed that, in the draw of the top card from a deck, the event Jack is independent of the event Club. That is, it showed $P(\mathsf{Jack} \mid \mathsf{Club}) = P(\mathsf{Jack})$.

1. Show directly that the event Club is independent of the event Jack. That is, show $P(\mathsf{Club} \mid \mathsf{Jack}) = P(\mathsf{Club})$. Show also that $P(\mathsf{Jack} \cap \mathsf{Club}) = P(\mathsf{Jack})P(\mathsf{Club})$.

2. Show, in general, that if $P(\mathsf{E}) \neq 0$ and $P(\mathsf{F}) \neq 0$, then $P(\mathsf{E} \mid \mathsf{F}) = P(\mathsf{E})$ if and only if $P(\mathsf{F} \mid \mathsf{E}) = P(\mathsf{F})$, and each of these holds if and only if $P(\mathsf{E} \cap \mathsf{F}) = P(\mathsf{E})P(\mathsf{F})$.

Exercise 2.4 The complement of a set E consists of all the elements in Ω that are not in E and is denoted by $\overline{\mathsf{E}}$.

1. Show that E is independent of F if and only if $\overline{\mathsf{E}}$ is independent of F, which is true if and only if $\overline{\mathsf{E}}$ is independent of $\overline{\mathsf{F}}$.

2. Example 2.6 showed that, for the objects in Figure 2.1, A and Square are conditionally independent given Black and given White. Let B be the set of all objects containing a B, and Circle be the set of all circular objects. Use the result just obtained to conclude that A and Circle, B and Square, and B and Circle are each conditionally independent given either Black or White.

Exercise 2.5 Show that in the draw of the top card from a deck, the event E $= \{kh, ks, qh\}$ and the event F $= \{kh, kc, qh\}$ are conditionally independent given the event G $= \{kh, ks, kc, kd\}$. Determine whether E and F are conditionally independent given $\overline{\mathsf{G}}$.

Exercise 2.6 Prove the Law of Total Probability, which says that if we have n mutually exclusive and exhaustive events $E_1, E_2, \ldots, E_n$, then for any other event F,

$$P(F) = P(F \cap E_1) + P(F \cap E_2) + \cdots + P(F \cap E_n).$$

Exercise 2.7 Let Ω be the set of all objects in Figure 2.1, and assign each object a probability of $1/13$. Let A be the set of all objects containing an A, and Square be the set of all square objects. Compute $P(A|\text{Square})$ directly and using Bayes' Theorem.

Section 2.2

Exercise 2.8 Consider the probability space and random variables given in Example 2.20.

1. Determine the joint distribution of S and W, the joint distribution of W and H, and the remaining values in the joint distribution of S, H, and W.

2. Show that the joint distribution of S and H can be obtained by summing the joint distribution of S, H, and W over all values of W.

Exercise 2.9 Let a joint probability distribution be given. Using the law of total probability, show that, in general, the probability distribution of any one of the random variables is obtained by summing over all values of the other variables.

Exercise 2.10 The chain rule says that for n random variables $X_1, X_2, \ldots, X_n$, defined on the same sample space Ω,

$$P(x_1, x_2, \ldots, x_n) = P(x_n | x_{n-1}, x_{n-2}, \ldots x_1) \cdots \times P(x_2|x_1) \times P(x_1)$$

whenever $P(x_1, x_2, \ldots, x_n) \neq 0$. Prove this rule.

Exercise 2.11 Use the results in Exercise 2.4 (1) to conclude that it was only necessary in Example 2.22 to show that $P(r, t) = P(r, t|s_1)$ for all values of r and t.

Exercise 2.12 Suppose we have two random variables X and Y with spaces $\{x_1, x_2\}$ and $\{y_1, y_2\}$, respectively.

1. Use the results in Exercise 2.4 (1) to conclude that we need only show $P(y_1|x_1) = P(y_1)$ to conclude $I_P(X, Y)$.

2. Develop an example showing that if X and Y both have spaces containing more than two values, then we need to check whether $P(y|x) = P(y)$ for all values of x and y to conclude $I_P(X, Y)$.

Exercise 2.13 Consider the probability space and random variables given in Example 2.20.

1. Are H and W independent?

2. Are H and W conditionally independent given S?

3. If this small sample is indicative of the probabilistic relationships among the variables in some population, what causal relationships might account for this dependency and conditional independency?

Exercise 2.14 In Example 2.25, it was left as an exercise to show for all values v of V, l of L, c of C, s of S, and f of F that

$$P(v, l | s, f, c) = P(v, l | c).$$

Show this.

Section 2.3

Exercise 2.15 Kerrich [1946] performed experiments such as tossing a coin many times, and he found that the relative frequency did appear to approach a limit. That is, for example, he found that after 100 tosses the relative frequency may have been .51, after 1000 tosses it may have been .508, after 10,000 tosses it may have been .5003, and after 100,000 tosses it may have been .50006. The pattern is that the 5 in the first place to the right of the decimal point remains in all relative frequencies after the first 100 tosses, the 0 in the second place remains in all relative frequencies after the first 1000 tosses, and so on. Toss a thumbtack at least 1000 times and see if you obtain similar results.

Exercise 2.16 Pick some upcoming event. It could be a sporting event or it could be the event that you will get an A in this course. Determine your probability of the event using Lindley's [1985] method of comparing the uncertain event to a draw of a ball from an urn. (See the discussion following Example 2.29.)

Section 2.4

Exercise 2.17 A forgetful nurse is supposed to give Mr. Nguyen a pill each day. The probability that the nurse will forget to give the pill on a given day is .3. If Mr. Nguyen receives the pill, the probability he will die is .1. If he does not receive the pill, the probability he will die is .8. Mr. Nguyen died today. Use Bayes' Theorem to compute the probability that the nurse forgot to give him the pill.

Exercise 2.18 An oil well might be drilled on Professor Neapolitan's farm in Texas. Based on what has happened on similar farms, we judge the probability of oil being present to be .5, the probability of only natural gas being present to be .2, and the probability of neither being present to be .3. If oil is present, a geological test will give a positive result with probability .9; if only natural gas is present, it will give a positive result with probability .3; and if neither is present, the test will be positive with probability .1. Suppose the test comes back positive. Use Bayes' Theorem to compute the probability that oil is present.

Chapter 3

Statistics Basics

This chapter reviews the statistics you need to read the remainder of this book. In Section 3.1 we present basic concepts in statistics such as expected value, variance, and maximum likelihood. Section 3.2 covers the Markov Chain Monte-Carlo (MCMC) approximation technique. Finally, Section 3.3 briefly reviews the normal distribution.

3.1 Basic Concepts

Next, we review several important statistical definitions.

3.1.1 Expected Value

Definition 3.1 *Suppose we have a discrete numeric random variable X, whose space is*

$$\{x_1, x_2, \ldots, x_n\}.$$

*Then the **expected value** $E(X)$ is given by*

$$E(X) = x_1 P(x_1) + x_2 P(x_2) + \cdots + x_n P(x_n).$$

Example 3.1 *Suppose we have a symmetric die such that the probability of each side showing is $1/6$. Let X be a random variable whose value is the number that shows when we toss the die. Then*

$$
\begin{aligned}
E(X) &= 1P(1) + 2P(2) + 3P(3) + 4P(4) + 5P(5) + 6P(6) \\
&= 1\left(\frac{1}{6}\right) + 2\left(\frac{1}{6}\right) + 3\left(\frac{1}{6}\right) + 4\left(\frac{1}{6}\right) + 5\left(\frac{1}{6}\right) + 6\left(\frac{1}{6}\right) \\
&= 3.5.
\end{aligned}
$$

If we threw the die many times, we would expect the average of the numbers showing to equal about 3.5.

Example 3.2 *Suppose we have the random variables S (Sex), H (height), and W (wage) discussed in Example 2.20. Recall that they were based on sex, height, and wage being distributed according to the following table:*

Case	Sex	Height (inches)	Wage ($)
1	*female*	64	30,000
2	*female*	64	30,000
3	*female*	64	40,000
4	*female*	64	40,000
5	*female*	68	30,000
6	*female*	68	40,000
7	*male*	64	40,000
8	*male*	64	50,000
9	*male*	68	40,000
10	*male*	68	50,000
11	*male*	70	40,000
12	*male*	70	50,000

Then

$$E(H) = 64\left(\frac{6}{12}\right) + 68\left(\frac{4}{12}\right) + 70\left(\frac{2}{12}\right) = 66.33$$

$$E(W) = 30{,}000\left(\frac{3}{12}\right) + 40{,}000\left(\frac{6}{12}\right) + 50{,}000\left(\frac{3}{12}\right) = 40{,}000.$$

The expected value of X is also called the **mean** and is sometimes denoted $\bar{X}$. In situations where probabilities are assigned according to how many times a value appears in a population (as in the previous example), it is simply the average of the values taken over all members of the population.

We have the following theorem, whose proof is left as an exercise, concerning two independent random variables:

Theorem 3.1 *If X and Y are independent random variables, then*

$$E(XY) = E(X)E(Y).$$

Example 3.3 *Let X be a random variable whose value is the number that shows when we toss a symmetric die, and let Y be a variable whose value is 1 if heads is the result of tossing a fair coin and 0 if tails is the result. Then X and Y are independent. We show next that $E(XY) = E(X)E(Y)$ as the previous theorem entails. To that end,*

$$
\begin{aligned}
E(XY) &= (1 \times 1)P(1,1) + (2 \times 1)P(2,1) + (3 \times 1)P(3,1) + \\
&\quad (4 \times 1)P(4,1) + (5 \times 1)P(5,1) + (6,1)P(6,1) + \\
&\quad (1 \times 0)P(1,0) + (2 \times 1)P(2,0) + (3 \times 0)P(3,1) + \\
&\quad (4 \times 0)P(4,0) + (5 \times 0)P(5,0) + (6,1)P(6,0) \\
&= 1\left(\frac{1}{6}\right)\left(\frac{1}{2}\right) + 2\left(\frac{1}{6}\right)\left(\frac{1}{2}\right) + 3\left(\frac{1}{6}\right)\left(\frac{1}{2}\right) + \\
&\quad 4\left(\frac{1}{6}\right)\left(\frac{1}{2}\right) + 5\left(\frac{1}{6}\right)\left(\frac{1}{2}\right) + 6\left(\frac{1}{6}\right)\left(\frac{1}{2}\right) \\
&= 1.75.
\end{aligned}
$$

Furthermore,

$$E(X)E(Y) = 3.5 \times .5 = 1.75.$$

3.1.2 Variance and Covariance

While the expected value is a summary statistic that often tells us approximately what values occur in a population, the variance tells us about how far the actual values are from the expected value. We discuss the variance next.

Definition 3.2 *Suppose we have a discrete numeric random variable X, whose space is*

$$\{x_1, x_2, \ldots, x_n\}.$$

*Then the **variance** $Var(X)$ is given by*

$$Var(X) = E\left([X - E(X)]^2\right).$$

The next theorem often enables us to compute the variance more easily. Its proof is left as an exercise.

Theorem 3.2 *We have*

$$Var(X) = E\left(X^2\right) - (E(X))^2.$$

Example 3.4 *Suppose again we have the die discussed in Example 3.1. Then*

$$
\begin{aligned}
E\left(X^2\right) &= 1^2 P(1) + 2^2 P(2) + 3^2 P(3) + 4^2 P(4) + 5^2 P(5) + 6^2 P(6) \\
&= 1^2\left(\frac{1}{6}\right) + 2^2\left(\frac{1}{6}\right) + 3^2\left(\frac{1}{6}\right) + 4^2\left(\frac{1}{6}\right) + 5^2\left(\frac{1}{6}\right) + 6^2\left(\frac{1}{6}\right) \\
&= 15.167.
\end{aligned}
$$

So

$$
\begin{aligned}
Var(X) &= E\left(X^2\right) - (E(X))^2 \\
&= 15.167 - (3.5)^2 = 2.917.
\end{aligned}
$$

In order to compare quantities of the same order of magnitude, the variance is often converted to the standard deviation σ_X, which is given by

$$\sigma_X = \sqrt{Var(X)}.$$

Example 3.5 *For the die discussed in the previous example,*

$$\sigma_X = \sqrt{2.917} = 1.708.$$

Example 3.6 *Suppose there are two neighboring communities, one community consists of "amazons," while the other consists of "pygmies." So, in the first community everyone is around 7 feet tall, while in the second community everyone is around 4 feet tall. Suppose further that there are the same number of people in both communities. Let H be a random variable representing the height of individuals in the two communities combined. Then it is left as an exercise to show*

$$
\begin{aligned}
E(H) &= 5.5 \\
\sigma_H &= 1.5.
\end{aligned}
$$

Example 3.7 *Suppose now there is a community in which everyone is about 5.5 feet tall. Then*

$$
\begin{aligned}
E(H) &= 5.5 \\
\sigma_H &= 0.
\end{aligned}
$$

The previous two examples illustrate that the expected value alone cannot tell us much about what we can expect concerning the actual values in the population. In the first example, we would never see individuals 5.5 feet tall, while in the second example that is all we would see. The standard deviation gives us an idea as to how far the values actually deviate from the expected value. However, it too is only a summary statistic, and the expected value and standard deviation together do not, in general, contain all information in the probability distribution. In the second example above they do indeed do this, whereas in the first example there could be many other distributions which yield the same expected value and standard deviation.

The next definition concerns the relationship between two random variables.

Definition 3.3 *Suppose we have two discrete numeric random variables X and Y. Then the* **covariance** *$Cov(X, Y)$ of X and Y is given by*

$$Cov(X, Y) = E\left([X - E(X)][Y - E(Y)]\right).$$

$Cov(X, Y)$ is often denoted σ_{XY}. It is left as an exercise to prove the following theorem.

Theorem 3.3 *We have the following:*

$$Cov(X, Y) = E(XY) - E(X)E(Y).$$

Clearly,

$$Cov(X, X) = Var(X),$$

and

$$Cov(X, Y) = Cov(Y, X).$$

Example 3.8 *Suppose we have the random variables in Example 3.2. Then*

$E(HW)$

$$
\begin{aligned}
= \ & (64 \times 30{,}000)P(64, 30{,}000) + (64 \times 40{,}000)P(64, 40{,}000) + \\
& (64 \times 50{,}000)P(64, 50{,}000) + (68 \times 30{,}000)P(68, 30{,}000) + \\
& (68 \times 40{,}000)P(68, 40{,}000) + (68 \times 50{,}000)P(68, 40{,}000) + \\
& (70 \times 40{,}000)P(70, 40{,}000) + (70 \times 50{,}000)P(70, 50{,}000) \\
= \ & (64 \times 30{,}000)\left(\frac{2}{12}\right) + (64 \times 40{,}000)\left(\frac{3}{12}\right) + (64 \times 50{,}000)\left(\frac{1}{12}\right) + \\
& (68 \times 30{,}000)\left(\frac{1}{12}\right) + (68 \times 40{,}000)\left(\frac{2}{12}\right) + (68 \times 50{,}000)\left(\frac{1}{12}\right) + \\
& (70 \times 40{,}000)\left(\frac{1}{12}\right) + (70 \times 50{,}000)\left(\frac{1}{12}\right) \\
= \ & 2{,}658{,}333.
\end{aligned}
$$

Therefore,

$$
\begin{aligned}
Cov(H, W) \ & = \ E(HW) - E(H)E(W) \\
& = \ 2{,}658{,}333 - 66.33 \times 40{,}000 \\
& = \ 5133.
\end{aligned}
$$

The covariance itself does not convey much meaning concerning the relationship between X and Y. To accomplish this we compute the correlation coefficient from it.

Suppose we have two discrete numeric random variables X and Y. Then the **correlation coefficient** $\rho(X, Y)$ of X and Y is given by

$$\rho(X, Y) = \frac{Cov(X, Y)}{\sigma_X \sigma_Y}.$$

Example 3.9 *Suppose we have the random variables in Example 3.2. It is left as an exercise to show*

$$\sigma_H = 2.516$$
$$\sigma_W = 7071.$$

We then have

$$\rho(H, W) = \frac{Cov(H, W)}{\sigma_H \sigma_W}$$

$$= \frac{5133}{2.516 \times 7071}$$

$$= .2885.$$

Notice the $\rho(H, W)$ is positive and less than 1. The correlation coefficient is between -1 and $+1$. A value greater than 0 indicates the variables are positively correlated, and a value less than 0 indicates they are negatively correlated. By positively correlated we mean as one increases the other increases, while by negatively correlated we mean as one increases the other decreases. In the example above as height increases wage tends to increase. That is, tall people tend to have larger wages. The next theorem states these results concretely.

Theorem 3.4 *For any two numeric random variables X and Y,*

$$|\rho(X, Y)| \leq 1.$$

Furthermore,

$$\rho(X, Y) = 1$$

if and only if there exists constants $a > 0$ and b such that

$$Y = aX + b,$$

and

$$\rho(X, Y) = -1$$

if and only if there exists constants $a < 0$ and b such that

$$Y = aX + b.$$

Finally, if X and Y are independent, then

$$\rho(X, Y) = 0.$$

Proof. *The proof can be found in [Feller, 1968].* ∎

Example 3.10 *Let X be a random variable whose value is the number that shows when we toss a symmetric die, and let Y be a random variable whose value is two times the number that shows. Then*

$$Y = 2X.$$

According to the previous theorem, $\rho(X, Y) = 1$. We show this directly.

$$
\begin{aligned}
Cov(XY) &= E(XY) - E(X)E(Y) \\
&\quad (1 \times 2)\left(\frac{1}{6}\right) + (2 \times 4)\left(\frac{1}{6}\right) + (3 \times 6)\left(\frac{1}{6}\right) + \\
&\quad (4 \times 8)\left(\frac{1}{6}\right) + (5 \times 10)\left(\frac{1}{6}\right) + (6 \times 12)\left(\frac{1}{6}\right) - (3.5)(7.0) \\
&= 5.833.
\end{aligned}
$$

Therefore,

$$
\begin{aligned}
\rho(X, Y) &= \frac{Cov(X, Y)}{\sigma_X \sigma_Y} \\
&= \frac{5.833}{1.708 \times 3.416} = 1.
\end{aligned}
$$

Example 3.11 *Let X be a random variable whose value is the number that shows when we toss a symmetric die, and let Y be a random variable whose value is the negative of the number that shows. Then*

$$Y = -X.$$

According to the previous theorem, $\rho(X, Y) = -1$. It is left as an exercise to show this directly.

Example 3.12 *As in Example 3.3, let X be a random variable whose value is the number that shows when we toss a symmetric die, and let Y be a variable whose value is 1 if heads is the result of tossing a fair coin and 0 if tails is the result. According to the previous theorem, $\rho(X, Y) = 0$. It is left as an exercise to show this directly.*

Notice that the theorem does not say that $\rho(X, Y) = 0$ if and only if X and Y are independent. Indeed, X and Y can be deterministically related and still have a zero correlation coefficient. The next example illustrates this concept.

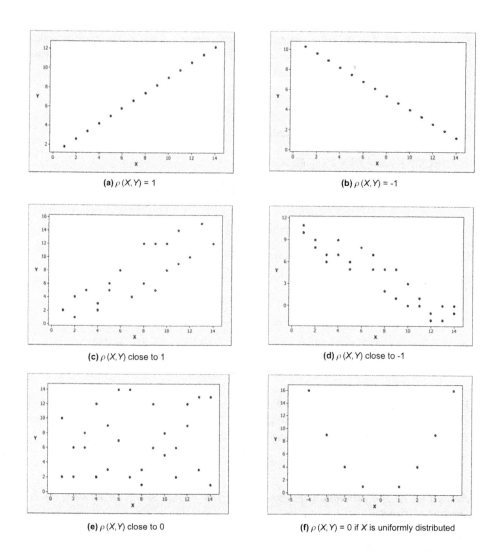

Figure 3.1: Several correlation coefficients of two random variables.

Example 3.13 *Let X's space be $\{-1, -2, 1, 2\}$, let each value have probability .25, let Y's space be $\{1, 4\}$, and let $Y = X^2$. Then $E(XY)$ is as follows*

$$
\begin{aligned}
E(XY) &= (-1)(1)\left(\frac{1}{4}\right) + (-1)(4)(0) + (-2)(1)(0) + (-2)(4)\left(\frac{1}{4}\right) + \\
&\quad (1)(1)\left(\frac{1}{4}\right) + (1)(4)(0) + 2(1)(0) + (2)(4)\left(\frac{1}{4}\right) \\
&= 0.
\end{aligned}
$$

Since clearly $E(X) = 0$, we conclude that $\rho(X, Y) = 0$.

We see that the correlation coefficient is not a general measure of the dependency between X and Y. Rather, it is a measure of the degree of linear dependence.

Figure 3.1 shows scatterplots illustrating correlation coefficients associated with some distributions. In each scatterplot, each point represents one (x, y) pair occurring in the joint distribution of X and Y.

We close with a theorem that gives us the variance of the sum of several random variables. Its proof is left as an exercise.

Theorem 3.5 *For random variables $X_1, X_2, \ldots, X_n$, we have*

$$
Var(X_1 + X_2 + \cdots + X_n) = \sum_{i=1}^{n} Var(X_i) + 2 \sum_{i \neq j} Cov(X_i, X_j).
$$

For two random variables X and Y, this equality is as follows:

$$
Var(X + Y) = Var(X) + Var(Y) + 2Cov(X, Y).
$$

Note that if X and Y are independent, then the variance of their sum is just the sum of their variances.

3.1.3 Maximum Likelihood Estimation

Recall that in Section 2.3.1 we discussed tossing a thumbtack. We noted that it could land on its flat end, which we will call *heads*, or it could land with the edge of the flat end and the point touching the ground, which we will call *tails*. We noted further that because the thumbtack is not symmetrical, we have no reason to apply the Principle of Indifference and assign probabilities of .5 to both outcomes. We then concluded that if we tossed it 10,000 times and it landed heads 3761 times, we should conclude that

$$
P(\text{Heads}) \approx \frac{3761}{10,000} = .3761.
$$

The value .3761 is an example of a maximum likelihood estimate. In general, if we have a probability distribution with an unknown parameter θ, and data is generated according to the distribution, the **maximum likelihood estimate**

(MLE) $\hat{\theta}$ of the parameter, based on the data, is the value of θ which makes the data most probable. That is, if D is our data, then $\hat{\theta}$ is the value of θ that maximizes

$$P(\mathsf{D}|\theta).$$

Example 3.14 *Suppose we toss a thumbtack n times and it lands heads k times. Our data D consists of the outcome of these n tosses. If we let $\theta = P(Heads)$ and we assume the tosses are independent, then*

$$P(\mathsf{D}|\theta) = \theta^k (1 - \theta)^{n-k}.$$

To find the maximum likelihood value of θ, we set the derivative of the logarithm of $P(\mathsf{D}|\theta)$ to 0, and solve for θ. To that end,

$$\ln\left(P(\mathsf{D}|\theta)\right) = k \ln \theta + (n - k) \ln(1 - \theta)$$

which means

$$\frac{d \ln\left(P(\mathsf{D}|\theta)\right)}{d\theta} = \frac{d\left(k \ln \theta + (n - k) \ln(1 - \theta)\right)}{d\theta}$$

$$= \frac{k}{\theta} - \frac{n - k}{1 - \theta}.$$

If we set this last expression to 0 and solve for θ, we obtain that

$$\theta = \frac{k}{n}.$$

So $\hat{\theta} = k/n$, which is what we would expect.

Example 3.15 *Suppose we toss a thumbtack four times and we observe the sequence $[heads, tails, heads, heads]$. Then the maximum likelihood estimate of the probability of heads is $3/4$. Let T_1, T_2, T_3, and T_4 be random variables such that the value of T_i is the outcome on the ith toss. If we estimate $P(T_i = heads)$ to be $3/4$ for all i and assume the trials are independent, then*

$$P(T_1 = heads, T_2 = tails, T_3 = heads, T_4 = heads)$$

$$= P(T_1 = heads)P(T_2 = tails)P(T_3 = heads)P(T_4 = heads)$$

$$= \frac{3}{4} \times \frac{1}{4} \times \frac{3}{4} \times \frac{3}{4} = .1055.$$

Any other estimate will make this sequence less probable. For example, if we estimate $P(T_i = heads)$ to be $1/2$ for all i, we have that

$$P(T_1 = heads, T_2 = tails, T_3 = heads, T_4 = heads)$$

$$= P(T_1 = heads)P(T_2 = tails)P(T_3 = heads)P(T_4 = heads)$$

$$= \frac{1}{2} \times \frac{1}{2} \times \frac{1}{2} \times \frac{1}{2} = .0625.$$

3.2 Markov Chain Monte Carlo

In Chapter 5, we will see that a Bayesian network can succinctly specify a joint probability distribution of n random variables. If each random variable only has a space size equal to 2 (That is, it has 2 possible values), there would still be 2^n values in the joint probability distribution. Suppose we have a function f defined on the values in the joint probability distribution, and we want to compute

$$I = \sum_{i=1}^{2^n} f(r_i) P(r_i),$$

where r_i for $1 \leq i \leq 2^n$ are the values in the joint probability distribution. For large n, this computation is intractable. However, we can estimate this sum using a technique called **Markov Chain Monte Carlo (MCMC)**. We discuss this technique shortly. First, we need to review Markov chains.

3.2.1 Markov Chains

This exposition is only for the purpose of review. If you are unfamiliar with Markov chains, you should consult a complete introduction as can be found in [Feller, 1968]. We start with the definition:

Definition 3.4 *A **Markov chain** consists of the following:*

1. *A set of outcomes (states) $e_1, e_2, \ldots$.*

2. *For each pair of states e_i and e_j a transition probability p_{ij} such that*

$$\sum_j p_{ij} = 1.$$

3. *A sequence of trials (random variables) $E^{(1)}, E^{(2)}, \ldots$ such that the outcome of each trial is one of the states, and*

$$P(E^{(h+1)} = e_j | E^{(h)} = e_i) = p_{ij}.$$

To completely specify a probability space we need to define initial probabilities $P(E^{(0)} = e_j) = p_j$, but these probabilities are not necessary to our theory and will not be discussed further.

Example 3.16 *Any Markov chain can be represented by an urn model. One such model is shown in Figure 3.2. The Markov chain is obtained by choosing an initial urn according to some probability distribution, picking a ball at random from that urn, moving to the urn indicated on the ball chosen, picking a ball at random from the new urn, and so on.*

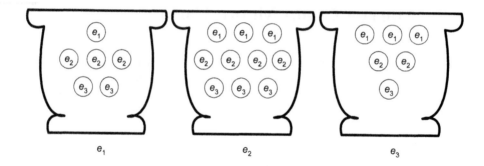

Figure 3.2: An urn model of a Markov chain.

The transition probabilities p_{ij} are arranged in a matrix of transition probabilities as follows:

$$\mathbf{P} = \begin{pmatrix} p_{11} & p_{12} & p_{13} & \cdots \\ p_{21} & p_{22} & p_{23} & \cdots \\ p_{31} & p_{32} & p_{33} & \cdots \\ \vdots & \vdots & \vdots & \ddots \end{pmatrix}.$$

This matrix is called the **transition matrix** for the chain.

Example 3.17 *For the Markov chain determined by the urns in Figure 3.2 the transition matrix is*

$$\mathbf{P} = \begin{pmatrix} 1/6 & 1/2 & 1/3 \\ 3/10 & 2/5 & 3/10 \\ 1/2 & 1/3 & 1/6 \end{pmatrix}.$$

A Markov chain is called **finite** if it has a finite number of states. Clearly the chain represented by the urns in Figure 3.2 is finite. We denote **the probability of a transition from e_i to e_j in exactly n trials** by $p_{ij}^{(n)}$. This is, $p_{ij}^{(n)}$ is the conditional probability of entering e_j at the nth trial given the initial state is e_i. We say e_j **is reachable from** e_i if there exists an $n \geq 0$ such that $p_{ij}^{(n)} > 0$. A Markov chain is called **irreducible** if every state is reachable from every other state.

Example 3.18 *Clearly, if $p_{ij} > 0$ for every i and j, the chain is irreducible.*

The state e_i has **period** $t > 1$ if $p_{ii}^{(n)} = 0$ unless $n = mt$ for some integer m, and t is the largest integer with this property. Such a state is called **periodic**. A state is **aperiodic** if no such $t > 1$ exists.

Example 3.19 *Clearly, if $p_{ii} > 0$, e_i is aperiodic.*

We denote by $f_{ij}^{(n)}$ the probability that starting from e_i the first entry to e_j occurs at the nth trial. Furthermore, we let

$$f_{ij} = \sum_{n=1}^{\infty} f_{ij}^{(n)}.$$

Clearly, $f_{ij} \leq 1$. When $f_{ij} = 1$, we call $P_{ij}(n) \equiv f_{ij}^{(n)}$ the **distribution of the first passage** for e_j starting at e_i. In particular, when $f_{ii} = 1$, we call $P_i(n) \equiv f_{ii}^{(n)}$ the **distribution of the recurrence times** for e_i, and we define the **mean recurrence time** for e_i to be

$$\mu_i = \sum_{n=1}^{\infty} n f_{ii}^{(n)}.$$

The state e_i is called **persistent** if $f_{ii} = 1$ and **transient** if $f_{ii} < 1$. A persistent state e_i is called **null** if its mean recurrence time $\mu_i = \infty$ and otherwise it is called **non-null**.

Example 3.20 *It can be shown that every state in a finite irreducible chain is persistent (See [Ash, 1970]), and that every persistent state in a finite chain is non-null (See [Feller, 1968]). Therefore every state in a finite irreducible chain is persistent and non-null.*

An aperiodic persistent non-null state is called **ergodic**. A Markov chain is called ergodic if all its states are ergodic.

Example 3.21 *Owing to Examples 3.18, 3.19, and 3.20, if in a finite chain we have $p_{ij} > 0$ for every i and j, the chain is an irreducible ergodic chain.*

We have the following theorem concerning irreducible ergodic chains:

Theorem 3.6 *In an irreducible ergodic chain the limits*

$$r_j = \lim_{n \to \infty} p_{ij}^{(n)} \tag{3.1}$$

exist and are independent of the initial state e_i. Furthermore, $r_j > 0$,

$$\sum_j r_j = 1, \tag{3.2}$$

$$r_j = \sum_i r_i p_{ij}, \tag{3.3}$$

and

$$r_j = \frac{1}{\mu_j},$$

where μ_j is the mean recurrence time of e_j.

The probability distribution

$$P(E = e_j) \equiv r_j$$

*is called the **stationary distribution** of the Markov chain.*

Conversely, suppose a chain is irreducible and aperiodic with transition matrix $\mathbf{P}$*, and there exists numbers* $r_j \geq 0$ *satisfying Equalities 3.2 and 3.3. Then the chain is ergodic, and the* r_j*s are given by Equality 3.1.*

Proof. *The proof can be found in [Feller, 1968].* ∎

We can write Equality 3.3 in the matrix/vector form

$$\mathbf{r}^T = \mathbf{r}^T \mathbf{P}. \tag{3.4}$$

That is,

$$
\begin{pmatrix} r_1 & r_2 & r_3 & \cdots \end{pmatrix} = \begin{pmatrix} r_1 & r_2 & r_3 & \cdots \end{pmatrix} \begin{pmatrix} p_{11} & p_{12} & p_{13} & \cdots \\ p_{21} & p_{22} & p_{23} & \cdots \\ p_{31} & p_{32} & p_{33} & \cdots \\ \vdots & \vdots & \vdots & \ddots \end{pmatrix}.
$$

Example 3.22 *Suppose we have the Markov chain determined by the urns in Figure 3.2. Then*

$$
\begin{pmatrix} r_1 & r_2 & r_3 \end{pmatrix} = \begin{pmatrix} r_1 & r_2 & r_3 \end{pmatrix} \begin{pmatrix} 1/6 & 1/2 & 1/3 \\ 3/10 & 2/5 & 3/10 \\ 1/2 & 1/3 & 1/6 \end{pmatrix}. \tag{3.5}
$$

Solving the system of equations determined by Equalities 3.2 and 3.5, we obtain

$$
\begin{pmatrix} r_1 & r_2 & r_3 \end{pmatrix} = \begin{pmatrix} 36/115 & 19/46 & 63/230 \end{pmatrix}.
$$

3.2.2   MCMC

Again our coverage is cursory. See [Hastings, 1970] for a more thorough introduction.

Suppose we have a finite set of states $\{e_1, e_2, \ldots e_s\}$, and a probability distribution $P(E = e_j) \equiv r_j$ defined on the states such that $r_j > 0$ for all j. Suppose further we have a function f defined on the states, and we wish to estimate

$$I = \sum_{j=1}^{s} f(e_j) r_j.$$

We can obtain an estimate as follows: Given we have a Markov chain with transition matrix $\mathbf{P}$ such that $\mathbf{r}^T = \begin{pmatrix} r_1 & r_2 & r_3 & \cdots \end{pmatrix}$ is its stationary distribution, we simulate the chain for trials $1, 2, \ldots M$. Then if k_i is the index of the state occupied at trial i, and

$$I' = \sum_{i=1}^{M} \frac{f(e_{k_i})}{M}, \tag{3.6}$$

the **ergodic theorem** says that $I' \to I$ with probability 1 (see [Tierney, 1996]). So, we can estimate I by I'. This approximation method is called **Markov chain Monte Carlo**. To obtain more rapid convergence, in practice a 'burn-in' number of iterations is used so that the probability of being in each state is approximately given by the stationary distribution. The sum in Equality 3.6 is then obtained over all iterations past the burn-in time. Methods for choosing a burn-in time and the number of iterations to use after burn-in are discussed in [Gilks et al, 1996].

It is not hard to see why the approximation converges. After a sufficient burn-in time, the chain will be in state e_j about r_j fraction of the time. So, if we do M iterations after burn in, we would have

$$\sum_{i=1}^{M} \frac{f(e_{k_i})}{M} \approx \sum_{j=1}^{s} \frac{f(e_j) r_j M}{M} = \sum_{j=1}^{s} f(e_j) r_j.$$

Example 3.23 *Suppose we have the Markov chain determined by the urns in Figure 3.2. Then, as shown in Example 3.22, the stationary distribution is*

$$\left(\begin{array}{ccc} r_1 & r_2 & r_3 \end{array} \right) = \left(\begin{array}{ccc} 36/115 & 19/46 & 63/230 \end{array} \right).$$

Suppose further we set $f(e_1) = 1$, $f(e_2) = 2$, and $f(e_3) = 3$. Then

$$\sum_{j=1}^{3} f(e_j) r_j = 1 \times \frac{36}{115} + 2 \times \frac{19}{46} + 3 \frac{63}{230} = 4.413.$$

We could approximate the value of 4.4286 by repeatedly sampling from the urns, and always choosing the next urn based on the ball sampled from the current urn. If, we take M such sample items after a burn-in period, then

$$\sum_{i=1}^{M} \frac{f(e_{k_i})}{M} \approx 4.413.$$

Of course, we would never do the approximation in the previous example because it is much easier just to compute the sum exactly. We would use an approximation when the number of states is forbiddingly large. Furthermore, in the previous example we started with a Markov chain and determined its stationary distribution. However, we must work in reverse. That is, we are starting with a probability distribution **r**, and we need to construct a Markov chain with transition matrix **P** such that **r** is its stationary distribution. Next we show two ways for doing this.

Metropolis-Hastings Method

The **Metropolis-Hastings method** proceeds as follows: We start with a probability distribution r and our goal is to create a Markov chain that has r as its stationary distribution. Owing to Theorem 3.6, we see from Equality 3.4

that we need only find an irreducible aperiodic chain such that its transition matrix $\mathbf{P}$ satisfies

$$\mathbf{r}^T = \mathbf{r}^T \mathbf{P}. \tag{3.7}$$

It is not hard to see that if we determine values p_{ij} such that for all i and j

$$r_i p_{ij} = r_j p_{ji} \tag{3.8}$$

the resultant $\mathbf{P}$ satisfies Equality 3.7. Towards determining such values, let $\mathbf{Q}$ be the transition matrix of an arbitrary Markov chain whose states are the members of our given finite set of states $\{e_1, e_2, \ldots e_s\}$, and let

$$\alpha_{ij} = \begin{cases} \dfrac{s_{ij}}{1 + \dfrac{r_i q_{ij}}{r_j q_{ji}}} & q_{ij} \neq 0,\ q_{ji} \neq 0 \\[4ex] 0 & q_{ij} = 0 \text{ or } q_{ji} = 0, \end{cases} \tag{3.9}$$

where s_{ij} is a symmetric function of i and j chosen so that $0 \leq \alpha_{ij} \leq 1$ for all i and j. We then take

$$\begin{aligned} p_{ij} &= \alpha_{ij} q_{ij} & i \neq j \\ p_{ii} &= 1 - \sum_{j \neq i} p_{ij}. \end{aligned} \tag{3.10}$$

Then for $i \neq j$

$$\begin{aligned} r_i p_{ij} &= r_i \alpha_{ij} q_{ij} \\[2ex] &= r_i \frac{s_{ij}}{1 + \dfrac{r_i q_{ij}}{r_j q_{ji}}} q_{ij} \\[2ex] &= r_i \frac{r_j q_{ji} s_{ij}}{r_j q_{ji} + r_i q_{ij}} q_{ij}, \end{aligned} \tag{3.11}$$

and

$$\begin{aligned} r_j p_{ji} &= r_j \alpha_{ji} q_{ji} \\[2ex] &= r_j \frac{s_{ji}}{1 + \dfrac{r_j q_{ji}}{r_i q_{ij}}} q_{ji} \\[2ex] &= r_j \frac{r_i q_{ij} s_{ij}}{r_i q_{ij} + r_j q_{ji}} q_{ij}. \end{aligned} \tag{3.12}$$

Equality 3.8 now follows from Equalities 3.11 and 3.12 because s_{ij} is symmetric. The irreducibility of $\mathbf{P}$ must be checked in each application.

Hastings [1970] suggests the following way of choosing $\mathbf{s}$: If q_{ij} and q_{ji} are both nonzero, set

$$s_{ij} = \begin{cases} 1 + \dfrac{r_i q_{ij}}{r_j q_{ji}} & \dfrac{r_j q_{ji}}{r_i q_{ij}} \geq 1 \\[3mm] 1 + \dfrac{r_j q_{ji}}{r_i q_{ij}} & \dfrac{r_j q_{ji}}{r_i q_{ij}} \leq 1 \end{cases} \quad . \tag{3.13}$$

Given this choice, we have

$$\alpha_{ij} = \begin{cases} 1 & q_{ij} \neq 0,\ q_{ji} \neq 0,\ \dfrac{r_j q_{ji}}{r_i q_{ij}} \geq 1 \\[3mm] \dfrac{r_j q_{ji}}{r_i q_{ij}} & q_{ij} \neq 0,\ q_{ji} \neq 0,\ \dfrac{r_j q_{ji}}{r_i q_{ij}} \leq 1 \\[3mm] 0 & q_{ij} = 0 \text{ or } q_{ji} = 0. \end{cases} \tag{3.14}$$

If we make $\mathbf{Q}$ symmetric (That is, $q_{ij} = q_{ji}$ for all i and j.), we have the method devised by Metropolis et al. [1953]. In this case

$$\alpha_{ij} = \begin{cases} 1 & q_{ij} \neq 0,\ r_j \geq r_i \\[2mm] r_j / r_i & q_{ij} \neq 0,\ r_j \leq r_i \\[2mm] 0 & q_{ij} = 0. \end{cases} \tag{3.15}$$

Note that with this choice if $\mathbf{Q}$ is irreducible so is $\mathbf{P}$.

Example 3.24 *Suppose* $\mathbf{r}^T = (\begin{array}{ccc} 1/8 & 3/8 & 1/2 \end{array})$. *Choose* $\mathbf{Q}$ *symmetric as follows:*

$$\mathbf{Q} = \begin{pmatrix} 1/3 & 1/3 & 1/3 \\ 1/3 & 1/3 & 1/3 \\ 1/3 & 1/3 & 1/3 \end{pmatrix}.$$

Choose $\mathbf{s}$ *according to Equality 3.13 so that* $\boldsymbol{\alpha}$ *has the values in Equality 3.15. We then have*

$$\boldsymbol{\alpha} = \begin{pmatrix} 1 & 1 & 1 \\ 1/3 & 1 & 1 \\ 1/4 & 3/4 & 1 \end{pmatrix}.$$

Using Equality 3.10 we have

$$\mathbf{P} = \begin{pmatrix} 1/3 & 1/3 & 1/3 \\ 1/9 & 5/9 & 1/3 \\ 1/12 & 1/4 & 2/3 \end{pmatrix}.$$

Notice that

$$\begin{aligned} \mathbf{r}^T\mathbf{P} &= (\begin{array}{ccc} 1/8 & 3/8 & 1/2 \end{array}) \begin{pmatrix} 1/3 & 1/3 & 1/3 \\ 1/9 & 5/9 & 1/3 \\ 1/12 & 1/4 & 2/3 \end{pmatrix} \\ &= (\begin{array}{ccc} 1/8 & 3/8 & 1/2 \end{array}) = \mathbf{r}^T \end{aligned}$$

as it should.

Once we have constructed matrices $\mathbf{Q}$ and $\boldsymbol{\alpha}$ as discussed above, we can conduct the simulation as follows:

1. Given the state occupied at the kth trial is e_i, choose a state using the probability distribution given by the ith row of $\mathbf{Q}$. Suppose that state is e_j.

2. Choose the state occupied at the $(k+1)$th trial to be e_j with probability α_{ij} and to be e_i with probability $1 - \alpha_{ij}$.

In this way, when state e_i is the current state, e_j will be chosen q_{ij} fraction of the time in Step (1), and of those times e_j will be chosen α_{ij} fraction of the time in Step (2). So, overall e_j will be chosen $\alpha_{ij}q_{ij} = p_{ij}$ fraction of the time (See Equality 3.10), which is what we want.

Gibb's Sampling Method

Next we show another method for creating a Markov chain whose stationary distribution is a particular distribution. The method is called **Gibb's sampling**, and it concerns the case where we have n random variables $X_1, X_2, \ldots X_n$ and a joint probability distribution P of the variables (as in Bayesian networks, introduced in Chapter 5). If we let $\mathbf{X} = \begin{pmatrix} X_1 & \cdots & X_n \end{pmatrix}^T$, we want to approximate

$$\sum_{\mathbf{x}} f(\mathbf{x}) P(\mathbf{x}).$$

To approximate this sum using MCMC, we need to create a Markov chain whose set of states is all possible values of $\mathbf{X}$, and whose stationary distribution is $P(\mathbf{x})$. We can accomplish this in the following way. We start with arbitrary values $x_1^{(1)}, x_2^{(1)}, \ldots x_n^{(1)}$ of the random variables $X_1, X_2, \ldots X_n$. Then we choose the values for mth iteration based on the values for the $(m-1)$th iteration as follows:

Sample $x_1^{(m)}$ using the distribution $P\left(x_1 | x_2^{(m-1)}, x_3^{(m-1)}, \ldots x_n^{(m-1)}\right)$.

Sample $x_2^{(m)}$ using the distribution $P\left(x_2 | x_1^{(m)}, x_3^{(m-1)}, \ldots x_n^{(m-1)}\right)$.

$$\vdots$$

Sample $x_k^{(m)}$ using the distribution $P\left(x_k | x_1^{(m)}, \ldots x_{k-1}^{(m)}, x_{k+1}^{(m-1)} \ldots x_n^{(m-1)}\right)$.

$$\vdots$$

Sample $x_n^{(m)}$ using the distribution $P\left(x_n | x_1^{(m)}, \ldots, x_{n-1}^{(m)}\right)$.

Notice that in the kth step, all variables except $x_k^{(m)}$ are unchanged, and the new value of $x_k^{(m)}$ is drawn from its distribution conditional on the current values of all the other variables.

As long as all conditional probabilities are nonzero, the chain is irreducible. Next we verify that $P(\mathbf{x})$ is the stationary distribution for the chain. If we let $p(\mathbf{x}; \hat{\mathbf{x}})$ denote the transition probability from $\mathbf{x}$ to $\hat{\mathbf{x}}$ in each iteration, we need to show

$$P(\hat{\mathbf{x}}) = \sum_{\mathbf{x}} P(\mathbf{x})p(\mathbf{x}; \hat{\mathbf{x}}). \qquad (3.16)$$

One can see that it suffices to show that Equality 3.16 holds for each iteration of each trial. To that end, for the kth step we have,

$$\sum_{\mathbf{x}} P(\mathbf{x})p_k(\mathbf{x}; \hat{\mathbf{x}})$$

$$= \sum_{x_1, \dots x_n} P(x_1, \dots x_n)p_k(x_1, \dots x_n; \hat{x}_1, \dots \hat{x}_n)$$

$$= \sum_{x_k} P(\hat{x}_1, \dots \hat{x}_{k-1}, x_k, \hat{x}_{k+1} \dots \hat{x}_n)P(\hat{x}_k | \hat{x}_1, \dots \hat{x}_{k-1}, \hat{x}_{k+1} \dots \hat{x}_n)$$

$$= P(\hat{x}_k | \hat{x}_1, \dots \hat{x}_{k-1}, \hat{x}_{k+1} \dots \hat{x}_n) \sum_{x_k} P(\hat{x}_1, \dots \hat{x}_{k-1}, x_k, \hat{x}_{k+1} \dots \hat{x}_n)$$

$$= P(\hat{x}_k | \hat{x}_1, \dots \hat{x}_{k-1}, \hat{x}_{k+1} \dots \hat{x}_n)P(\hat{x}_1, \dots \hat{x}_{k-1}, \hat{x}_{k+1} \dots \hat{x}_n)$$

$$= P(\hat{x}_1, \dots \hat{x}_{k-1}, \hat{x}_k, \hat{x}_{k+1} \dots \hat{x}_n)$$

$$= P(\hat{\mathbf{x}}).$$

The second step follows because $p_k(\mathbf{x}; \hat{\mathbf{x}}) = 0$ unless $\hat{x}_j = x_j$ for all $j \neq k$.

See [Geman and Geman, 1984] for more on Gibb's sampling.

3.3 The Normal Distribution

So far all of our probability distributions have been discrete. We will need to use the normal distribution in this text, which is a continuous probability distributions. Although we don't have space to review continuous distributions, we do briefly review properties of the normal distribution. You are referred to an introductory probability or statistics text such as [Anderson et al., 2007] if you are not familiar with continuous probability distributions.

The normal distribution is defined as follows:

Definition 3.5 *The **normal density function with parameters** μ and σ, where $-\infty < \mu < \infty$ and $\sigma > 0$, is*

$$\rho(x) = \frac{1}{\sqrt{2\pi}\sigma}e^{-\frac{(x-\mu)^2}{2\sigma^2}} \qquad -\infty < x < \infty, \qquad (3.17)$$

and is denoted $\text{NormalDen}(x; \mu, \sigma^2)$.

*A random variables X that has this density function is said to have a **normal distribution**.*

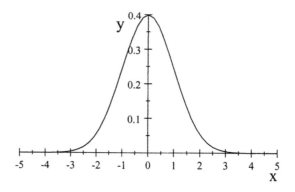

Figure 3.3: The standard normal density function.

If the random variable X has the normal density function, then

$$E(X) = \mu \quad \text{and} \quad V(X) = \sigma^2.$$

The density function NormalDen$(x; 0, 1^2)$ is called the **standard normal density function** . Figure 3.3 shows this density function.

EXERCISES

Section 3.1

Exercise 3.1 Suppose a fair six-sided die is to be rolled. The player will receive a number of dollars equal to the number of dots that show, except when five or six dots show, in which case the player will lose $5 or $6 respectively.

1. Compute the expected value of the amount of money the player will win or lose.

2. Compute the variance and standard deviation of the amount of money the player will win or lose.

3. If the game is repeated 100 times, compute the most the player could win, the most the player could lose, and the amount the player can expect to win or lose.

Exercise 3.2 Prove Theorem 3.1.

Exercise 3.3 Prove Theorem 3.2.

Exercise 3.4 Prove Theorem 3.3.

Exercise 3.5 Prove Theorem 3.5.

Exercise 3.6 Suppose we have the following data on 10 college students concerning their grade point averages (GPA) and scores on the graduate record exam (GRE).

Student	GPA	GRE
1	2.2	1400
2	2.4	1300
3	2.8	1550
4	3.1	1600
5	3.3	1400
6	3.3	1700
7	3.4	1650
8	3.7	1800
9	3.9	1700
10	4.0	1800

1. Develop a scatterplot of GRE versus GPA.

2. Determine the sample covariance.

3. Determine the sample correlation coefficient. What does this value tell us about the relationship between the two variables?

Exercise 3.7 Suppose a basketball player successfully makes 712 free-throws in 1048 attempts during a basketball season. Determine the maximum likelihood estimate of the probability that this player will successfully make a free-throw.

Section 3.2

Exercise 3.8 Suppose we have a Markov chain with the following transition matrix:

$$\begin{pmatrix} 1/5 & 2/5 & 2/5 \\ 1/7 & 4/7 & 2/7 \\ 3/8 & 1/8 & 1/2 \end{pmatrix}.$$

Determine the stationary distribution for the chain.

Exercise 3.9 Suppose we have the distribution $\mathbf{r}^T = \begin{pmatrix} 1/9 & 2/3 & 2/9 \end{pmatrix}$. Using the Metropolis-Hastings method, find a transition matrix for a Markov chain that has this as its stationary distribution. Do it both with a matrix $\mathbf{Q}$ that is symmetric and with one that is not.

Section 3.3

Exercise 3.10 Suppose we have the following normal density function:

$$\rho(x) = \frac{1}{\sqrt{2\pi}(3)} e^{-\frac{(x-8)^2}{2(3)^2}} \qquad -\infty < x < \infty.$$

What are the $E(X)$ and $Var(X)$?

Chapter 4

Genetics Basics

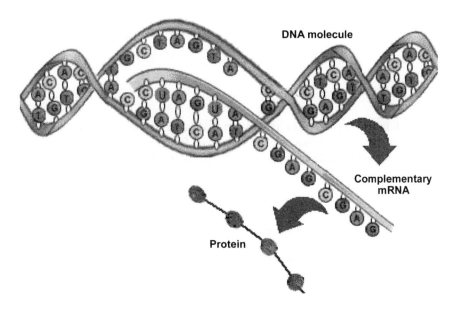

In this chapter we review the basics of genetics necessary to an understanding of the rest of this text. This is only a brief overview. You are referred to [Griffiths et al., 2007] for a thorough treatment.

4.1 Organisms and Cells

We start by discussing fundamental definitions concerning living entities.

4.1.1 Basic Definitions

An **organism** is an individual form of life such as a plant, animal, bacterium, protist, or fungus. A **cell** is the basic structural and functional unit of all organisms. The **cell nucleus** is a membrane-bound structure inside the cell.

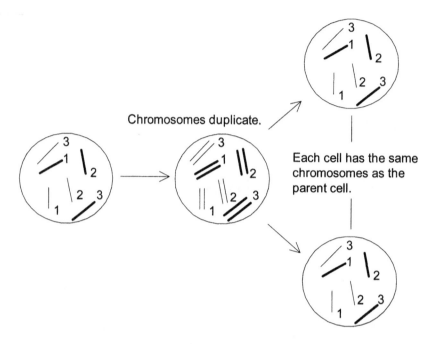

Figure 4.1: A depiction of mitosis in a diploid cell of an organism whose genome has three chromosomes. For the sake of illustration, we depict homologous chromosomes as straight segments. This is not how they actually exist in the nucleus.

Organisms that have cells that do not contain a nucleus are called **prokaryotes**. Most prokaryotes (e.g., bacteria) are unicellular. That is, the organism consists of a single cell. However, some (e.g., cyanobacteria) are multicellular. Organisms whose cells contain a nucleus are called **eukaryotes**. Some eukaryotes, such as the yeast schizosaccharomyces and protozoa such as the amoeba, are unicellular; others, such as the human, are multicellular.

A **chromosome** is a long threadlike macromolecule consisting of deoxyribonucleic acid (DNA). Chromosomes are the carriers of biologically expressed hereditary characteristics. A **locus** on a chromosome is a section of the chromosome that we distinguish. A locus that encodes a particular biologically expressed hereditary characteristic is called a **gene**, and we say the chromosomes carry the **genetic content** of the organism. (We discuss genes much more in Section 4.2.) In eukaryotes, chromosomes are found inside the nucleus. A **genome** is a complete set of chromosomes in an organism. The human genome contains 23 chromosomes. A **haploid cell** contains one genome; that is, it contains one set of chromosomes. So a human haploid cell contains 23 chromosomes. A **diploid cell** contains two genomes; that is, it contains two sets of chromosomes. Each chromosome in one set is matched with a chromosome in the other set. This pair of chromosomes is called a **homologous pair**.

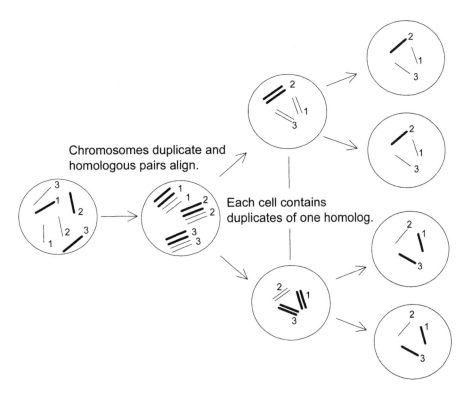

Chromosomes duplicate and homologous pairs align.

Each cell contains duplicates of one homolog.

Figure 4.2: A depiction of meiosis.

Each chromosome in the pair is called a **homolog**. So a human diploid cell contains $2 \times 23 = 46$ chromosomes.

A **somatic cell** is one of the cells in the body of the organism. A **haploid organism** is an organism whose somatic cells are haploid. Most prokaryotes are haploid; however, some bacteria are polypoid, meaning that they have more than two genomes. Some eukaryotes are haploid. For example, fungi and protozoa such as the amoeba are haploid eukaryotes. A **diploid organism** is an organism whose somatic cells are diploid. Humans are diploid organisms.

A **gamete** is a mature sexual reproductive cell that unites with another gamete to become a **zygote**, which eventually grows into a new organism. A gamete is always haploid. The gamete produced by a male is called a **sperm**, whereas the gamete that is produced by the female is called an **egg**. **Germ cells** are precursors of gametes. They are diploid.

4.1.2 Cell Division in Eukaryotes

Mitosis is the process by which somatic cells (i.e., cells not destined to become gametes) divide to produce new cells. Figure 4.1 illustrates mitosis in the case of a diploid cell in an organism whose genome has three chromosomes, labeled

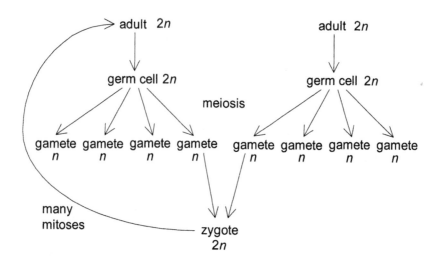

Figure 4.3: The life cycle for a diploid species whose genome contains n chromosomes. The number of chromosomes in each cell is shown.

1, 2, and 3. In that figure, the chromosomes in one of the genomes are depicted as thick lines, whereas the chromosomes in the other genome are depicted as thin lines. (Since we are only interested in computational biology, in this and all figures in this text there is no effort to make the figures look true to life or to show details of a process that are not relevant to our purposes.) In mitosis each chromosome in the parent cell first divides into two copies of itself. Initially these copies are actually attached and are called **chromatids**. One copy goes to each child cell. So the child cell has the exact same chromosomes as the parent cell. Technically, the stage in which the genetic material duplicates is not part of mitosis, but for our computational purposes we need not make that distinction.

Meiosis is the process by which germ cells divide to become gametes. Figure 4.2 illustrates meiosis in an organism whose genome has three chromosomes, labeled 1, 2, and 3. First each chromosome duplicates itself and aligns with its homolog. As is the case in mitosis, at this stage the duplicates are actually attached and are called **chromatids**. After the first cell division, each child cell contains duplicates of one of the homologs in the parent cell. After the second cell division, each cell (gamete) contains one of the homologs in the parent cell. Each gamete obtains one member of each homologous pair of chromosomes. Since there are three homologous pairs, this means there are $2^3 = 8$ different gametes that could possibly be produced. Note that exactly two of these possibilities will be realized in the four gametes that are actually produced. That is, one pair of gametes will have one genetic content, while the other pair will have a second genetic content. Since the human genome contains 23 homologous pairs, there are $2^{23} = 8,388,608$ different gametes a given human could produce. Owing to crossing-over, which will be discussed shortly, the number

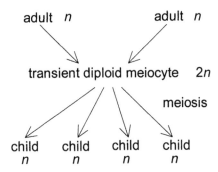

Figure 4.4: Haploid sexual reproduction by binary fusion. The genome contains n chromosomes. The adults combine to produce a meiocyte.

is even larger than that.

4.1.3 Reproduction

A **species** is a fundamental category of biological classification of organisms. Organisms in the same species have similar appearance and genetic content. For organisms that engage in sexual reproduction, ordinarily members of the same species are capable of breeding among themselves but are not capable of breeding with members of another species. Next we discuss reproduction and the life cycle for diploid and haploid species.

Diploid Species

Figure 4.3 shows the life cycle for a diploid species whose genome has size n. Each adult produces a gamete, the two gametes combine to form a zygote, and the zygote grows to become a new adult. This process is called **sexual reproduction**.

Haploid Species

Some unicellular haploid organisms such as the amoeba reproduce asexually by a process called **binary fission**. The unicellular organism splits by mitosis into two new organisms. So, each new organism has the exact same genetic content as the original organism.

Some unicellular haploid organisms (e.g., some of the protozoa) reproduce sexually by a process called **fusion**. This process is illustrated in Figure 4.4. Two adult cells themselves first combine to form a transient diploid meiocyte. The transient diploid meiocyte contains a homologous pair of chromosomes, one from each parent. A given child can obtain a given homolog from each parent. So the children are not genetic copies of the parents. For example, if the genome size is 3, there are $2^3 = 8$ different chromosome combinations that

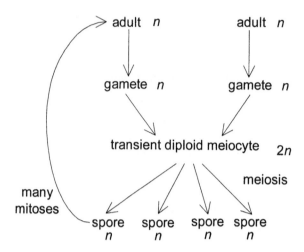

Figure 4.5: Sexual reproduction in multicellular haploid organisms. The genome contains n chromosomes. The adults produce gametes, which then combine to produce a meiocyte.

a child could have. Note that exactly two of these possibilities will be realized in the four children. That is, one pair of children will have one genetic content, whereas the other pair will have a second genetic content. Look again at Figure 4.2 to see why this is the case.

Multicellular haploid organisms such as the pink bread mold *Neurospora* undergo sexual reproduction, as illustrated in Figure 4.5. The difference between this form of reproduction and fusion (as discussed previously) is that the haploid adult produces a haploid gamete which combines with another gamete to become a meiocyte. In the case of fusion the two adult cells themselves combine.

4.2   Genes

A **gene** is a locus (section of a chromosome) that is functional in the sense that it is responsible for some biological expression in the organism. Genes are responsible for both the structure and the processes of the organism. Many organisms have tens of thousands of genes on each chromosome. Here we discuss the behavior of genes from a high-level perspective; then we review their chemical structure and low-level behavior.

4.2.1   High-Level Gene Behavior

An **allele** is any of several forms of a gene, usually arising through mutation. Alleles are responsible for hereditary variation. The following classic example illustrates this concept.

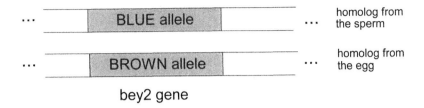

Figure 4.6: Sections of two homologous chromosomes. The allele for blue eyes comes from the father, and the allele for brown eyes comes from the mother.

Example 4.1 *The bey2 gene on chromosome 15 is responsible for eye color in humans. There is one allele for blue eyes, which we call BLUE, and one for brown eyes, which we call BROWN. As is the case for all genes, an individual gets one allele from each parent. Figure 4.6 shows an example in which the BLUE allele comes from the father and the BROWN allele comes from the mother. The BLUE allele is* **recessive**. *This means that if an individual receives one BLUE allele and one BROWN allele, that individual will have brown eyes. The only way the individual could have blue eyes would be for the individual to have two BLUE alleles. We also say that the brown allele is* **dominant**.

The previous example is over-simplified for the sake of illustration. There are actually several genes related to eye color. For example, a second gene for eye color, the gey gene located on chromosome 19 has green and blue alleles, with green dominant over blue. An individual with at least one bey2 brown allele will have brown eyes regardless of the gey gene. An individual with two bey2 blue alleles and at least one gey green allele will have green eyes.

The **genotype** of an organism is its genetic makeup; the **phenotype** of an organism is its appearance resulting from the interaction of the genotype and the environment. So if one individual had a BLUE allele and a BROWN allele, and another individual had two BROWN alleles, they would have different genotypes relative to the bey2 gene. However, for both of them the phenotype relative to eye color would be brown eyes.

Mammals have a pair of chromosomes, labeled X and Y, that share a small region of homology. The human, for example, has 22 homologous pairs of chromosomes plus either two X chromosomes or an X and a Y chromosome. These chromosomes determine the sex of the individual. Females have two X chromosomes; males have an X chromosome and a Y chromosome. Thus a sperm cell, which is produced by the male, can carry either an X or a Y chromosome, whereas an egg cell, which is produced by the female, always carries an X chromosome. We see that the male determines the sex of the offspring. The X chromosome carries many genes with function unrelated to sexual determination, whereas the Y chromosome carries few genes other than those related to male determination. A characteristic determined by a gene that is located on the X chromosome is called a **sex-linked characteristic**.

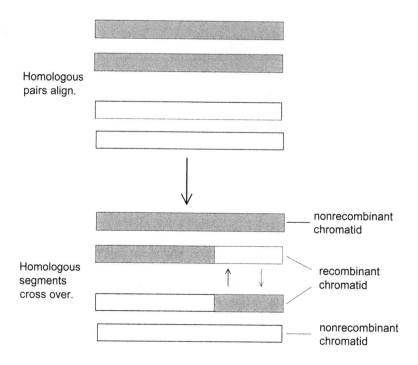

Figure 4.7: An illustration of crossing-over.

Example 4.2 *The cones in the retina of the human eye are sensitive to blue, red, and green light. These are our primary colors. All other colors are perceived as mixtures of these three. The gene for blue sensitivity is located on chromosome 7; the genes for red and green sensitivity are located on the X chromosome. So red-green color vision is a sex-linked characteristic. The mechanism by which the genes for red and green sensitivity result in red-green color blindness is a bit complex. See [Hartl and Jones, 2006] for a complete discussion. For our purposes, let NORMAL denote that the X chromosome carries normal red-green color vision and DEFECT denote that the X chromosome carries a defect resulting in red-green color blindness. DEFECT is recessive to NORMAL. So for a woman to be red-green color blind, she must have the DEFECT condition on both her X chromosomes. For a man to be red-green color blind, he need only have the DEFECT condition on his single X chromosome. (The Y chromosome does not carry the genes for red and green sensitivity.) For this reason, about 8% of human males are red-green color blind, while only about 0.5% of females are.*

We mentioned previously that, since a human gamete has 23 chromosomes and each of these chromosomes can come from either genome, there are $2^{23} = 8,388,608$ different genetic combinations a parent could pass on to his or her offspring. Actually, there are many more than this, for the following reason.

Figure 4.8: A section of DNA.

Recall that during meiosis homologous pairs of chromosomes line up. Often the chromatids from the two homologs that are next to each other exchange corresponding segments of genetic material. This exchange, which is called **crossing-over**, is illustrated in Figure 4.7. At a computational level, crossing-over occurs because each recombinant chromatid breaks at the same point and then recombines with the loose segment from the other recombinant chromatid. So, every allele on one of these segments moves to a chromatid of the other member of the homologous pair.

Gene conversion (nonreciprocal recombination) is similar to crossing-over except that one of the homologs has a segment of chromosome replaced by the homologous segment from the other homolog, whereas the other homolog is unchanged.

4.2.2 Low-Level Gene Behavior

Next we discuss the chemical structure of genes and how they behave at the molecular level.

DNA

Chromosomes consist of the compound **deoxyribonucleic acid (DNA)**. DNA is composed of four basic molecules called **nucleotides**. Each nucleotide contains a pentose sugar (deoxyribose), a phosphate group, and a purine or pyrimidine base. The **purines**, adenine (A) and guanine (G), are similar in structure, as are the **pyrimidines**, cytosine (C) and thymine (T). DNA is a macromolecule composed of two complementary strands, each strand consisting of a sequence of nucleotides. The strands are joined together by hydrogen bonds between pairs of nucleotides. Adenine always pairs with thymine, and guanine always pairs with cytosine. Each such pair is called a **canonical base pair (bp)**, and A, G, C, and T are called **bases**. Authors often used the terms *nucleotide* and *base* interchangeably.

A section of DNA is depicted in Figure 4.8. You may recall from your biology course that the strands twist around each other to form a right-handed double helix. However, for our computational purposes, we need only consider them as character strings as shown in Figure 4.8. It is believed that a chromosome is just one long DNA molecule. As illustrated in Figures 4.1 and 4.2, when a cell divides, each chromosome first duplicates itself. This is accomplished as follows: The two strands of DNA separate, and then each strand serves as a template

··· U U C A G G C ···

Figure 4.9: A section of RNA.

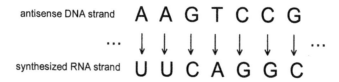

Figure 4.10: The transcription process.

for synthesis of a new complementary strand. This is called **semiconservative replication** because each new chromosome consists of an original parent strand and a newly synthesized child strand.

Ribonucleic acid (RNA) differs from DNA by having ribose instead of deoxyribose as the sugar, having the base uracil (U) instead of thymine, and being single-stranded. A section of RNA is depicted in Figure 4.9.

Genes and DNA

A **gene** is a section of a chromosome, often consisting of thousands of base pairs, but the size of genes varies a great deal. Most genes, called **protein-coding genes**, produce a **protein**, which is a macromolecule composed of **amino acids**. For example, the human genome has approximately 30,000 genes, and over 26,000 are protein-coding. It is the protein that then actually does the work in the cell. The coding regions of a gene, which are the regions that determine the resultant protein, are called **exons**; any space between two exons is called an **intron**.

Let's look at a simplified description of the means by which a gene produces a protein. First, the gene produces **messenger RNA (mRNA)** by a process called **transcription**. The transcription process starts when as many as 50 different **protein transcription factors** bind to **promoter** sites on the DNA. An enzyme, called **RNA polymerase**, then binds to the complex of transcription factors. Working together, they open the DNA double helix. One of the resultant strands of DNA is called the **antisense strand**; the complementary strand is called the **sense strand**. The **antisense strand** synthesizes a single strand of RNA called **precursor messenger RNA (pre-mRNA)**. The original DNA duplex is then rebuilt by reconnecting the hydrogen bonds. Finally, the intron space is removed from the pre-mRNA (along with some other tasks), resulting in **messenger RNA (mRNA)**. The result of the transcription process is illustrated in Figure 4.10. In the transcription process, multiple RNA polymerases can bind on a single DNA template, resulting in multiple

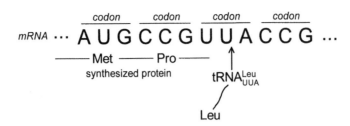

Figure 4.11: A depiction of translation.

rounds of transcription.

The section of mRNA corresponding to a given gene then produces the protein corresponding to that gene. The process by which the mRNA strand synthesizes the protein is called **translation**. Translation takes place as follows: A mRNA molecule consists of triplets of sequential nucleotides called **codons**. Each codon codes for a unique amino acid. For example, the codon UUA codes for the amino acid leucine (Leu). During translation, in sequence each codon synthesizes one amino acid in the protein. This is accomplished via the work of a **transfer RNA (tRNA)** molecule. A tRNA molecule is a small molecule that has associated with it both a codon and a protein. Its job is to deliver the protein to the codon. A given tRNA molecule is denoted with the protein as a superscript and the codon as a subscript. For example, the tRNA molecule that delivers the protein leucine to the codon UUA is denoted $tRNA_{UUA}^{Leu}$. The entire protein is synthesized by the sequential delivery of amino acids to the codons by tRNA molecules. This process is illustrated in Figure 4.11.

The means by which a gene produces a protein, which we just described, is summarized in the figure at the beginning of this chapter.

Since there are four different nucleotides, the number of different codons is $4 \times 4 \times 4 = 64$. Three of these codons, namely UUA, UAG, and UGA, are **stop codons**, which signal a termination to translation. The remaining 61 codons code for amino acids and are called **sense codons**. Since there are only 20 amino acids, a given amino acid can have several codons that code for it. For example, GUU, GUC, GUA, and GUG all code for valine (Val). Codons that code for the same amino acid are called **synonymous**. With a few exceptions, the correspondence between the codons and the amino acids is the same in all organisms. For that reason it is called the **universal genetic code** (in spite of the few exceptions). Tables showing this code can be found in many texts (e.g., [Li, 1997]).

Since all cells in an organism contain the same genetic code, you might be wondering how we obtain differentiation in structure and processes. The level of mRNA produced by a gene is called the **gene expression level**. Recall that the transcription process starts with protein transcription factors binding to promoter sites on the DNA. Although all cells contain the same genetic code, their protein composition is quite different. Thus different cells have different protein transcription factors, which means that the same gene in one cell may

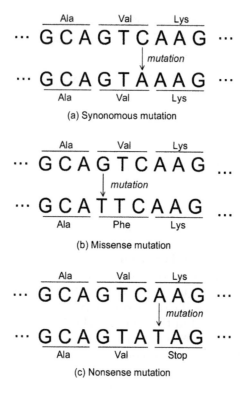

Figure 4.12: Types of substitution mutations.

be expressed at a different level than it is in another cell. We see that the protein produced by one gene can have a causal effect on the level of mRNA of another gene. You might still be wondering how all this starts with the zygote. That is, when the zygote, which is a single cell, first divides into two cells, how do these two cells obtain different protein content? This occurs due to asymmetric localization of cytoplasmic molecules (proteins and mRNA) within the cell before it divides. For example, in the fruit fly (*Drosophila melanogaster*) bicoid mRNA from the mother is translated into bicoid protein in the zygote. The protein diffuses through the egg, forming a gradient. The result is a high concentration of bicoid protein at one end of the egg and a low concentration at the other end. Bicoid is a transcription factor that turns on different genes at different levels.

4.3 Mutations

Recall that during cell division each chromosome replicates itself, with each new chromosome containing one strand from the parent chromosome. Sometimes errors occur during the DNA replication process. These errors are called

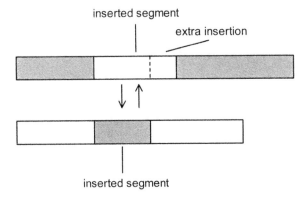

inserted segment

extra insertion

inserted segment

Figure 4.13: Unequal crossing-over.

mutations. Mutations can occur in either somatic cells or germ cells. It is believed that mutations in germ cells are the source of all variation in evolution. On the other hand, mutations in a somatic cell could affect the organism (e.g., cause cancer) but would have no effect on the offspring. Note that in the case of reproduction in haploids by binary fission and by fusion (see Section 4.1.3) there is no distinction between somatic cells and germ cells. Next we review a few of the various types of mutations.

4.3.1 Substitution Mutations

In a substitution mutation, one nucleotide is simply replaced by another. A **transition** substitution mutation is a change between an A and a G (purines) or a C and a T (pyrimidines). A **transversion** substitution is a change between a purine and a pyrimidine. A mutation is called **synonymous** if it causes no change in the amino acid produced. That is, the mutated codon codes for the same amino acid as the original codon. Otherwise, it is called **nonsynonymous**. A **missense** nonsynonymous mutation changes the codon to one that codes for a different amino acid. A **nonsense** nonsynonymous mutation changes the codon into a stop codon. These different types of mutations are illustrated in Figure 4.12.

4.3.2 Insertions and Deletions

An **insertion mutation** occurs when a section of DNA is added to a chromosome; a **deletion mutation** occurs when a section of DNA is removed from a chromosome. One cause of both types of mutation is **unequal crossing-over**. Recall that crossing-over occurs when two homologous chromosomes exchange segments of genetic material. If one homolog gives a longer segment than it receives, we have unequal crossing-over. This is illustrated in Figure 4.13. In that figure the bottom chromosome gives a longer segment than it

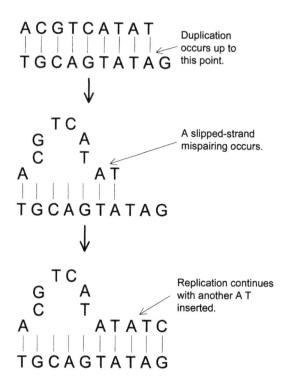

Figure 4.14: An insertion due to slip-strand mispairing.

receives, resulting in a deletion mutation for that chromosome and an insertion mutation for the top chromosome.

Another source of insertions and deletions is **replication slippage** (also called **slip-strand mispairing**) during DNA replication. An example is illustrated in Figure 4.14. This type of mutation occurs in DNA regions containing contiguous short repeats. In Figure 4.14 the top strand is being synthesized by the bottom strand. The bottom strand contains the sequence TATA which consists of a TA repeat. After the ATAT sequence is synthesized, the top strand "slips", and the second AT in the top strand becomes paired with the first TA in the bottom strand. Then the second TA in the bottom strand synthesizes another AT, resulting in an insertion in the top strand. Depending on the direction of the slippage, we can have either an insertion or a deletion.

Some DNA sequences, called **transposable elements**, possess the intrinsic ability to move from one genomic location, called the **donor site**, to another genomic location, called the **target site**. This type of insertion mutation is called **transposition**. When the sequence is retained at the donor site, we call the transposition **conservative**, and when it is excised from the donor site, we call the transposition **duplicative**. So, duplicative transposition increases the number of copies of the transposable element.

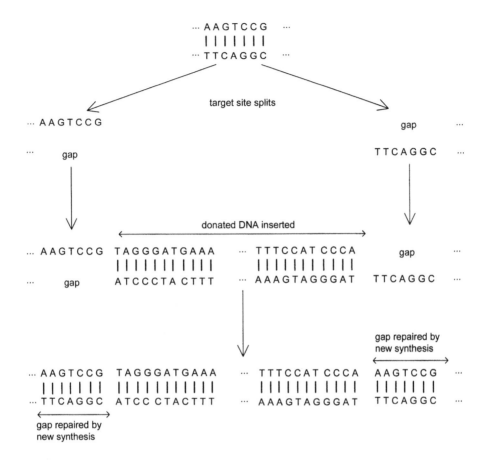

Figure 4.15: An illustration of transposition.

Figure 4.15 illustrates the process of transposition. First the target site splits open, resulting in a sequence of unpaired nucleotides at each end of the two pieces. That is, there is a gap in one of the strands where there are no nucleotides. Then the transposable element is inserted, and finally the synthesis of new nucleotides repairs the gaps. Ordinarily, transposable elements have **inverted repeats** at each of their ends. Figure 4.15 shows the transposable element called *Ds* in maize (corn). At one end we have the sequence TAGGGATGAAA on the top from left to right; on the other end we have the same sequence from right to left on the bottom. Transposable elements are not uncommon. About 50% of the human genome consists of transposable elements, most of them being evolutionary remnants no longer capable of transposing.

Horizontal gene transfer is a form of transposition in which genetic material is transferred from one species to another. A common form of horizontal gene transfer in bacteria is **transformation**, which consists of the following steps:

1. DNA is released to the environment.

2. The recipient cell enters a competent state for receipt of the DNA.

3. The recipient cell interacts with the DNA.

4. The DNA enters the recipient cell.

5. The DNA is integrated into the recipient cell's genome.

Another type of horizontal gene transfer is **conjugation**, in which the donor organism and recipient organism must physically interact. In bacterium-to-bacterium conjugation, a plasmid transports the DNA from the host to the recipient. **Transduction** is similar to conjugation in that a vehicle transports the DNA, but the host and recipient need not be in physical contact. For example, in bacteria, a bacteriophage (bacteria virus) encapsulates DNA from the host, and when the phage attaches to the recipient, the DNA is injected into the recipient. Horizontal gene transfer from eukaryotes to prokaryotes ordinarily occurs by transduction, whereas horizontal gene transfer from prokaryotes to eukaryotes ordinarily occurs by conjugation. Horizontal gene transfer can occur from eukaryotes to eukaryotes by a form of transduction in which a retrovirus is the vehicle. A retrovirus is a virus consisting of RNA rather than DNA, which has the ability to perform a reversal of genetic transcription. That is, it can transcribe from RNA to DNA rather than the usual DNA to RNA. The newly transcribed DNA is incorporated into the host cell's genome. The virus then replicates as part of the cell's DNA.

Some researchers believe transposition, in particular horizontal gene transfer, has played a significant role in evolution. See [Li, 1997] for a discussion.

4.3.3   Mutation Rates

It is estimated that the average rate of mutation in mammalian DNA is about 3 to 5×10^{-9} nucleotide substitutions per nucleotide site per year.

In the Rous sarcoma virus the rate is estimated to be about 1.4×10^{-4}. The human genome has about 3×10^9 nucleotides, so a given egg cell in a human undergoes about one mutation a year.

EXERCISES

Section 4.1

Exercise 4.1 Suppose a diploid organism has 15 chromosomes in one of its genomes.

1. How many chromosomes are in each of its somatic cells?

2. How many chromosomes are in each of its gametes?

3. Suppose one of its somatic cells undergoes mitosis.

   (a) How many cells will be produced?

   (b) Will these cells all have the same genetic content?

4. Suppose one of its germ cells undergoes meiosis.

   (a) How many germ cells will be produced?

   (b) Will these germs cells all have the same genetic content?

Exercise 4.2 Suppose two adult haploid organisms reproduce by fusion.

1. How many children will be produced?

2. Will the genetic content of the children all be the same?

Section 4.2

Exercise 4.3 Consider the eye color of a human being as determined by the bey2 gene. Recall that the allele for brown eyes is dominant. For each of the following parent allele combinations, determine the eye color of the individual.

Father	Mother
BLUE	BLUE
BLUE	BROWN
BROWN	BLUE
BROWN	BROWN

Exercise 4.4 Consider red-green color blindness in human beings. Assume the simplified model in which there is a single gene for red-green color sensitivity, the gene has two alleles, NORMAL and DEFECT, and DEFECT is recessive to NORMAL.

1. For each of the following parent allele combinations, determine whether a female is color-blind.

X Chromosome from Father	X Chromosome from Mother
NORMAL	NORMAL
NORMAL	DEFECT
DEFECT	NORMAL
DEFECT	DEFECT

2. For each of the following alleles on the single X chromosome received from the mother, determine whether a male is color-blind (recall that the father gives him a Y chromosome rather than an X chromosome).

| X Chromosome from Mother |
| NORMAL |
| DEFECT |

Exercise 4.5 Suppose that there is p fraction of DEFECT alleles in the population.

1. What fraction of females will be color-blind?

2. What fraction of males will be color-blind?

Exercise 4.6 During cell division, how many double-stranded DNA molecules result from each double-stranded DNA molecule in the original cell?

Exercise 4.7 During transcription, how many single-stranded RNA molecules are produced from each double-stranded DNA molecule?

Exercise 4.8 Do the following:

1. State the number of different codons.

2. State the number of codons that are not stop codons.

3. State the number of different amino acids.

4. Explain why we have more codons than amino acids.

Section 4.3

Exercise 4.9 Define a mutation.

Exercise 4.10 If a mutation occurs in a somatic cell of an individual, can the genotype of its offspring be affected?

Exercise 4.11 If a mutation occurs in a germ cell of an individual, can the genotype of its offspring be affected? State whether the phenotype of its offspring could be affected for each of the following types of mutations in a germ cell:

1. Synonymous mutation.

2. Missense mutation.

3. Nonsense mutation.

Exercise 4.12 Determine the expected value of the number of mutations at a given site in $10,000$ years for the following mutation rates:

1. 5×10^{-9} nucleotide substitutions per nucleotide site per year (typical in mammals).

2. 1.4×10^{-4} nucleotide substitutions per nucleotide site per year (typical in the Rous sarcoma virus).

Part II

Bayesian Networks

Chapter 5

Foundations of Bayesian Networks

The Reverend Thomas Bayes (1702–1761) developed Bayes' Theorem in the eighteenth century. Since that time the theorem has had a great impact on statistical inference because it enables us to infer the probability of a cause when its effect is observed. In the 1980s, the method was extended to model the probabilistic relationships among many causally related variables. The graphical structures that describe these relationships have come to be known as *Bayesian networks*. This chapter introduces these networks. (Applications of Bayesian networks to bioinformatics appear in Part III.) In Sections 5.1 and 5.2 we define Bayesian networks and discuss their properties. Section 5.3 shows how causal graphs often yield Bayesian networks. In Section 5.4 we discuss doing probabilistic inference using Bayesian networks. Section 5.5 introduces Bayesian networks containing continuous variables.

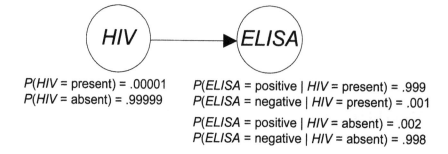

$P(HIV = \text{present}) = .00001$
$P(HIV = \text{absent}) = .99999$

$P(ELISA = \text{positive} \mid HIV = \text{present}) = .999$
$P(ELISA = \text{negative} \mid HIV = \text{present}) = .001$

$P(ELISA = \text{positive} \mid HIV = \text{absent}) = .002$
$P(ELISA = \text{negative} \mid HIV = \text{absent}) = .998$

Figure 5.1: A two-node Bayesian network.

5.1   What Is a Bayesian Network?

Recall that in Example 2.30, we computed the probability of Joe having the HIV virus, given that he tested positive for it using the ELISA test. Specifically, we knew that

$$P(ELISA = positive | HIV = present) = .999$$

$$P(ELISA = positive | HIV = absent) = .002,$$

and

$$P(HIV = present) = .00001.$$

We then employed Bayes' Theorem to compute

$P(present|positive)$

$$= \frac{P(positive|present)P(present)}{P(positive|present)P(present) + P(positive|absent)P(absent)}$$

$$= \frac{(.999)(.00001)}{(.999)(.00001) + (.002)(.99999)}$$

$$= .00497.$$

We summarize the information used in this computation in Figure 5.1, which is a two-node/variable Bayesian network. Notice that it represents the random variables HIV and $ELISA$ by nodes in a directed acyclic graph (DAG) and the causal relationship between these variables with an edge from HIV to $ELISA$. That is, the presence of HIV has a causal effect on whether the test result is positive; so there is an edge from HIV to $ELISA$. Besides showing a DAG representing the causal relationships, Figure 5.1 shows the prior probability distribution of HIV and the conditional probability distribution of $ELISA$ given each value of its parent HIV. In general, Bayesian networks consist of a DAG, whose edges represent relationships among random variables that are often (but not always) causal; the prior probability distribution of every variable that is a root in the DAG; and the conditional probability distribution of every

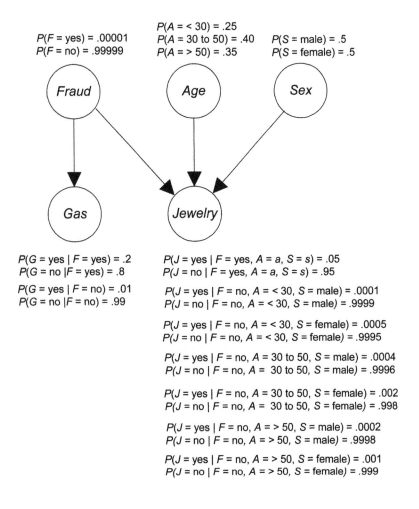

Figure 5.2: Bayesian network for detecting credit card fraud.

non-root variable given each set of values of its parents. We use the terms *node* and *variable* interchangeably in discussing Bayesian networks.

Let's illustrate a more complex Bayesian network by considering the problem of detecting credit card fraud (taken from [Heckerman, 1996]). Suppose that we have identified the following variables as being relevant to the problem:

Variable	What the Variable Represents
Fraud (F)	Whether the current purchase is fraudulent
Gas (G)	Whether gas has been purchased in the last 24 hours
Jewelry (J)	Whether jewelry has been purchased in the last 24 hours
Age (A)	Age of the card holder
Sex (S)	Sex of the card holder

These variables are all causally related. That is, a credit card thief is likely to buy gas and jewelry, and middle-aged women are most likely to buy jewelry, whereas young men are least likely to buy jewelry. Figure 5.2 shows a DAG representing these causal relationships. Notice that it also shows the conditional probability distribution of every non-root variable given each set of values of its parents. The Jewelry variable has three parents, and there is a conditional probability distribution for every combination of values of those parents. The DAG and the conditional distributions together constitute a Bayesian network.

You could have a few questions concerning this Bayesian network. First you might ask, "What value does it have?" That is, what useful information can we obtain from it? Recall how we used Bayes' Theorem to compute $P(HIV = present|ELISA = positive)$ from the information in the Bayesian network in Figure 5.1. Similarly, we can compute the probability of credit card fraud given values of the other variables in this Bayesian network. For example, we can compute $P(F = yes|G = yes, J = yes, A = \; < 30, S = female)$. If this probability is sufficiently high, we can deny the current purchase or require additional identification. The computation is not a simple application of Bayes' Theorem as was the case for the two-node Bayesian network in Figure 5.1. Rather it is done using sophisticated algorithms.

Second, you might ask how we obtained the probabilities in the network. They can either be obtained from the subjective judgements of an expert in the area or be learned from data. (In Chapter 8 we discuss techniques for learning them from data.)

Finally, you could ask why we are bothering to include the variables for age and sex in the network when the age and sex of the card holder has nothing to do with whether the card has been stolen (fraud). That is, fraud has no causal effect on the card holder's age or sex, and vice versa. The reason we include these variables is quite subtle. It is because fraud, age, and sex all have a common effect, namely the purchasing of jewelry. So, when we know jewelry has been purchased, the three variables are rendered probabilistically dependent owing to what psychologists call **discounting**. For example, if jewelry has been purchased in the last 24 hours, it increases the likelihood of fraud. However, if the card holder is a middle-aged woman, the likelihood of fraud is lessened (discounted) because such women are prone to buying jewelry. That is, the fact that the card holder is a middle-aged woman explains the jewelry purchase. On the other hand, if the card holder is a young man, the likelihood of fraud is increased because such men are unlikely to purchase jewelry.

We have informally introduced Bayesian networks, their properties, and their usefulness. Next we formally develop their mathematical properties.

5.2 Properties of Bayesian Networks

After defining Bayesian networks, we show how they are ordinarily represented.

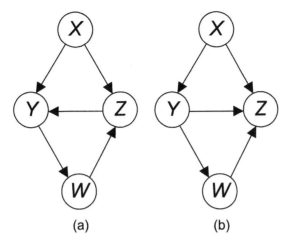

(a) (b)

Figure 5.3: Both graphs are directed graphs; only the one in (b) is a directed acyclic graph.

5.2.1 Definition of a Bayesian Network

First, let's review some graph theory. A **directed graph** is a pair (V, E), where V is a finite, nonempty set whose elements are called **nodes** (or vertices), and E is a set of ordered pairs of distinct elements of V. Elements of E are called **directed edges**, and if $(X, Y) \in \mathsf{E}$, we say there is an edge from X to Y. The graph in Figure 5.3 (a) is a directed graph. The set of nodes in that figure is

$$\mathsf{V} = \{X, Y, Z, W\},$$

and the set of edges is

$$\mathsf{E} = \{(X, Y), (X, Z), (Y, W), (W, Z), (Z, Y)\}.$$

A **path** in a directed graph is a sequence of nodes $[X_1, X_2, \ldots, X_k]$ such that $(X_{i-1}, X_i) \in \mathsf{E}$ for $2 \leq i \leq k$. For example, $[X, Y, W, Z]$ is a path in the directed graph in Figure 5.3 (a). A **chain** in a directed graph is a sequence of nodes $[X_1, X_2, \ldots, X_k]$ such that $(X_{i-1}, X_i) \in \mathsf{E}$ or $(X_i, X_{i-1}) \in \mathsf{E}$ for $2 \leq i \leq k$. For example, $[Y, W, Z, X]$ is a chain in the directed graph in Figure 5.3 (b), but it is not a path. A **cycle** in a directed graph is a path from a node to itself. In Figure 5.3 (a) $[Y, W, Z, Y]$ is a cycle from Y to Y. However, in Figure 5.3 (b) $[Y, W, Z, Y]$ is not a cycle because it is not a path. A directed graph $\mathbb{G}$ is called a **directed acyclic graph** (DAG) if it contains no cycles. The directed graph in Figure 5.3 (b) is a DAG, whereas the one in Figure 5.3 (a) is not.

Given a DAG $\mathbb{G} = (\mathsf{V}, \mathsf{E})$ and nodes X and Y in V, Y is called a **parent** of X if there is an edge from Y to X, Y is called a **descendent** of X and X is called an **ancestor** of Y if there is a path from X to Y, and Y is called a

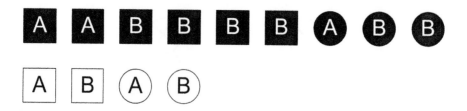

Figure 5.4: The random variables L and S are not independent, but they are conditionally independent given C.

nondescendent of X if Y is not a descendent of X and Y is not a parent of X.[1]

We can now state the following definition.

Definition 5.1 *Suppose we have a joint probability distribution P of the random variables in some set V and a DAG $\mathbb{G} = (V, E)$. We say that $(\mathbb{G}, P)$ satisfies the **Markov condition** if for each variable $X \in V$, X is conditionally independent of the set of all its nondescendents given the set of all its parents. Using the notation established in Chapter 2, Section 2.2.2, this means that if we denote the sets of parents and nondescendents of X by PA_X and ND_X, respectively, then*

$$I_P(X, ND_X | PA_X).$$

*If $(\mathbb{G}, P)$ satisfies the Markov condition, $(\mathbb{G}, P)$ is called a **Bayesian network**.*

Example 5.1 *Recall Chapter 2, Figure 2.1, which appears again as Figure 5.4. In Chapter 2, Example 2.23 we let P assign $1/13$ to each object in the figure, and we defined these random variables on the set containing the objects.*

Variable	Value	Outcomes Mapped to This Value
L	l_1	All objects containing an A
	l_2	All objects containing a B
S	s_1	All square objects
	s_2	All circular objects
C	c_1	All black objects
	c_2	All white objects

We then showed that L and S are conditionally independent given C. That is, using the notation established in Chapter 2, Section 2.2.2, we showed

$$I_P(L, S | C).$$

Consider the DAG $\mathbb{G}$ in Figure 5.5. For that DAG we have the following.

---
[1] It is not standard to exclude a node's parents from its nondescendents, but this definition better serves our purposes.

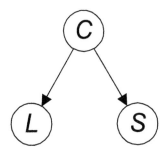

Figure 5.5: The joint probability distribution of L, S, and C constitutes a Bayesian network with this DAG.

Node	Parents	Nondescendents
L	C	S
S	C	L
C	$\varnothing$	$\varnothing$

For $(\mathbb{G}, P)$ to satisfy the Markov condition, we need to have

$$I_P(L, S|C)$$
$$I_P(S, L|C).$$

Note that since C has no nondescendents, we do not have a conditional independency for C. Since independence is symmetric, $I_P(L, S|C)$ implies $I_P(L, S|C)$. Therefore, all the conditional independencies required by the Markov condition are satisfied, and $(\mathbb{G}, P)$ is a Bayesian network.

Next we further illustrate the Markov condition with a more complex DAG.

Example 5.2 *Consider the DAG $\mathbb{G}$ in Figure 5.6. If $(\mathbb{G}, P)$ satisfied the Markov condition with some probability distribution P of X, Y, Z, W, and V, we would have the following conditional independencies.*

Node	Parents	Nondescendents	Conditional Independency
X	$\varnothing$	$\varnothing$	None
Y	X	Z, V	$I_P(Y, \{Z, V\}|X)$
Z	X	Y	$I_P(Z, Y|X)$
W	Y, Z	X, V	$I_P(W, \{X, V\}|\{Y, Z\})$
V	Z	X, Y, W	$I_P(V, \{X, Y, W\}|Z)$

5.2.2 Representation of a Bayesian Network

A Bayesian network $(\mathbb{G}, P)$, by definition, is a DAG $\mathbb{G}$ and joint probability distribution P that together satisfy the Markov condition. Then why in Figures 5.1 and 5.2 do we show a Bayesian network as a DAG and a set of conditional

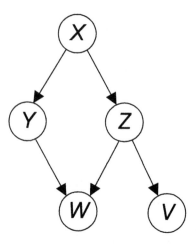

Figure 5.6: A DAG.

probability distributions? The reason is that $(\mathbb{G}, P)$ satisfies the Markov condition if and only if P is equal to the product of its conditional distributions in $\mathbb{G}$. Specifically, we have the following theorem.

Theorem 5.1 $(\mathbb{G}, P)$ *satisfies the Markov condition (and thus is a Bayesian network) if and only if P is equal to the product of its conditional distributions of all nodes given their parents in G, whenever these conditional distributions exist.*

Proof. *The proof can be found in [Neapolitan, 2004].* ∎

Example 5.3 *We showed that the joint probability distribution P of the random variables L, S, and C defined on the set of objects in Figure 5.4 constitutes a Bayesian network with the DAG $\mathbb{G}$ in Figure 5.5. Next we illustrate that the preceding theorem is correct by showing that P is equal to the product of its conditional distributions in $\mathbb{G}$. Figure 5.7 shows those conditional distributions. We computed them directly from Figure 5.4. For example, since there are nine black objects (c_1) and six of them are squares (s_1), we compute*

$$P(s_1|c_1) = \frac{6}{9} = \frac{2}{3}.$$

The other conditional distributions are computed in the same way. To show that the joint distribution is the product of the conditional distributions, we need to show for all values of i, j, and k that

$$P(s_i, l_j, c_k) = P(s_i|c_k)P(l_j|c_k)P(c_k).$$

There are a total of eight combinations. We show that the equality holds for one of them. It is left as an exercise to show that it holds for the others. To

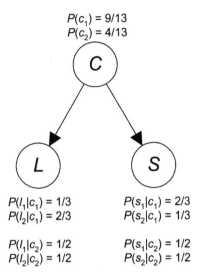

$P(c_1) = 9/13$
$P(c_2) = 4/13$

C

L S

$P(l_1|c_1) = 1/3$ $\quad$ $P(s_1|c_1) = 2/3$
$P(l_2|c_1) = 2/3$ $\quad$ $P(s_2|c_1) = 1/3$

$P(l_1|c_2) = 1/2$ $\quad$ $P(s_1|c_2) = 1/2$
$P(l_2|c_2) = 1/2$ $\quad$ $P(s_2|c_2) = 1/2$

Figure 5.7: A Bayesian network representing the probability distribution P of the random variables L, S, and C defined on the set of objects in Figure 5.4.

that end, we have directly from Figure 5.4 that

$$P(s_1, l_1, c_1) = \frac{2}{13}.$$

From Figure 5.7 we have

$$P(s_1|c_1)P(l_1|c_1)P(c_1) = \frac{2}{3} \times \frac{1}{3} \times \frac{9}{13} = \frac{2}{13}.$$

Owing to Theorem 5.1, we can represent a Bayesian network $(\mathbb{G}, P)$ using the DAG $\mathbb{G}$ and the conditional distributions. We don't need to show every value in the joint distributions. These values can all be computed from the conditional distributions. So we always show a Bayesian network as the DAG and the conditional distributions as we did in Figures 5.1, 5.2, and 5.7. Herein lies the representational power of Bayesian networks. If there is a large number of variables, there are many values in the joint distribution. However, if the DAG is sparse, there are relatively few values in the conditional distributions. For example, suppose all variables are binary, and a joint distribution satisfies the Markov condition with the DAG in Figure 5.8. Then there are $2^{10} = 1024$ values in the joint distribution, but only $2 + 2 + 8 \times 8 = 68$ values in the conditional distributions. Note that we are not even including redundant parameters in this count. For example, in the Bayesian network in Figure 5.7 it is not necessary to show $P(c_2) = 4/13$ because $P(c_2) = 1 - P(c_1)$. So we need only show $P(c_1) = 9/13$. If we eliminate redundant parameters, there are only 34 values in the conditional distributions for the DAG in Figure 5.8 but still 1023 in the joint distribution. We see then that a Bayesian network is a structure for representing a joint probability distribution succinctly.

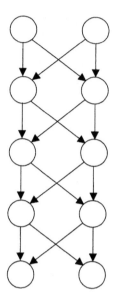

Figure 5.8: If all variables are binary and a joint distribution satisifes the Markov condition with this DAG, there are 1024 values in the joint distribution, but only 68 values in the conditional distributions.

It is important to realize that we can't take just any DAG and expect a joint distribution to equal the product of its conditional distributions in the DAG. This is only true if the Markov condition is satisfied. The next example illustrates this idea.

Example 5.4 *Consider again the joint probability distribution P of the random variables L, S, and C defined on the set of objects in Figure 5.4. Figure 5.9 shows its conditional distributions for the DAG in that figure. Note that we no longer show redundant parameters in our figures. If P satisfied the Markov condition with this DAG, we would have to have $I_P(L, S)$ because L has no parents and S is the sole nondescendent of L. It is left as an exercise to show that this independency does not hold. Furthermore, P is not equal to the product of its conditional distributions in this DAG. For example, we have directly from Figure 5.4 that*

$$P(s_1, l_1, c_1) = \frac{2}{13} = .15385.$$

From Figure 5.9 we have

$$P(c_1|l_1, s_1)P(l_1)P(s_1) = \frac{2}{3} \times \frac{5}{13} \times \frac{8}{13} = .15779.$$

It seems that we are left with a conundrum. That is, our goal is to succinctly represent a joint probability distribution using a DAG and conditional

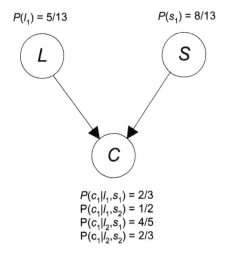

$P(l_1) = 5/13$ $P(s_1) = 8/13$

$P(c_1|l_1,s_1) = 2/3$
$P(c_1|l_1,s_2) = 1/2$
$P(c_1|l_2,s_1) = 4/5$
$P(c_1|l_2,s_2) = 2/3$

Figure 5.9: The joint probability distribution P of the random variables L, S, and C defined on the set of objects in Figure 5.4 does not satisfy the Markov condition with this DAG.

distributions for the DAG (a Bayesian network) rather than enumerating every value in the joint distribution. However, we don't know which DAG to use until we check whether the Markov condition is satisfied, and, in general, we would need to have the joint distribution to check this. A common way out of this predicament is to construct a **causal DAG**, which is a DAG in which there is an edge from X to Y if X causes Y. The DAGs in Figures 5.1 and 5.2 are causal; other DAGs shown so far in this chapter are not causal.

Next we discuss why a causal DAG should satisfy the Markov condition with the probability distribution of the variables in the DAG. A second way of obtaining the DAG is to learn it from data. This second way is discussed in Chapter 8.

5.3 Causal Networks as Bayesian Networks

Before discussing why a causal DAG should often satisfy the Markov condition with the probability distribution of the variables in the DAG, we formalize the notion of causality.

5.3.1 Causality

After providing an operational definition of a cause, we show a comprehensive example of identifying a cause according to this definition.

An Operational Definition of a Cause

One dictionary definition of a cause is "the one, such as a person, an event, or a condition, that is responsible for an action or a result." Although useful, this definition is certainly not the last word on the concept of causation, which has been investigated for centuries (see, e.g., [Hume, 1748]; [Piaget, 1966]; [Eells, 1991]; [Salmon, 1997]; [Spirtes et al., 1993; 2000]; [Pearl, 2000]). This definition does, however, shed light on an operational method for identifying causal relationships. That is, if the action of making variable X take some value sometimes changes the value taken by variable Y, then we assume X is responsible for sometimes changing Y's value, and we conclude X is a cause[2] of Y. More formally, we say we **manipulate** X when we force X to take some value, and we say X **causes** Y if there is some manipulation of X that leads to a change in the probability distribution of Y. We assume that if manipulating X leads to a change in the probability distribution of Y, then X obtaining a value by any means whatsoever also leads to a change in the probability distribution of Y. So we assume that causes and their effects are statistically correlated. However, as we shall discuss soon, variables can be correlated without one causing the other.

A manipulation consists of a **randomized controlled experiment (RCE)** using some specific population of entities (e.g., individuals with chest pain) in some specific context (e.g., they currently receive no chest pain medication and they live in a particular geographical area). The causal relationship discovered is then relative to this population and this context.

Let's discuss how the manipulation proceeds. We first identify the population of entities we want to consider. Our random variables are features of these entities. Next we ascertain the causal relationship we want to investigate. Suppose we are trying to determine if variable X is a cause of variable Y. We then sample a number of entities from the population. For every entity selected, we manipulate the value of X so that each of its possible values is given to the same number of entities (if X is continuous, we choose the values of X according to a uniform distribution). After the value of X is set for a given entity, we measure the value of Y for that entity. The more the resultant data show a dependency between X and Y, the more the data support that X causes Y. The manipulation of X can be represented by a variable M that is external to the system being studied. There is one value m_i of M for each value x_i of X; the probabilities of all values of M are the same; and when M equals m_i, X equals x_i. That is, the relationship between M and X is deterministic. The data support that X causes Y to the extent that the data indicate $P(y_i|m_j) \neq P(y_i|m_k)$ for $j \neq k$. Manipulation is actually a special kind of causal relationship that we assume exists primordially and is within our control so that we can define and discover other causal relationships.

---

[2]This notion of causality does not pertain to *token* causality, which concerns individual, causal events rather than probabilistic relationships among variables.

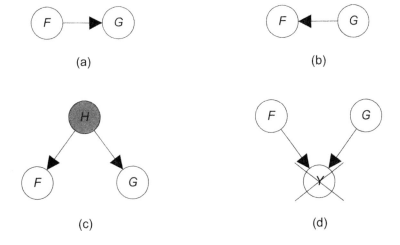

Figure 5.10: The edges in these graphs represent causal influences. All four causal relationships could account for F and G being correlated.

An Illustration of Manipulation

We demonstrate these ideas with a comprehensive example concerning recent headline news. The pharmaceutical company Merck had been marketing its drug finasteride as medication for men with benign prostatic hyperplasia (BPH). Based on anecdotal evidence, it seemed that there was a correlation between use of the drug and regrowth of scalp hair. Let's assume that Merck took a random sample from the population of interest and, based on that sample, determined there is a correlation between finasteride use and hair regrowth. Should Merck conclude that finasteride causes hair regrowth and therefore market it as a cure for baldness? Not necessarily. There are quite a few causal explanations for the correlation of two variables. We discuss these relationships next.

Possible Causal Relationships Let F be a variable representing finasteride use and G be a variable representing scalp hair growth. The actual values of F and G are unimportant to the present discussion. We could use either continuous or discrete values. If F caused G, then indeed they would be statistically correlated, but this would also be the case if G caused F or if they had some hidden common cause H. If we represent a causal influence by a directed edge, Figure 5.10 shows these three possibilities plus one more. Figure 5.10 (a) shows the conjecture that F causes G, which we already suspect might be the case. However, it could be that G causes F (Figure 5.10 (b)). You may argue that, based on domain knowledge, this does not seem reasonable. However, we do not, in general, have domain knowledge when doing a statistical analysis. So, from the correlation alone, the causal relationships in Figures 5.10 (a) and 5.10 (b) are equally reasonable. Even in this domain, G causing F seems possible. A man might have used some other hair regrowth product such as minoxidil that caused him to regrow hair, become excited about the regrowth, and decide to try other products such as finasteride, which he heard might cause regrowth.

A third possibility, shown in Figure 5.10 (c), is that F and G have some hidden common cause H that accounts for their statistical correlation. For example, a man concerned about hair loss might try both finasteride and minoxidil in his effort to regrow hair. The minoxidil might cause hair regrowth, whereas the finasteride might not. In this case the man's concern is a cause of finasteride use and hair regrowth (indirectly through minoxidil use), whereas the two are not causally related.

A fourth possibility is that our sample (or even our entire population) consists of individuals who have some (possibly hidden) effect of both F and G. For example, suppose finasteride and apprehension about lack of hair regrowth both cause hypertension,[3] and our sample consists of individuals who have hypertension Y. We say a node is **instantiated** when we know its value for the entity currently being modeled. So we are saying the variable Y is instantiated to the same value for every entity in our sample. This situation is depicted in Figure 5.10 (d), where the cross through Y means that the variable is instantiated. Usually, the instantiation of a common effect creates a dependency between its causes because each cause explains the occurrence of the effect, thereby making the other cause less likely. As noted earlier, psychologists call this **discounting**. So, if this were the case, discounting would explain the correlation between F and G. This type of dependency is called **selection bias**.[4]

A final possibility (not shown in Figure 5.10) is that F and G are not causally related at all. A notable example of this situation occurs when our entities are points in time and our random variables are values of properties at these different points in time. Such variables are often correlated without having any apparent causal connection. For example, if our population consists of points in time, J is the Dow Jones Average at a given time, and L is Professor Neapolitan's hairline at a given time, then J and L are correlated.[5] Yet they do not seem to be causally connected. Some argue that there are hidden common causes beyond our ability to measure. We will not discuss this issue further here. We only want to note the difficulty with such correlations. In light of the factors we've discussed, we see then that we cannot deduce a causal relationship between two variables from the mere fact that they are statistically correlated.

Any of the causal relationships shown in Figure 5.10 could occur in combination, resulting in F and G being correlated. For example, it could be both that finasteride causes hair regrowth and that excitement about regrowth may cause the use of finasteride, meaning we could have a causal loop or feedback. So, we could have the causal relationships in both Figures 5.10 (a) and 5.10 (b).

---

[3] There is no evidence that either finasteride or apprehension about the lack of hair regrowth causes hypertension. This example is only for the sake of illustration.

[4] This could happen if our sample is a **convenience sample**, which is a sample in which the participants are selected at the convenience of the researcher. The researcher makes no attempt to ensure that the sample is an accurate representation of the larger population. In the context of the current example, this might be the case if it is convenient for the researcher to observe males hospitalized for hypertension.

[5] Unfortunately, his hairline did not go back down in fall, 2008.

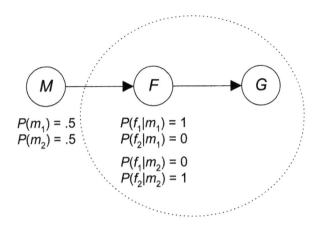

$P(m_1) = .5$ $P(f_1|m_1) = 1$
$P(m_2) = .5$ $P(f_2|m_1) = 0$

$P(f_1|m_2) = 0$
$P(f_2|m_2) = 1$

Figure 5.11: An RCE investigating whether F causes G.

It might not be obvious why two variables with a common cause would be correlated. Consider the present example. Suppose H is a common cause of F and G and neither F nor G caused the other. Then H and F are correlated, because H causes F, and H and G are correlated because H causes G, which implies F and G are correlated transitively through H. Here is a more detailed explanation. For the sake of example, suppose h_1 is a value of H that has a causal influence on F taking value f_1 and on G taking value g_1. Then if F had value f_1, each of its causes would become more probable because one of them should be responsible. So, $P(h_1|f_1) > P(h_1)$. Now, since the probability of h_1 has gone up, the probability of g_1 would also go up because h_1 causes g_1. Therefore, $P(g_1|f_1) > P(g_1)$, which means F and G are correlated.

Merck's Manipulation Study Since Merck could not conclude that finasteride causes hair regrowth from their mere correlation alone, they did a manipulation study to test this conjecture. The study was done on 1879 men aged 18 to 41 with mild to moderate hair loss of the vertex and anterior mid-scalp areas. Half of the men were given 1 mg. of finasteride, whereas the other half were given 1 mg. of a placebo. The following table shows the possible values of the variables in the study, including the manipulation variable M.

Variable	Value	When the Variable Takes This Value
F	f_1	Subject takes 1 mg. of finasteride
	f_2	Subject takes 1 mg. of a placebo
G	g_1	Subject has significant hair regrowth
	g_2	Subject does not have significant hair regrowth
M	m_1	Subject is chosen to take 1 mg. of finasteride
	m_2	Subject is chosen to take 1 mg. of a placebo

An RCE used to test the conjecture that F causes G is shown in Figure 5.11. There is an oval around the system being studied (F and G and their possible causal relationship) to indicate that the manipulation comes from outside the system. The edges in Figure 5.11 represent causal influences. The RCE supports the conjecture that F causes G to the extent that the data support $P(g_1|m_1) \neq P(g_1|m_2)$. Merck decided that "significant hair regrowth" would be judged according to the opinion of independent dermatologists. A panel of independent dermatologists evaluated photos of the men after 24 months of treatment. The panel judged that significant hair regrowth was demonstrated in 66% of men treated with finasteride, compared to 7% of men treated with placebo.

Basing our probability on these results, we have that $P(g_1|m_1) \approx .67$ and $P(g_1|m_2) \approx .07$. In a more analytical analysis, only 17% of men treated with finasteride demonstrated hair loss (defined as any decrease in hair count from baseline). In contrast, 72% of the placebo group lost hair, as measured by hair count. Merck concluded that finasteride does indeed cause hair regrowth and on December 22, 1997, announced that the U.S. Food and Drug Administration granted marketing clearance to Propecia™ (finasteride 1 mg.) for treatment of male pattern hair loss (androgenetic alopecia), for use in men only (see [McClennan and Markham, 1999] for more on this topic).

5.3.2   Causality and the Markov Condition

First we more rigorously define a causal DAG. After that we state the causal Markov assumption and argue why it should be satisfied.

Causal DAGs

We say X is a **cause** of Y if a manipulation of X results in a change in the probability distribution of Y. A **causal graph** is a directed graph containing a set of causally related random variables V such that for every $X, Y \in$ V there is an edge from X to Y if and only if X is a cause of Y, and there is no subset of variables W_{XY} of V such that if we knew the values of the variables in W_{XY}, a manipulation of X would no longer change the probability distribution of Y. If there is an edge from X to Y, we call X a **direct cause** of Y. Note that whether or not X is a direct cause of Y depends on the variables included in V. A causal graph is a **causal DAG** if the causal graph is acyclic (i.e., there are no causal feedback loops).

Example 5.5 *Testosterone (T) is known to convert to dihydro-testosterone (D), and dihydro-testosterone is believed to be the hormone necessary for erectile function (E). A study in [Lugg et al., 1995] tested the causal relationship among these variables in rats. They manipulated testosterone to low levels and found that both dihydro-testosterone and erectile function declined. They then held dihydro-testosterone fixed at low levels and found that erectile function was low regardless of the manipulated value of testosterone. Finally, they held dihydro-testosterone fixed at high levels and found that erectile function was high regardless of the manipulated value of testosterone. So they learned that, in a causal graph containing only the variables T, D, and E, T is a direct cause*

testosterone   dihydro-testosterone   erectile function

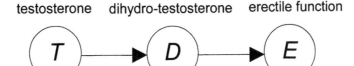

Figure 5.12: A causal DAG.

of D, and D is a direct cause of E, but, although T is a cause of E, it is not a direct cause. So the causal graph (DAG) is the one in Figure 5.12.

Notice that if the variable D were not in the DAG in Figure 5.12, T would be called a direct cause of E, and there would be an edge from T directly into E instead of the directed path through D. In general, our edges always represent only the relationships among the identified variables. It seems we can usually conceive of intermediate, unidentified variables along each edge. Consider the following example taken from [Spirtes et al., 1993; 2000], p. 42:

> If C is the event of striking a match, and A is the event of the match catching on fire, and no other events are considered, then C is a direct cause of A. If, however, we added B, the sulfur on the match tip achieved sufficient heat to combine with the oxygen, then we could no longer say that C directly caused A, but rather C directly caused B and B directly caused A. Accordingly, we say that B is a causal mediary between C and A if C causes B and B causes A.

Note that, in this intuitive explanation, a variable name is used to also stand for a value of the variable. For example, A is a variable whose value is *on-fire* or *not-on-fire*, and A is also used to represent that the match is on fire. Clearly, we can add more causal mediaries. For example, we could add the variable D, representing whether the match tip is abraded by a rough surface. C would then cause D, which would cause B, and so on. We could go much further and describe the chemical reaction that occurs when sulfur combines with oxygen.

Indeed, it seems we can conceive of a continuum of events in any causal description of a process. We see then that the set of observable variables is observer dependent. Apparently an individual, given myriad sensory input, selectively records discernible events and develops cause/effect relationships among them. Therefore, rather than assuming that there is a set of causally related variables out there, it seems more appropriate to only assume that, in a given context or application, we identify certain variables and develop a set of causal relationships among them.

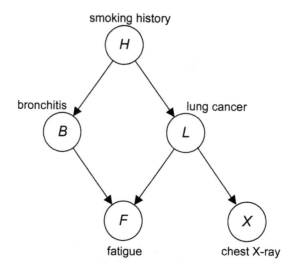

Figure 5.13: A causal DAG.

The Causal Markov Assumption

If we assume that the observed probability distribution P of a set of random variables V satisfies the Markov condition with the causal DAG G containing the variables, we say we are making the **causal Markov assumption**, and we call $(\mathbb{G}, P)$ a **causal network**. Why should we make the causal Markov assumption? To answer this question we show several examples.

Example 5.6 *Consider again the situation involving testosterone (T), dihydro-testosterone (D), and erectile function (E). Recall the manipulation study in [Lugg et al., 1995], which we discussed in Example 5.5. This study showed that if we instantiate D, the value of E is independent of the value of T. So there is experimental evidence that the Markov condition is satisfied for a three-variable causal chain.*

Example 5.7 *A history of smoking (H) is known to cause both bronchitis (B) and lung cancer (L). Lung cancer and bronchitis both cause fatigue (F), but only lung cancer can cause a chest X-ray (X) to be positive. There are no other causal relationships among the variables. Figure 5.13 shows a causal DAG containing these variables. The causal Markov assumption for that DAG entails the following conditional independencies.*

Node	Parents	Nondescendents	Conditional Independency
H	$\varnothing$	$\varnothing$	None
B	H	L, X	$I_P(B, \{L, X\}|H)$
L	H	B	$I_P(L, B|H)$
F	B, L	H, X	$I_P(F, \{H, X\}|\{B, L\})$
X	L	H, B, F	$I_P(X, \{H, B, F\}|L)$

Given the causal relationship in Figure 5.13, we would not expect bronchitis and lung cancer to be independent, because if someone had lung cancer it would make it more probable that the individual smoked (since smoking can cause lung cancer), which would make it more probable that another effect of smoking, namely bronchitis, was present. However, if we knew someone smoked, it would already be more probable that the person had bronchitis. Learning that the individual had lung cancer could no longer increase the probability of smoking (which is now 1), which means it cannot change the probability of bronchitis. That is, the variable H shields B from the influence of L, which is what the causal Markov condition says. Similarly, a positive chest X-ray increases the probability of lung cancer, which in turn increases the probability of smoking, which in turn increases the probability of bronchitis. So, a chest X-ray and bronchitis are not independent. However, if we knew the person had lung cancer, the chest X-ray could not change the probability of lung cancer and thereby change the probability of bronchitis. So B is independent of X conditional on L, which is what the causal Markov condition says.

In summary, if we create a causal graph containing the variables X and Y, if X and Y do not have a hidden common cause (i.e., a cause that is not in our graph), if there are no causal paths from Y back to X (i.e., our graph is a DAG), and if we do not have selection bias (i.e., our probability distribution is not obtained from a population in which a common effect is instantiated to the same value for all members of the population), then we feel X and Y are independent if we condition on a set of variables including at least one variable in each of the causal paths from X to Y. Since the set of all parents of Y is such a set, we feel that the Markov condition holds relative to X and Y. So we conclude that the causal Markov assumption is justified for a causal graph if the following conditions are satisfied:

1. There are no hidden common causes. That is, all common causes are represented in the graph.

2. There are no causal feedback loops. That is, our graph is a DAG.

3. Selection bias is not present.

Note that, for the Markov condition to hold, there must be an edge from X to Y whenever there is a causal path from X to Y besides the ones containing variables in our graph. However, we need not stipulate this requirement because it is entailed by the definition of a causal graph. Recall that in a causal graph there is an edge from X to Y if X is a direct cause of Y.

Perhaps the condition that is most frequently violated is that there can be no hidden common causes. We discuss this condition further with a final example.

Example 5.8 Suppose we wanted to create a causal DAG containing the variables cold (C), sneezing (S), and runny nose (R). Since a cold can cause both sneezing and a runny nose and neither of these conditions can cause each other,

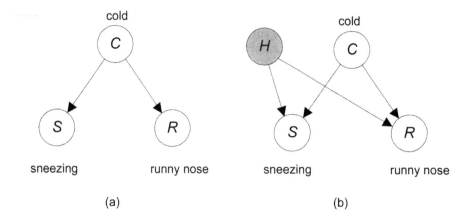

Figure 5.14: The causal Markov assumption would not hold for the DAG in (a) if there is a hidden common cause as depicted in (b).

we would create the DAG in Figure 5.14 (a). The causal Markov condition for that DAG would entail $I_P(S, R|C)$. However, if there were a hidden common cause of S and R as depicted in Figure 5.14 (b), this conditional independency would not hold because even if the value of C were known, S would change the probability of H, which in turn would change the probability of R. Indeed, there is at least one other cause of sneezing and runny nose, namely hay fever. So when making the causal Markov assumption, we must be certain that we have identified all common causes.

5.3.3   The Markov Condition without Causality

We have argued that a causal DAG often satisfies the Markov condition with the joint probability distribution of the random variables in the DAG. This does not mean that the edges in a DAG in a Bayesian network must be causal. That is, a DAG can satisfy the Markov condition with the probability distribution of the variables in the DAG without the edges being causal. For example, we showed that the joint probability distribution P of the random variables L, S, and C defined on the set of objects in Figure 5.4 satisfies the Markov condition with the DAG $\mathbb{G}$ in Figure 5.5. However, we would not argue that the color of the objects causes their shape or the letter that is on them. As another example, if we reversed the edges in the DAG in Figure 5.12 to obtain the DAG $E \rightarrow DHT \rightarrow T$, the new DAG would also satisfy the Markov condition with the probability distribution of the variables, yet the edges would not be causal.

5.4   Inference in Bayesian Networks

As noted previously, a standard application of Bayes' Theorem is inference in a two-node Bayesian network. Larger Bayesian networks address the problem

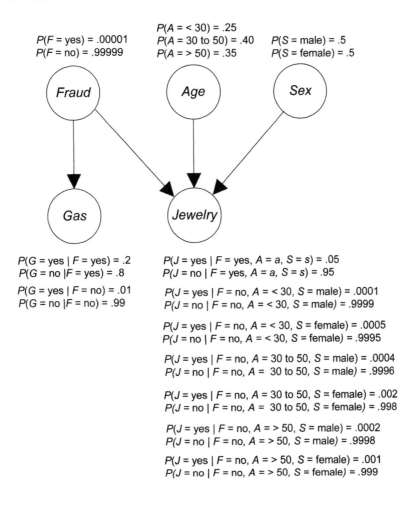

$P(F = \text{yes}) = .00001$
$P(F = \text{no}) = .99999$

$P(A = < 30) = .25$
$P(A = 30 \text{ to } 50) = .40$
$P(A = > 50) = .35$

$P(S = \text{male}) = .5$
$P(S = \text{female}) = .5$

$P(G = \text{yes} \mid F = \text{yes}) = .2$
$P(G = \text{no} \mid F = \text{yes}) = .8$

$P(G = \text{yes} \mid F = \text{no}) = .01$
$P(G = \text{no} \mid F = \text{no}) = .99$

$P(J = \text{yes} \mid F = \text{yes}, A = a, S = s) = .05$
$P(J = \text{no} \mid F = \text{yes}, A = a, S = s) = .95$

$P(J = \text{yes} \mid F = \text{no}, A = < 30, S = \text{male}) = .0001$
$P(J = \text{no} \mid F = \text{no}, A = < 30, S = \text{male}) = .9999$

$P(J = \text{yes} \mid F = \text{no}, A = < 30, S = \text{female}) = .0005$
$P(J = \text{no} \mid F = \text{no}, A = < 30, S = \text{female}) = .9995$

$P(J = \text{yes} \mid F = \text{no}, A = 30 \text{ to } 50, S = \text{male}) = .0004$
$P(J = \text{no} \mid F = \text{no}, A = 30 \text{ to } 50, S = \text{male}) = .9996$

$P(J = \text{yes} \mid F = \text{no}, A = 30 \text{ to } 50, S = \text{female}) = .002$
$P(J = \text{no} \mid F = \text{no}, A = 30 \text{ to } 50, S = \text{female}) = .998$

$P(J = \text{yes} \mid F = \text{no}, A = > 50, S = \text{male}) = .0002$
$P(J = \text{no} \mid F = \text{no}, A = > 50, S = \text{male}) = .9998$

$P(J = \text{yes} \mid F = \text{no}, A = > 50, S = \text{female}) = .001$
$P(J = \text{no} \mid F = \text{no}, A = > 50, S = \text{female}) = .999$

Figure 5.15: Bayesian network for detecting credit card fraud.

of representing the joint probability distribution of a large number of variables. For example, Figure 5.2, which appears again as Figure 5.15, represents the joint probability distribution of variables related to credit card fraud. Inference in this network consists of computing the conditional probability of some variable (or set of variables), given that other variables are instantiated to certain values. For example, we might want to compute the probability of credit card fraud, given that gas has been purchased, jewelry has been purchased, and the card holder is male. To accomplish this inference we need sophisticated algorithms. First, we show simple examples illustrating how one of these algorithms uses the Markov condition and Bayes' Theorem to do inference. Then we reference papers describing some of the algorithms. Finally we show examples using the algorithms to do inference.

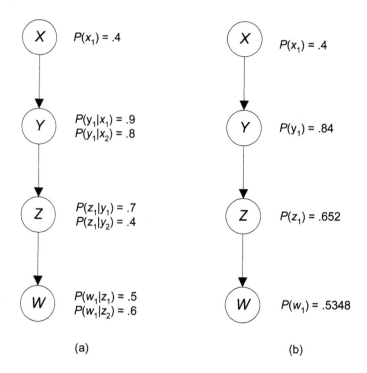

Figure 5.16: A Bayesian network appears in (a), and the prior probabilities of the variables in that network are shown in (b). Each variable has only two values, so only the probability of one is shown in (a).

5.4.1   Examples of Inference

Next we present some examples illustrating how the conditional independencies entailed by the Markov condition can be exploited to accomplish inference in a Bayesian network.

Example 5.9 *Consider the Bayesian network in Figure 5.16 (a). The prior probabilities of all variables can be computed using the Law of Total Probability:*

$$P(y_1) = P(y_1|x_1)P(x_1) + P(y_1|x_2)P(x_2) = (.9)(.4) + (.8)(.6) = .84$$

$$P(z_1) = P(z_1|y_1)P(y_1) + P(z_1|y_2)P(y_2) = (.7)(.84) + (.4)(.16) = .652$$

$$P(w_1) = P(w_1|z_1)P(z_1) + P(w_1|z_2)P(z_2) = (.5)(.652) + (.6)(.348) = .5348.$$

These probabilities are shown in Figure 5.16 (b). Note that the computation for each variable requires information determined for its parent. We can therefore consider this method a message-passing algorithm in which each node passes its child a message needed to compute the child's probabilities. Clearly, this algorithm applies to an arbitrarily long linked list and to trees.

Example 5.10 *Suppose now that X is instantiated for x_1. Since the Markov condition entails that each variable is conditionally independent of X given its parent, we can compute the conditional probabilities of the remaining variables by again using the Law of Total Probability (however, now with the background information that $X = x_1$) and passing messages down as follows:*

$$P(y_1|x_1) = .9$$

$$
\begin{aligned}
P(z_1|x_1) &= P(z_1|y_1,x_1)P(y_1|x_1) + P(z_1|y_2,x_1)P(y_2|x_1) \\
&= P(z_1|y_1)P(y_1|x_1) + P(z_1|y_2)P(y_2|x_1) \quad // \text{ Markov condition} \\
&= (.7)(.9) + (.4)(.1) = .67
\end{aligned}
$$

$$
\begin{aligned}
P(w_1|x_1) &= P(w_1|z_1,x_1)P(z_1|x_1) + P(w_1|z_2,x_1)P(z_2|x_1) \\
&= P(w_1|z_1)P(z_1|x_1) + P(w_1|z_2)P(z_2|x_1) \\
&= (.5)(.67) + (.6)(1 - .67) = .533.
\end{aligned}
$$

Clearly, this algorithm also applies to an arbitrarily long linked list and to trees.

The preceding example shows how we can use downward propagation of messages to compute the conditional probabilities of variables below the instantiated variable. Next we illustrate how to compute conditional probabilities of variables above the instantiated variable.

Example 5.11 *Suppose W is instantiated for w_1 (and no other variable is instantiated). We can use upward propagation of messages to compute the conditional probabilities of the remaining variables. First, we use Bayes' Theorem to compute $P(z_1|w_1)$:*

$$P(z_1|w_1) = \frac{P(w_1|z_1)P(z_1)}{P(w_1)} = \frac{(.5)(.652)}{.5348} = .6096.$$

Then, to compute $P(y_1|w_1)$, we again apply Bayes' Theorem:

$$P(y_1|w_1) = \frac{P(w_1|y_1)P(y_1)}{P(w_1)}.$$

We cannot yet complete this computation because we do not know $P(w_1|y_1)$. We can obtain this value using downward propagation as follows:

$$P(w_1|y_1) = (P(w_1|z_1)P(z_1|y_1) + P(w_1|z_2)P(z_2|y_1)).$$

After doing this computation, also computing $P(w_1|y_2)$ (because X will need this value) and then determining $P(y_1|w_1)$, we pass $P(w_1|y_1)$ and $P(w_1|y_2)$ to X. We then compute $P(w_1|x_1)$ and $P(x_1|w_1)$ in sequence:

$$P(w_1|x_1) = (P(w_1|y_1)P(y_1|x_1) + P(w_1|y_2)P(y_2|x_1)$$

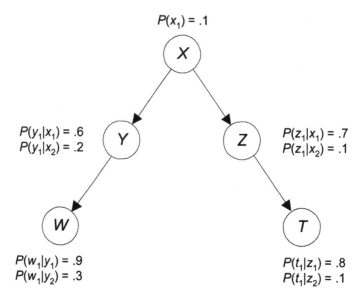

Figure 5.17: A Bayesian network. Each variable has only two possible values, so only the probability of one is shown.

$$P(x_1|w_1) = \frac{P(w_1|x_1)P(x_1)}{P(w_1)}.$$

It is left as an exercise to perform these computations. Clearly, this upward propagation scheme applies to an arbitrarily long linked list.

The next example shows how to turn corners in a tree.

Example 5.12 *Consider the Bayesian network in Figure 5.17. Suppose W is instantiated for w_1. We compute $P(y_1|w_1)$ followed by $P(x_1|w_1)$ using the upward propagation algorithm just described. Then we proceed to compute $P(z_1|w_1)$ followed by $P(t_1|w_1)$ using the downward propagation algorithm. This is left as an exercise.*

5.4.2   Inference Algorithms and Packages

By exploiting local independencies as we did in the previous subsection, Pearl [1986, 1988] developed a message-passing algorithm for inference in Bayesian networks. Based on a method originated in [Lauritzen and Spiegelhalter, 1988], Jensen et al. [1990] developed an inference algorithm that involves the extraction of an undirected triangulated graph from the DAG in a Bayesian network and the creation of a tree, whose vertices are the cliques of this triangulated graph. Such a tree is called a **junction tree**. Conditional probabilities are then computed by passing messages in the junction tree. Li and D'Ambrosio

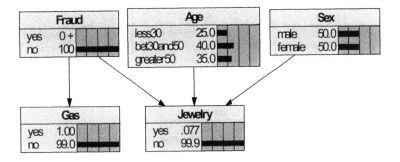

Figure 5.18: The fraud detection Bayesian network in Figure 5.15, implemented using Netica.

[1994] took a different approach. They developed an algorithm that approximates finding the optimal way to compute marginal distributions of interest from the joint probability distribution. They call this **symbolic probabilistic inference** (SPI).

All these algorithms are worst-case nonpolynomial time. This is not surprising, since the problem of inference in Bayesian networks has been shown to be NP-hard [Cooper, 1990]. In light of this result, approximation algorithms for inference in Bayesian networks have been developed. One such algorithm, likelihood weighting, was developed independently in [Fung and Chang, 1990] and [Shachter and Peot, 1990]. It is proven in [Dagum and Luby, 1993] that the problem of approximate inference in Bayesian networks is also NP-hard. However, there are restricted classes of Bayesian networks that are provably amenable to a polynomial-time solution (see [Dagum and Chavez, 1993]). Indeed, a variant of the likelihood weighting algorithm, which is worst-case polynomial time as long as the network does not contain extreme conditional probabilities, appears in [Pradhan and Dagum, 1996].

Practitioners need not concern themselves with all these algorithms since a number of packages for doing inference in Bayesian networks have been developed. A few of them are shown here:

1. Netica (www.norsys.com/)

2. GeNIe (genie.sis.pitt.edu/)

3. HUGIN (/www.hugin.com/)

4. Elvira (www.ia.uned.es/~elvira/)

5. BUGS (www.mrc-bsu.cam.ac.uk/bugs/)

In this book we ordinarily use Netica to illustrate inference. Figure 5.18 shows the fraud detection network in Figure 5.15 implemented using Netica.

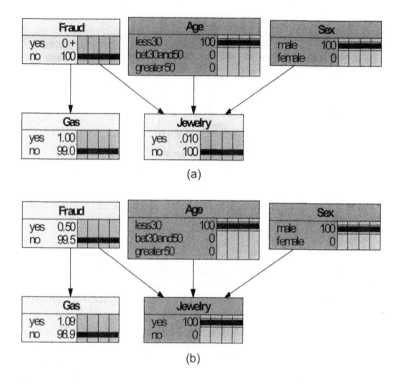

Figure 5.19: In (a) *Age* has been instantiated to *less30* and *Sex* has been instantiated to *male*. In (b) *Age* has been instantiated to *less30*, *Sex* has been instantiated to *male*, and *Jewelry* has been instantiated to *yes*.

5.4.3   Inference Using Netica

Next we illustrate inference in a Bayesian network using Netica. Notice from Figure 5.18 that Netica computes and shows the prior probabilities of the variables rather than showing the conditional probability distributions. Probabilities are shown as percentages. For example, the fact that there is a .077 next to *yes* in the *Jewelry* node means

$$P(Jewelry = yes) = .00077.$$

This is the prior probability of a jewelry purchase in the past 24 hours being charged to any particular credit card.

After variables are instantiated, Netica shows the conditional probabilities of the other variables given these instantiations. In Figure 5.19 (a) we instantiated *Age* to *less30* and *Sex* to *male*. So the fact that there is .010 next to *yes* in the *Jewelry* node means

$$P(Jewelry = yes|Age = less30, Sex = male) = .00010.$$

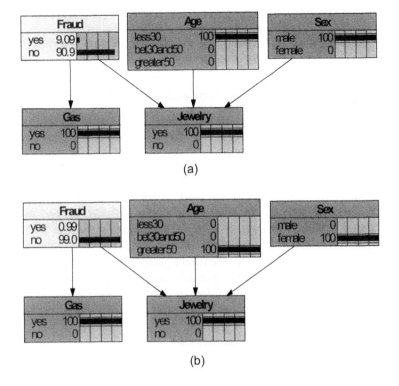

Figure 5.20: *Sex* and *Jewelry* have both been instantiated to *yes* in both (a) and (b). However, in (a) the card holder is a young man, whereas in (b) it is an older woman.

Notice that the probability of *Fraud* has not changed. This is what we would expect. First the Markov condition says that *Fraud* should be independent of *Age* and *Sex*. Second, it seems they should be independent. That is, the fact that the card holder is a young man should not make it more or less likely that the card is being used fraudulently. Figure 5.19 (b) has the same instantiations as Figure 5.19 (a) except that we have also instantiated *Jewelry* to *yes*. Notice that the probability of *Fraud* has now changed. First, the jewelry purchase makes *Fraud* more likely to be *yes*. Second, the fact that the card holder is a young man means it is less likely the card holder would make the purchase, thereby making *Fraud* even more likely to be *yes*.

In Figures 5.20 (a) and 5.20 (b), *Gas* and *Jewelry* have both been instantiated to *yes*. However, in Figure 5.20 (a) the card holder is a young man, whereas in Figure 5.20 (b) it is an older woman. This illustrates discounting of the jewelry purchase. When the card holder is a young man, the probability of *Fraud* being *yes* is high (.0909). However, when it is an older woman, it is still low (.0099) because the fact that the card holder is an older woman explains the jewelry purchase.

5.5   Networks with Continuous Variables

So far in all our Bayesian networks the variables have been discrete. Next we discuss Bayesian networks that contain continuous variables.

5.5.1   Gaussian Bayesian Networks

Gaussian Bayesian networks contain variables that are normally distributed. (The normal distribution is reviewed in Section 3.3.) We motivate such networks with the following example.

Example 5.13 *Suppose you are considering taking a job that pays $10 an hour and you expect to work 40 hours per week. However, you are not guaranteed 40 hours, and you estimate the number of hours actually worked in a week to be normally distributed with mean 40 and standard deviation 5. You have not yet fully investigated the benefits such as bonus pay and nontaxable deductions (e.g., contributions to a retirement program). However, you estimate these other influences on your gross taxable weekly income to also be normally distributed with mean 0 (that is, you feel they about offset) and standard deviation 30. Furthermore, you assume that these other influences are independent of your hours worked.*

We define the following random variables.

Variable	What the Variable Represents
X	Hours worked in the week
Y	Salary obtained in the week

Based on the preceding discussion, X is distributed as follows:

$$\rho(x) = \text{NormalDen}(x; 40, 5^2).$$

A portion of your salary Y is a deterministic function of X. That is, you will receive $10x$ dollars if you work x hours. However, your gross salary may be greater or less than this based on the other influences we discussed. That is,

$$y = 10x + \varepsilon_Y,$$

where

$$\rho(\varepsilon_Y) = \text{NormalDen}(\varepsilon_Y; 0, 30^2).$$

Since the expected value of those other influences is 0,

$$E(Y|x) = 10x.$$

and since the variance of those other influences is 30^2,

$$V(Y|x) = 30^2.$$

So Y is distributed conditionally as follows:

$$\rho(y|x) = \text{NormalDen}(y; 10x, 30^2).$$

Therefore, the relationship between X and Y is represented by the Bayesian network in Figure 5.21.

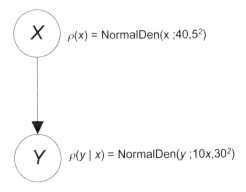

Figure 5.21: A Gaussian Bayesian network.

The Bayesian network we've just developed is an example of a Gaussian Bayesian network. In general, in a Gaussian Bayesian network, the root is normally distributed, and each non-root Y is a linear function of its parents plus an error term ε_Y that is normally distributed with mean 0 and variance σ_Y^2. So if $X_1, X_2, \ldots$ and X_k are the parents of Y, then

$$Y = b_1 x_1 + b_2 x_2 + \cdots b_k x_k + \varepsilon_Y, \tag{5.1}$$

where

$$\rho(\varepsilon_Y) = \text{NormalDen}(\varepsilon_Y; 0, \sigma_Y^2),$$

and Y is distributed conditionally as follows:

$$\rho(y|x) = \text{NormalDen}(y; b_1 x_1 + b_2 x_2 + \cdots b_k x_k, \sigma_Y^2).$$

The linear relationship in Equality 5.1 has been used in causal models in economics [Joereskog, 1982], in structural equations in psychology [Bentler, 1980], and in path analysis in sociology and genetics [Kenny, 1979], [Wright, 1921].

Pearl [1988] developed an exact inference algorithm for Gaussian Bayesian networks. It is described in [Neapolitan, 2004]. Most Bayesian network inference algorithms handle Gaussian Bayesian networks. Some use the exact algorithm; others discretize the continuous distribution and then do inference using discrete variables. Next we show two examples.

Example 5.14 *HUGIN (www.hugin.com/) does exact inference in Gaussian Bayesian networks. Figure 5.22 shows the network in Figure 5.21 developed using HUGIN. The prior means and variances are shown under the DAG. Suppose now you just got your paycheck and it is only $300. Your spouse becomes suspicious that you did not work very many hours. So your spouse instantiates Y to 300 in the network. The updated mean and variance of X are shown in Figure 5.22 under the priors. It turns out that the expected value of the hours you worked is only about 32.64.*

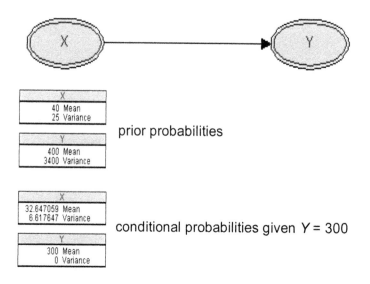

Figure 5.22: The Bayesian network in Figure 5.21, implemented in HUGIN.

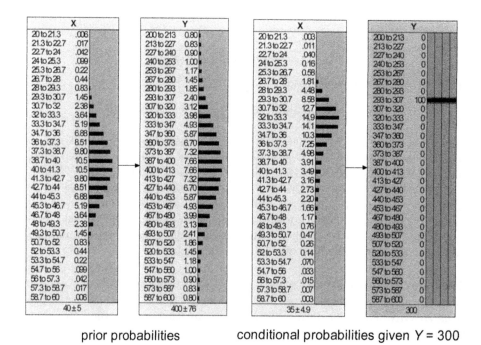

Figure 5.23: The Bayesian network in Figure 5.21, implemented in Netica.

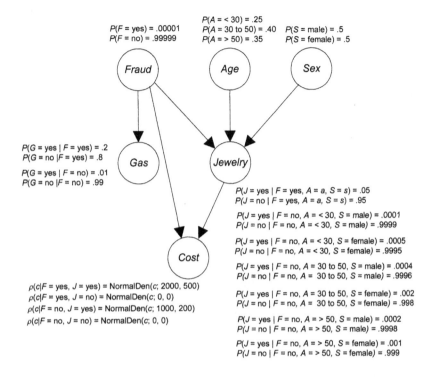

Figure 5.24: A hybrid Bayesian network modeling the situation in which the cost of the jewelry is likely to be higher if the purchase was fraudulent.

Netica (www.norsys.com/) requires that we discretize continuous variables. On the left in Figure 5.23 we show the network in Figure 5.21 with no nodes instantiated; on the right Y is instantiated to 300. Notice that, owing to the approximation, we do not obtain the value of 32.64 for the updated mean of X. Rather, we obtain the value of 35. If we had a finer discretization, we should have obtained better results.

5.5.2 Hybrid Networks

Hybrid Bayesian networks contain both discrete and continuous variables. Figure 5.24 shows a hybrid network, which will be discussed shortly. Methods for exact inference in hybrid Bayesian networks have been developed. For example, Shenoy [2006] develops a method that approximates general hybrid Bayesian networks by a mixture of Gaussian Bayesian networks. However, packages often deal with hybrid networks by discretizing the continuous distributions. HUGIN allows Gaussian variables to have discrete parents while still doing exact inference. It could, therefore, handle the Bayesian network in the following example.

Example 5.15 *Recall the Bayesian network in Figure 5.2, which models fraudulent use of a credit card. Suppose that if jewelry is purchased, the cost of the*

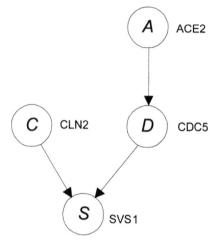

$\rho(s \mid C{=}\text{low}, D{=}\text{low}) = \text{NormalDen}(s; .6, .1)$

$\rho(s \mid C{=}\text{low}, D{=}\text{high}) = \text{NormalDen}(s; 1.3, .3)$

$\rho(s \mid C{=}\text{high}, D{=}\text{low}) = \text{NormalDen}(s; 1.1, .2)$

$\rho(s \mid C{=}\text{high}, D{=}\text{high}) = \text{NormalDen}(s; 1.7, .4)$

Figure 5.25: A Bayesian network showing possible causal relationships among the expression levels of genes. Only the conditional probability distribution of the leaf is shown.

jewelry is likely to be greater if the purchase was due to fraudulent use. We could model this situation using the hybrid Bayesian network in Figure 5.24. The variable Cost is normally distributed given each set of values of its discrete parents. Note that if $J = no$, the distribution is NormalDen$(s; 0, 0)$. This is the same as stating that

$$P(C = 0 | F = yes, J = no) = 0.$$

However, we showed the conditional probability distribution as a normal distribution to be consistent with the other distributions of C.

Example 5.16 *Recall from Section 4.2.2 that the protein transcription factor produced by one gene can have a causal effect on the level of mRNA (called the gene expression level) of another gene. In Chapter 12 we discuss methods that use Bayesian networks to learn these causal effects from data. Gene expression level is often set as the ratio of measured expression to a control level. So, values greater than 1 would indicate a relatively high expression level, whereas values less than 1 would indicate a relatively low expression level. Since gene expression levels are continuous, we could try learning a Gaussian Bayesian network. Another approach taken in [Segal et al., 2005] is to learn a network in* which each variable is normally distributed given values of its parents. However, each parent has only two values, namely *high* and *low*, which determine the conditional distribution of the child. The value *high* represents all expression levels greater than 1, and the value *low* represents all expression levels less than or equal to 1. Such a network appears in Figure 5.25. The nodes in the network represent genes. This network is not exactly hybrid, because every variable is continuous. However, the conditional distributions are based on discrete values.

5.6 How Do We Obtain the Probabilities?

So far we have simply shown the conditional probability distributions in the Bayesian networks we have presented. We have not been concerned with how we obtained them. For example, in the credit card fraud example we simply stated that $P(Age = less30) = .25$. However, how did we obtain this and other probabilities? As mentioned at the beginning of this chapter, they can either be obtained from the subjective judgements of an expert in the area, or they can be learned from data. In Chapter 8 we discuss techniques for learning them from data. Here, we show two techniques for simplifying the process of ascertaining them. The first technique concerns the case where a node has multiple parents, while the second technique concerns nodes that represent continuous random variables.

5.6.1 The Noisy OR-Gate Model

After discussing a problem in obtaining the conditional probabilities when a node has multiple parents, we present models that address this problem.

Difficulty Inherent in Multiple Parents

Suppose lung cancer, bronchitis, and tuberculosis all cause fatigue, and we need to model this relationship as part of a system for medical diagnosis. The portion of the DAG concerning only these four variables appears in Figure 5.26. We need to assess eight conditional probabilities for node F, one for each of the eight combinations of that node's parents. That is, we need to assess the following:

$$P(F = yes|B = no, T = no, L = no)$$

$$P(F = yes|B = no, T = no, L = yes)$$

$$\cdots$$

$$P(F = yes|B = yes, T = yes, L = yes).$$

It would be quite difficult to obtain these values either from data or from an expert physician. For example, to obtain the value of $P(F = yes|B = yes, T = yes, L = no)$ directly from data, we would need a sufficiently large population of individuals who are known to have both bronchitis and tuberculosis, but not lung cancer. To obtain this value directly from an expert, the expert would

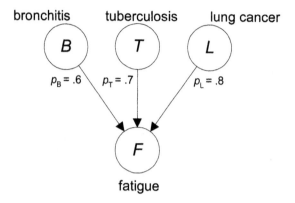

Figure 5.26: We need to assess eight conditional probabilities for node F.

have to be familiar with the likelihood of being fatigued when two diseases are present and the third is not. Next, we show a method for obtaining these conditional probabilities in an indirect way.

The Basic Noisy OR-Gate Model

The noisy OR-gate model concerns the case where the relationships between variables ordinarily represent causal influences, and each variable has only two values. The situation shown in Figure 5.26 is a typical example. Rather than assessing all eight probabilities, we assess the causal strength of each cause for its effect. The **causal strength** is the probability of the cause resulting in the effect whenever the cause is present. In Figure 5.26 we have shown the causal strength p_B of bronchitis for fatigue to be .6. *The assumption is that bronchitis will always result in fatigue unless some unknown mechanism inhibits this from taking place,* and this inhibition takes place 40% of the time. So 60% of the time bronchitis will result in fatigue. Presently, *we assume that all causes of the effect are articulated in the DAG,* and the effect cannot occur unless at least one of its causes is present. In this case, mathematically we have

$$p_B = P(F = yes | B = yes, T = no, L = no).$$

The causal strengths of tuberculosis and lung cancer for fatigue are also shown in Figure 5.26. These three causal strengths should not be as difficult to ascertain as all eight conditional probabilities. For example, to obtain p_B from data we only need a population of individuals who have lung bronchitis and do not have the other diseases. To obtain p_B from an expert, the expert need only ascertain the frequency with which bronchitis gives rise to fatigue.

We can obtain the eight conditional probabilities we need from the three causal strengths if we make one additional assumption. *We need to assume that the mechanisms that inhibit the causes act independently from each other.* For example, the mechanism that inhibits bronchitis from resulting in fatigue acts independently from the mechanism that inhibits tuberculosis from resulting in fatigue. Mathematically, this assumption is as follows:

$$
\begin{aligned}
P(F = no|B = yes, T = yes, L = no) &= (1 - p_B)(1 - p_T) \\
&= (1 - .6)(1 - .7) = .12.
\end{aligned}
$$

Note that in the previous equality we are conditioning on bronchitis and tuberculosis both being present and lung cancer being absent. In this case, fatigue should occur unless the causal effects of bronchitis and tuberculosis are both inhibited. Since we have assumed these inhibitions act independently, the probability that both effects are inhibited is the product of the probabilities that each is inhibited, which is $(1 - p_B)(1 - p_T)$.

In this same way, if all three causes are present, we have

$$
\begin{aligned}
P(F = no|B = yes, T = yes, L = yes) &= (1 - p_B)(1 - p_T)(1 - p_L) \\
&= (1 - .6)(1 - .7)(1 - .8) = .024.
\end{aligned}
$$

Notice that when more causes are present, it is less probable that fatigue will be absent. This is what we would expect. In the following example we compute all eight conditional probabilities needed for node F in Figure 5.26.

Example 5.17 *Suppose we make the assumptions in the noisy OR-gate model, and the causal strengths of bronchitis, tuberculosis, and lung cancer for fatigue are the ones shown in Figure 5.26. Then*

$$
P(F = no|B = no, T = no, L = no) = 1
$$

$$
\begin{aligned}
P(F = no|B = no, T = no, L = yes) &= (1 - p_L) \\
&= (1 - .8) = .2
\end{aligned}
$$

$$
\begin{aligned}
P(F = no|B = no, T = yes, L = no) &= (1 - p_T) \\
&= (1 - .7) = .3
\end{aligned}
$$

$$
\begin{aligned}
P(F = no|B = no, T = yes, L = yes) &= (1 - p_T)(1 - p_L) \\
&= (1 - .7)(1 - .8) = .06
\end{aligned}
$$

$$
\begin{aligned}
P(F = no|B = yes, T = no, L = no) &= (1 - p_B) \\
&= (1 - .6) = .4
\end{aligned}
$$

$$
\begin{aligned}
P(F = no|B = yes, T = no, L = yes) &= (1 - p_B)(1 - p_L) \\
&= (1 - .6)(1 - .8) = .08
\end{aligned}
$$

$$
\begin{aligned}
P(F = no|B = yes, T = yes, L = no) &= (1 - p_B)(1 - p_T) \\
&= (1 - .6)(1 - .7) = .12
\end{aligned}
$$

$$P(F = no|B = yes, T = yes, L = yes) = (1 - p_B)(1 - p_T)(1 - p_L)$$
$$= (1 - .6)(1 - .7)(1 - .8) = .024.$$

Note that since the variables are binary, these are the only values we need to ascertain. The remaining probabilities are uniquely determined by these. For example,

$$P(F = yes|B = yes, T = yes, L = yes) = 1 - .024 = .976.$$

Although we illustrated the model for three causes, it clearly extends to an arbitrary number of causes. We showed the assumptions in the model in italics when we introduced them. Next, we summarize them and show the general formula.

The **noisy OR-gate model** makes the following three assumptions:

1. **Causal inhibition:** This assumption entails that there is some mechanism which inhibits a cause from bringing about its effect, and the presence of the cause results in the presence of the effect if and only if this mechanism is disabled (turned off).

2. **Exception independence:** This assumption entails that the mechanism that inhibits one cause is independent of the mechanism that inhibits other causes.

3. **Accountability:** This assumption entails that an effect can happen only if at least one of its causes is present and is not being inhibited.

The **general formula for the noisy OR-gate model** is as follows: Suppose Y has n causes $X_1, X_2, \ldots, X_n$, all variables are binary, and we assume the noisy OR-gate model. Let p_i be the causal strength of X_i for Y. That is,

$$p_i = P(Y = yes|X_1 = no, X_2 = no, \ldots X_i = yes, \ldots X_n = no).$$

Then if X is a set of nodes that are instantiated to yes,

$$P(Y = no|\mathsf{X}) = \prod_{i \text{ such that } X_i \in \mathsf{X}} (1 - p_i).$$

The Leaky Noisy OR-Gate Model

Of the three assumptions in the noisy OR-gate model, the assumption of accountability seems to be justified least often. For example, in the case of fatigue there are certainly other causes of fatigue such as listening to a lecture by Professor Neapolitan. So the model in Figure 5.26 does not contain all causes of fatigue, and the assumption of accountability is not justified. It seems in many, if not most, situations we would not be certain that we have elaborated all known causes of an effect. Next, we show a version of the model that does not assume accountability. The derivation of the formula for this model is not simple and intuitive like the one for the basic noisy OR-gate model. So we first present the model without deriving it and then show the derivation.

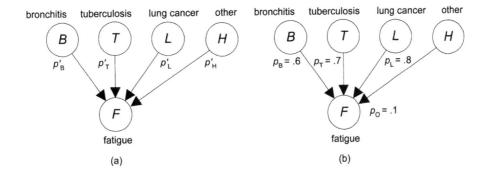

Figure 5.27: The probabilities in (a) are the causal strengths in the noisy OR-gate model. The probabilities in (b) are the ones we ascertain.

The Leaky Noisy OR-Gate Formula The leaky noisy OR-gate model assumes that all causes that have not been articulated can be grouped into one other cause H and that the articulated causes, along with H, satisfy the three assumptions in the noisy OR-gate model. This is illustrated for the fatigue example in Figure 5.27 (a). The probabilities in that figure are the causal strengths in the noisy OR-gate model. For example,

$$p'_B = P(F = yes | B = yes, T = no, L = no, H = no).$$

We could not ascertain these values because we do know whether or not H is present. The probabilities in Figure 5.27 (b) are the ones we could ascertain. For each of the three articulated causes, the probability shown is the probability the effect is present given the remaining two articulated causes are not present. For example,

$$p_B = P(F = yes | B = yes, T = no, L = no).$$

Note the difference in the probabilities p'_B and p_B. The latter one does not condition on a value of H, while the former one does. The probability p_0 is different from the other probabilities. It is the probability that the effect will be present given none of the articulated causes are present. That is,

$$p_0 = P(F = yes | B = no, T = no, L = no).$$

Note again that we are not conditioning on a value of H.

Our goal is to develop conditional probability distributions for a Bayesian network containing the nodes B, T, L, and F from the probabilities ascertained in Figure 5.27 (b). We show an example that realizes this goal after presenting the formula necessary to the task.

The **general formula for the leaky noisy OR-gate model** is as follows (a derivation appears in the next subsection): Suppose Y has n causes $X_1, X_2, \ldots, X_n$, all variables are binary, and we assume the leaky noisy OR-gate model. Let

$$p_i = P(Y = yes | X_1 = no, X_2 = no, \ldots X_i = yes, \ldots X_n = no) \qquad (5.2)$$

$$p_0 = P(Y = yes | X_1 = no, X_2 = no, \ldots X_n = no). \qquad (5.3)$$

Then if X is a set of nodes that are instantiated to yes,

$$P(Y = no | \mathsf{X}) = (1 - p_0) \prod_{i \text{ such that } X_i \in \mathsf{X}} \frac{1 - p_i}{1 - p_0}.$$

Example 5.18 *Let's compute the conditional probabilities for a Bayesian network containing the nodes B, T, L, and F from the probabilities ascertained in Figure 5.27 (b). We have*

$$
\begin{aligned}
P(F = no | B = no, T = no, L = no) &= 1 - p_0 \\
&= 1 - .1 = .9
\end{aligned}
$$

$$
\begin{aligned}
P(F = no | B = no, T = no, L = yes) &= (1 - p_0) \frac{1 - p_L}{1 - p_0} \\
&= 1 - .8 = .2
\end{aligned}
$$

$$
\begin{aligned}
P(F = no | B = no, T = yes, L = no) &= (1 - p_0) \frac{1 - p_T}{1 - p_0} \\
&= 1 - .7 = .3
\end{aligned}
$$

$$
\begin{aligned}
P(F = no | B = no, T = yes, L = yes) &= (1 - p_0) \frac{1 - p_T}{1 - p_0} \frac{1 - p_L}{1 - p_0} \\
&= \frac{(1 - .7)(1 - .8)}{1 - .1} = .067
\end{aligned}
$$

$$
\begin{aligned}
P(F = no | B = yes, T = no, L = no) &= (1 - p_0) \frac{1 - p_B}{1 - p_0} \\
&= 1 - .6 = .4
\end{aligned}
$$

$$
\begin{aligned}
P(F = no | B = yes, T = no, L = yes) &= (1 - p_0) \frac{1 - p_B}{1 - p_0} \frac{1 - p_L}{1 - p_0} \\
&= \frac{(1 - .6)(1 - .8)}{1 - .1} = .089
\end{aligned}
$$

$$
\begin{aligned}
P(F = no | B = yes, T = yes, L = no) &= (1 - p_0) \frac{1 - p_B}{1 - p_0} \frac{1 - p_T}{1 - p_0} \\
&= \frac{(1 - .6)(1 - .7)}{1 - .1} = .133
\end{aligned}
$$

$$
\begin{aligned}
P(F = no | B = yes, T = yes, L = yes) &= (1 - p_0) \frac{1 - p_B}{1 - p_0} \frac{1 - p_T}{1 - p_0} \frac{1 - p_L}{1 - p_0} \\
&= \frac{(1 - .6)(1 - 7)(1 - .8)}{(1 - .1)(1 - .1)} = .030.
\end{aligned}
$$

A Derivation of the Formula The following lemmas and theorem derive the formula in the leaky noisy OR-gate model.

Lemma 5.1 *Given the assumptions and notation shown above for the leaky noisy OR-gate model,*

$$p_0 = p'_H \times P(H = yes).$$

Proof. *Owing to Equality 5.3, we have*

$$
\begin{aligned}
p_0 &= P(Y = yes | X_1 = no, X_2 = no, \ldots X_n = no) \\
&= P(Y = yes | X_1 = no, X_2 = no, \ldots X_n = no, H = yes)P(H = yes) + \\
&\quad P(Y = yes | X_1 = no, X_2 = no, \ldots X_n = no, H = no)P(H = no) \\
&= p'_H \times P(H = yes) + 0 \times P(H = no).
\end{aligned}
$$

This completes the proof. ∎

Lemma 5.2 *Given the assumptions and notation shown above for the leaky noisy OR-gate model,*

$$1 - p'_i = \frac{1 - p_i}{1 - p_0}.$$

Proof. *Owing to Equality 5.2, we have* ∎

$$1 - p_i$$

$$
\begin{aligned}
&= P(Y = no | X_1 = no, \ldots X_i = yes, \ldots X_n = no) \\
&= P(Y = no | X_1 = no, \ldots X_i = yes, \ldots X_n = no, H = yes)P(H = yes) + \\
&\quad P(Y = no | X_1 = no, \ldots X_i = yes, \ldots X_n = no, H = no)P(H = no) \\
&= (1 - p'_i)(1 - p'_H)P(H = yes) + (1 - p'_i)P(H = no) \\
&= (1 - p'_i)(1 - p'_H \times P(H)) \\
&= (1 - p'_i)(1 - p_0).
\end{aligned}
$$

The last equality is due to Lemma 5.1. This completes the proof.

Theorem 5.2 *Given the assumptions and notation shown above for the leaky noisy OR-gate model,*

$$P(Y = no | \mathsf{X}) = (1 - p_0) \prod_{i \text{ such that } X_i \in \mathsf{X}} \frac{1 - p_i}{1 - p_0}.$$

Proof. We have

$$
\begin{aligned}
P(Y = no | \mathsf{X}) &= P(Y = no | \mathsf{X}, H = yes)P(H = yes) + \\
&\quad\; P(Y = no | \mathsf{X}, H = yes)P(H = yes) \\
&= P(H = yes)(1 - p'_H) \prod_{i\ such\ that\ X_i \in \mathsf{X}} (1 - p'_i) + \\
&\quad\; P(H = no) \prod_{i\ such\ that\ X_i \in \mathsf{X}} (1 - p'_i) \\
&= (1 - p_0) \prod_{i\ such\ that\ X_i \in \mathsf{X}} (1 - p'_i) \\
&= (1 - p_0) \prod_{i\ such\ that\ X_i \in \mathsf{X}} \frac{1 - p_i}{1 - p_0}.
\end{aligned}
$$

The second to the last equality is due to Lemma 5.1, and the last is due to Lemma 5.2. ∎

Further Models

A generalization of the noisy OR-gate model to the case of more than two values appears in [Srinivas, 1993]. Diez and Druzdzel [2002] propose a general framework for canonical models, classifying them into three categories: deterministic, noisy, and leaky. They then analyze the most common families of canonical models, namely the noisy OR/MAX, the noisy AND/MIN, and the noisy XOR. Other models for succinctly representing the conditional distributions use the **sigmoid** function [Neal, 1992] and the **logit** function [McLachlan and Krishnan, 1997]. Another approach to reducing the number of parameter estimates is the use of **embedded Bayesian networks**, which is discussed in [Heckerman and Meek, 1997].

5.6.2   Methods for Discretizing Continuous Variables

Often, a Bayesian network contains both discrete and continuous random variables. For example, the Bayesian network in Figure 5.2 contains four random variables that are discrete and one, namely *Age*, that is continuous.[6] However, notice that a continuous probability density function for node *Age* does not appear in the network. Rather, the possible values of the node are three ranges for ages, and the probability of each of these ranges is specified in the network. This is called **discretizing** the continuous variables. Although many Bayesian network inference packages allow the user to specify both continuous variables and discrete variables in the same network, we can sometimes obtain simpler and better inference results by representing the variables as discrete. One reason for this is that, if we discretize the variables, we do not need to assume any particular continuous probability density function. Next, we present two of the most popular methods for discretizing continuous random variables.

---

[6]Technically, if we count age only by years it is discrete. However, even in this case, it is usually represented by a continuous distribution because there are so many values.

Bracket Medians Method

In the **Bracket Medians Method** the mass in a continuous probability distribution function $F(x) = P(X \leq x)$ is divided into n equally spaced intervals. The method proceeds as follows ($n = 5$ in this explanation):

1. Determine n equally spaced intervals in the interval $[0, 1]$. If $n = 5$, the intervals are $[0, .2]$, $[.2, .4]$, $[.4, .6]$, $[.6, .8]$, and $[.8, 1.0]$.

2. Determine points x_1, x_2, x_3, x_4, x_5, and x_6 such that

$$
\begin{aligned}
P(X \leq x_1) &= .0 \\
P(X \leq x_2) &= .2 \\
P(X \leq x_3) &= .4 \\
P(X \leq x_4) &= .6 \\
P(X \leq x_5) &= .8 \\
P(X \leq x_6) &= 1.0,
\end{aligned}
$$

where the values on the right in these equalities are the endpoints of the five intervals.

3. For each interval $[x_i, x_{i+1}]$ compute the bracket median d_i, which is the value such that

$$
P(x_i \leq X \leq d_i) = P(d_i \leq X \leq x_{i+1}).
$$

4. Define the discrete variable D with the following probabilities:

$$
\begin{aligned}
P(D = d_1) &= .2 \\
P(D = d_2) &= .2 \\
P(D = d_3) &= .2 \\
P(D = d_4) &= .2 \\
P(D = d_5) &= .2.
\end{aligned}
$$

Example 5.19 *Recall that the normal density function is given by*

$$
f(x) = \frac{1}{\sqrt{2\pi}\sigma} e^{-\frac{(x - \mu)^2}{2\sigma^2}} \qquad -\infty < x < \infty,
$$

where

$$
E(X) = \mu \qquad and \qquad Var(X) = \sigma^2,
$$

and the cumulative distribution function for this density function is given by

$$
F(x) = P(X \leq x) = \frac{1}{\sqrt{2\pi}\sigma} \int_{-\infty}^{x} e^{-\frac{(t - \mu)^2}{2\sigma^2}} dt \qquad -\infty < x < \infty.
$$

These functions for $\mu = 50$ and $\sigma = 15$ are shown in Figure 5.28. This might be the distribution of age for some particular population. Next we use the Bracket Medians Method to discretize it into three ranges. Then $n = 3$ and our four steps are as follows:

1. Since there is essentially no mass < 0 or > 100, our three intervals are $[0, .333]$, $[.333, .666]$, and $[.666, 1]$.

2. We need to find points x_1, x_2, x_3, and x_4 such that

$$
\begin{aligned}
P(X \le x_1) &= .0 \\
P(X \le x_2) &= .333 \\
P(X \le x_3) &= .666 \\
P(X \le x_4) &= 1.
\end{aligned}
$$

Clearly, $x_1 = 0$ and $x_4 = 100$. To determine x_2 we need to determine

$$x_2 = F^{-1}(.333).$$

Using the mathematics package Maple, we have

$$x_2 = \text{NormalInv}(.333; 50, 15) = 43.5.$$

Similarly,

$$x_3 = \text{NormalInv}(.666; 50, 15) = 56.4.$$

In summary, we have

$$x_1 = 0 \qquad x_2 = 43.5 \qquad x_3 = 56.4 \qquad x_4 = 1.$$

3. Compute the bracket medians. We compute them using Maple by solving the following equations:

$$\text{NormalDist}(d_1; 50, 15)$$

$$= \text{NormalDist}(43.5; 50, 15) - \text{NormalDist}(d_1; 50, 15)$$

Solution is $d_1 = 35.5$.

$$\text{NormalDist}(d_2; 50, 15) - \text{NormalDist}(43.5; 50, 15)$$

$$= \text{NormalDist}(56.4; 50, 15) - \text{NormalDist}(d_2; 50, 15)$$

Solution is $d_2 = 50.0$.

$$\text{NormalDist}(d_3; 50, 15) - \text{NormalDist}(56.4; 50, 15)$$

$$= 1 - \text{NormalDist}(d_3; 50, 15)$$

Solution is $d_3 = 64.5$.

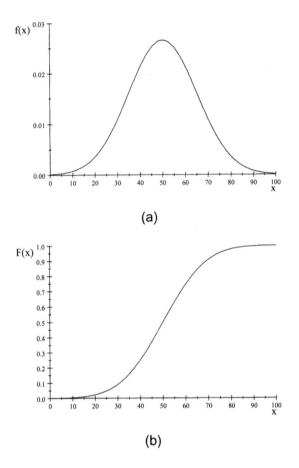

Figure 5.28: The normal density function with $\mu = 50$ and $\sigma = 15$ appears in (a), while the corresponding normal cumulative distribution function appears in (b).

4. *Finally, we set*

$$P(D = 35.5) \quad = \quad .333$$
$$P(D = 50.0) \quad = \quad .333$$
$$P(D = 64.5) \quad = \quad .333.$$

If, for example, a data item's continuous value is between 0 and 43.5, we assign the data item a discrete value of 35.5.

The variable D requires a numeric value if we need to perform computations using it. However, if the variable does not require a numeric value for computational purposes, we need not perform Step 3 in the Bracket Medians Method.

Rather, we just show ranges as the values of D. In the previous example, we would set

$$P(D = \; < 43.5) \;=\; .333$$
$$P(D = 43.5 \text{ to } 56.4) \;=\; .333$$
$$P(D = \; > 56.4) \;=\; .333.$$

Recall that this is what we did for *Age* in the Bayesian network in Figure 5.2. In this case, if a data item's continuous value is between 0 and 43.5, we simply assign the data item that range.

Pearson-Tukey Method

In some applications we want to give special attention to the case when a data item falls in the tail of a density function. For example, if we are trying to predict whether a company will go bankrupt, then unusually low cash flow is indicative they will, while unusually high cash flow is indicative they will not [McKee and Lensberg, 2002]. Values in the middle are not indicative one way or the other. In such cases we want to group the values in each tail together. The Bracket Medians Method does not do this. However, the Pearson-Tukey Method [Keefer, 1983], which we describe next, does. Neapolitan and Jiang [2007] discuss the bankruptcy prediction application in detail.

In the **Pearson-Tukey Method** the mass in a continuous probability distribution function $F(x) = P(X \le x)$ is divided into three intervals. The method proceeds as follows:

1. Determine points x_1, x_2, and x_3 such that

$$P(X \le x_1) \;=\; .05$$
$$P(X \le x_2) \;=\; .50$$
$$P(X \le x_3) \;=\; .95.$$

2. Define the discrete variable D with the following probabilities:

$$P(D = x_1) \;=\; .185$$
$$P(D = x_2) \;=\; .63$$
$$P(D = x_3) \;=\; .185.$$

Example 5.20 *Suppose we have the normal distribution discussed in Example 5.19. Next, we apply the Pearson-Tukey Method to that distribution.*

1. *Using Maple, we have*

$$x_1 = \text{NormalInv}(.05; 50, 15) = 25.3$$

$$x_2 = \text{NormalInv}(.50; 50, 15) = 50$$

$$x_3 = \text{NormalInv}(.95; 50, 15) = 74.7.$$

2. We set

$$P(D = 25.3) = .185$$
$$P(D = 50.0) = .63$$
$$P(D = 74.7) = .185.$$

To assign data items discrete values, we need to determine the range of values corresponding to each of the cutoff points. That is, we compute the following:

$$\text{NormalInv}(.185; 50, 15) = 36.6$$

$$\text{NormalInv}(1 - .185; 50, 15) = 63.4.$$

If a data item's continuous value is < 36.6, we assign the data item the value 25.3; if the value is in [36.6, 63.4], we assign the value 50; and if the value is > 63.4, we assign the value 74.7.

Notice that when we used the Pearson-Tukey Method, the middle discrete value represented numbers in the interval [36.6, 63.4], while when we used the Bracket's Median Method, the middle discrete value represented numbers in the interval [43.5, 56.4]. The interval for the Pearson-Tukey Method is larger, meaning more numbers in the middle are treated as the same discrete value, and the other two discrete values represent values only in the tails.

If the variable does not require a numeric value for computational purposes, we need not perform Steps 1 and 2, but rather just determine the range of values corresponding to each of the cutoff points and just show ranges as the values of D. In the previous example, we would set

$$P(D = \ < 36.6) = .185$$
$$P(D = 36.6 \text{ to } 63.4) = .63$$
$$P(D = \ > 63.4) = .185.$$

In this case if a data item's continuous value is between 0 and 36.6, we simply assign the data item that range.

EXERCISES

Section 5.2

Exercise 5.1 In Example 5.3 it was left as an exercise to show for all values of s, l, and c that

$$P(s, l, c) = P(s|c)P(l|c)P(c).$$

Show this.

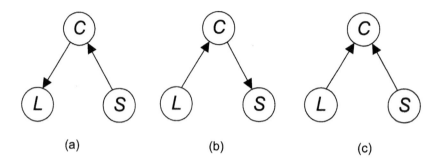

Figure 5.29: The probability distribution discussed in Example 5.1 satisfies the Markov condition with the DAGs in (a) and (b), but not with the DAG in (c).

Exercise 5.2 Consider the joint probability distribution P in Example 5.1.

1. Show that probability distribution P satisfies the Markov condition with the DAG in Figure 5.29 (a) and that P is equal to the product of its conditional distributions in that DAG.

2. Show that P probability distribution satisfies the Markov condition with the DAG in Figure 5.29 (b) and that P is equal to the product of its conditional distributions in that DAG.

3. Show that P probability distribution does not satisfy the Markov condition with the DAG in Figure 5.29 (c) and that P is not equal to the product of its conditional distributions in that DAG.

Section 5.3

Exercise 5.3 Professor Morris investigated gender bias in hiring in the following way. He gave hiring personnel equal numbers of male and female résumés to review, and then he investigated whether their evaluations were correlated with gender. When he submitted a paper summarizing his results to a psychology journal, the reviewers rejected the paper because they said this was an example of fat-hand manipulation. Investigate the concept of fat-hand manipulation, and explain why the journal reviewers might have thought this.

Exercise 5.4 Consider the following piece of medical knowledge: tuberculosis and lung cancer can each cause shortness of breath (dyspnea) and a positive chest X-ray. Bronchitis is another cause of dyspnea. A recent visit to Asia could increase the probability of tuberculosis. Smoking can cause both lung cancer and bronchitis. Create a DAG representing the causal relationships among these variables. Complete the construction of a Bayesian network by determining values for the conditional probability distributions in this DAG, either based on your own subjective judgement or from data.

Exercise 5.5 Explain why, if we reverse the edges in the DAG in Figure 5.12 to obtain the DAG $E \rightarrow D \rightarrow T$, the new DAG also satisfies the Markov condition with the probability distribution of the variables.

Section 5.4

Exercise 5.6 Compute $P(x_1|w_1)$, assuming the Bayesian network in Figure 5.16.

Exercise 5.7 Compute $P(t_1|w_1)$, assuming the Bayesian network in Figure 5.17.

Exercise 5.8 Compute $P(x_1|t_2, w_1)$, assuming the Bayesian network in Figure 5.17.

Exercise 5.9 Using Netica, develop the Bayesian network in Figure 5.2, and use that network to determine the following conditional probabilities.

1. $P(F = yes|Sex = male)$. Is this conditional probability different from $P(F = yes)$? Explain why it is or is not.

2. $P(F = yes|J = yes)$. Is this conditional probability different from $P(F = yes)$? Explain why it is or is not.

3. $P(F = yes|Sex = male, J = yes)$. Is this conditional probability different from $P(F = yes|J = yes)$? Explain why it is or is not.

4. $P(G = yes|F = yes)$. Is this conditional probability different from $P(G = yes)$? Explain why it is or is not.

5. $P(G = yes|J = yes)$. Is this conditional probability different from $P(G = yes)$? Explain why it is or is not.

6. $P(G = yes|J = yes, F = yes)$. Is this conditional probability different from $P(G = yes|F = yes)$? Explain why it is or is not.

7. $P(G = yes|A = < 30)$. Is this conditional probability different from $P(G = yes)$? Explain why it is or is not.

8. $P(G = yes|A = < 30, J = yes)$. Is this conditional probability different from $P(G = yes|J = yes)$? Explain why it is or is not.

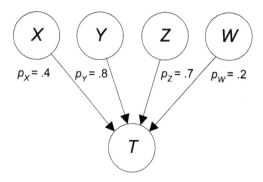

Figure 5.30: The noisy OR-gate model is assumed.

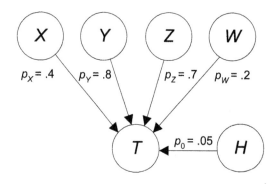

Figure 5.31: The leaky noisy OR-gate model is assumed.

Section 5.5

Exercise 5.10 Using Netica, HUGIN, or some other Bayesian network software package, implement the Bayesian network in Figure 5.24. Using that network compute the conditional probabilities in Exercise 5.9. Compare your answers to those obtained in Exercise 5.9.

Section 5.6

Exercise 5.11 Assume the noisy OR-gate model and the causal strengths are those shown in Figure 5.30. Compute the probability $T = yes$ for all combinations of values of the parents.

Exercise 5.12 Assume the leaky noisy OR-gate model and the relevant probabilities are those shown in Figure 5.31. Compute the probability $T = yes$ for all combinations of values of the parents. Compare the results to those obtained in Exercise 5.11.

Exercise 5.13 Suppose we have the normal density function with $\mu = 100$ and $\sigma = 20$.

1. Discretize this function into four ranges using the Brackets Median Method.

2. Discretize this function using the Pearson-Tukey Method.

Chapter 6

Further Properties of Bayesian Networks

The previous chapter introduced only one relationship between probability distributions and DAGs, namely the Markov condition. However, the Markov condition entails only independencies; it does not entail any dependencies. That is, when we only know that $(\mathbb{G}, P)$ satisfies the Markov condition, we know that the absence of an edge between X and Y entails there is no direct dependency between X and Y, but the presence of an edge between X and Y does not mean there is a direct dependency. In general, we would want an edge to mean there is a direct dependency. In Section 6.2, we discuss another condition, namely the faithfulness condition, which does entail this. Certain DAGs are equivalent in the sense that they entail the same conditional independencies. This is discussed in Section 6.3. Finally, in Section 6.4 we discuss Markov blankets and Markov boundaries, which are sets of variables that render a given variable conditionally independent of all other variables. Before any of this, in Section 6.1 we discuss the conditional independencies entailed by the Markov condition.

6.1 Entailed Conditional Independencies

If $(\mathbb{G}, P)$ satisfies the Markov condition, then each node in $\mathbb{G}$ is conditionally independent of the set of all its nondescendents given its parents. Do these conditional independencies entail any other conditional independencies? That is, if $(\mathbb{G}, P)$ satisfies the Markov condition, are there any other conditional independencies that P must satisfy other than the one based on a node's parents? The answer is yes. Such conditional independencies are called *entailed conditional independencies*. Specifically, we say a DAG **entails a conditional independency** if every probability distribution that satisfies the Markov condition with the DAG must have the conditional independency. Before explicitly showing all the entailed conditional independencies, we illustrate that one would expect these conditional independencies.

6.1.1 Examples of Entailed Conditional Independencies

Suppose some distribution P satisfies the Markov condition with the DAG in Figure 6.1 (a). Then we know $I_P(C, \{F, G\}|, B)$, because B is the parent of C, and F and G are nondescendents of C. Furthermore, we know $I_P(B, G|F)$ because F is the parent of B, and G is a nondescendent of B. These are the only conditional independencies, according to the statement of the Markov condition. However, can any other conditional independencies be deduced from them? For example, can we conclude $I_P(C, G|F)$? Let's first give the variables meaning and the DAG a causal interpretation to see if we would expect this conditional independency.

Suppose we are investigating the way that professors obtain citations, and the variables represent the following:

G:	Graduate program quality
F:	First job quality
B:	Number of publications
C:	Number of citations

Furthermore, suppose the DAG in Figure 6.1 (a) represents the causal relationships among these variables, and there are no hidden common causes.[1] Then it is reasonable to make the causal Markov assumption, and we would feel that

---

[1] We make no claim that this model accurately represents the causal relationships among the variables. See [Spirtes et al., 1993; 2000] for a detailed discussion of this problem.

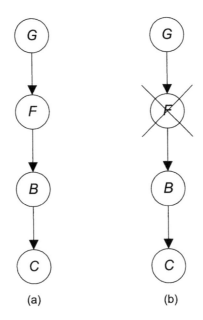

(a)   (b)

Figure 6.1: A causal DAG is shown in (a). The variable F is instantiated in (b).

the probability distribution of the variables satisfies the Markov condition with the DAG. Suppose we learn that Professor La Budde attended a graduate program (G) of high quality. We would now expect that his first job (F) may well be of high quality, which means that he should have a large number of publications (B), which in turn implies he should have a large number of citations (C). Therefore, we would not expect $I_P(C, G)$.

Suppose we next learn that Professor Pellegrini's first job (F) was of high quality. That is, we instantiate F to "high quality." The cross through the node F in Figure 6.1 (b) indicates it is instantiated. We would now expect his number of publications (B) to be large and, in turn, his number of citations (C) to be large. If Professor Pellegrini then tells us he attended a graduate program (G) of high quality, would we expect the number of citations to be even higher than we previously thought? It seems not. The graduate program's high quality implies that the number of citations is probably large, because it implies that the first job is probably of high quality. Once we already know that the first job is of high quality, the information concerning the graduate program should be irrelevant to our beliefs concerning the number of citations. Therefore, we would expect C to not only be conditionally independent of G given its parent B, but also its grandparent F. Either one seems to block the dependency between G and C that exists through the chain $[G, F, B, C]$. So, we would expect $I_P(C, G|F)$.

It is straightforward to show that the Markov condition does indeed entail

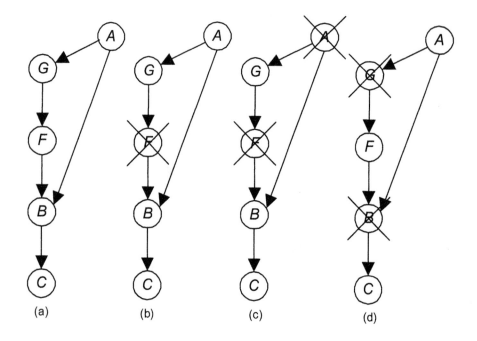

Figure 6.2: A causal DAG is shown in (a). The variable F is instantiated in (b). The variables A and F are instantiated in (c). The variables B and G are instantiated in (d).

$I_P(C, G | F)$ for the DAG $\mathbb{G}$ in Figure 6.1. We show this next. If $(\mathbb{G}, P)$ satisfies the Markov condition,

$$
\begin{aligned}
P(c|g, f) &= \sum_b P(c|b, g, f) P(b|g, f) \\
&= \sum_b P(c|b, f) P(b|f) \\
&= P(c|f).
\end{aligned}
$$

The first equality is due to the Law of Total Probability (in the background space that we know the values of g and f), the second equality is due to the Markov condition, and the last equality is again due to the Law of Total Probability.

If we have an arbitrarily long directed linked list of variables, and P satisfies the Markov condition with that list, in the same way as just illustrated we can show that for any variable in the list, the set of variables above it is conditionally independent of the set of variables below it given that variable. That is, the variable blocks the dependency transmitted through the chain.

Suppose now that P does not satisfy the Markov condition with the DAG in Figure 6.1 (a) because there is a common cause A of G and B. For the sake of illustration, let's say that A represents the following in the current example:

A: Ability

Further, suppose there are no other hidden common causes, so we would now expect P to satisfy the Markov condition with the DAG in Figure 6.2 (a). Would we still expect $I_P(C, G|F)$? It seems not. For example, as before, suppose that we initially learn Professor Pellegrini's first job (F) was of high quality. This instantiation is shown in Figure 6.2 (b). We learn next that his graduate program (G) was of high quality. Given the current model, the fact that G is of high quality is indicative of his having high ability (A), which can directly affect his publication rate (B) and therefore his citation rate (C). So, we now would feel his citation rate (C) could be even higher than what we thought when we only knew his first job (F) was of high quality. This means we would not feel $I_P(C, G|F)$, as we did with the previous model. Suppose next that we know Professor Pellegrini's first job (F) was of high quality and that he has high ability (A). These instantiations appear in Figure 6.2 (c). In this case, his attendance at a high-quality graduate program (G) can no longer be indicative of his ability (A), and therefore, it cannot affect our belief concerning his citation rate (C) through the chain $[G, A, B, C]$. That is, this chain is blocked at A. So, we would expect $I_P(C, G|\{A, F\})$. Indeed, it is possible to prove that the Markov condition does entail $I_P(C, G|\{A, F\})$.

Finally, consider the conditional independency $I_P(F, A|G)$. This independency is obtained directly by applying the Markov condition to the DAG in Figure 6.2 (a). So, we will not offer an intuitive explanation for it. Rather, we discuss whether we would expect the conditional independency to still exist if we also knew the state of B. Suppose we first learn that Professor Georgakis has a high publication rate (B) and attended a high-quality graduate program (G). These instantiations are shown in Figure 6.2 (d). We later learn she also has high ability (A). In this case, her high ability (A) could explain her high publication rate (B), thereby making it less probable she had a high-quality first job (F). As mentioned in Section 5.1, psychologists call this *discounting*. So, the chain $[A, B, F]$ is opened by instantiating B, and we would not expect $I_P(F, A|\{B, G\})$. Indeed, the Markov condition does not entail $I_P(F, A|\{B, G\})$. Note that the instantiation of C should also open the chain $[A, B, F]$. That is, if we know the citation rate (C) is high, then it is probable the publication rate (B) is high, and each of the causes of B can explain this high probability. Indeed, the Markov condition does not entail $I_P(F, A|\{C, G\})$, either.

6.1.2   d-Separation

Figure 6.3 shows the chains that can transmit a dependency between variables X and Y in a Bayesian network. To discuss these dependencies intuitively, we give the edges in that figure causal interpretations as follows:

1. The chain $[X, B, C, D, Y]$ is a causal path from X to Y. In general, there is a dependency between X and Y on this chain, and the instantiation of any intermediate cause on the chain blocks the dependency.

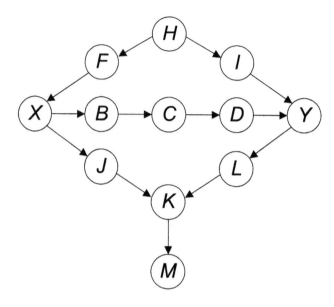

Figure 6.3: This DAG illustrates the chains that can transmit a dependency between X and Y.

2. The chain $[X, F, H, I, Y]$ is a chain in which H is a common cause of X and Y. In general, there is a dependency between any X and Y on this chain, and the instantiation of the common cause H or either of the intermediate causes F and I blocks the dependency.

3. The chain $[X, J, K, L, Y]$ is a chain in which X and Y both cause K. There is no dependency between X and Y on this chain. However, if we instantiate K or M, in general a dependency would be created. We would then need to also instantiate J or L.

To render X and Y conditionally independent, we need to instantiate at least one variable on all the chains that transmit a dependency between X and Y. So, we would need to instantiate at least one variable on the chain $[X, B, C, D, Y]$, at least one variable on the chain $[X, F, H, I, Y]$, and, if K or M are instantiated, at least one other variable on the chain $[X, J, K, L, Y]$.

Now that we've discussed intuitively how dependencies can be transmitted and blocked in a DAG, we show precisely what conditional independencies are entailed by the Markov condition. To do this, we need the notion of *d-separation*, which we define shortly. First we present some preliminary concepts. We say there is a **head-to-head meeting** at X on a chain in a DAG $\mathbb{G}$ if the edges incident to X both have their arrows into X. For example, the chain $Y \leftarrow W \rightarrow X \leftarrow V$ has a head-to-head meeting at X. We say there is a **head-to-tail meeting** at X on a chain in a DAG $\mathbb{G}$ if precisely one of the edges incident to X has its arrows into X. For example, the chain $Y \leftarrow W \leftarrow X \leftarrow V$

has a head-to-tail meeting at X. We say there is a **tail-to-tail meeting** at X on a chain in a DAG $\mathbb{G}$ if neither of the edges incident to X has its arrows into X. For example, the chain $Y \leftarrow W \leftarrow X \rightarrow V$ has a tail-to-tail meeting at X. We now have the following definition:

Definition 6.1 *Suppose we have a DAG $\mathbb{G} = (\mathsf{V}, \mathsf{E})$, a chain ρ in the DAG connecting two nodes X and Y, and a subset of nodes $\mathsf{W} \subseteq \mathsf{V}$. We say that the chain ρ is **blocked** by W if at least one of the following is true:*

1. *There is a node $Z \in \mathsf{W}$ that has a head-to-tail meeting on ρ.*

2. *There is a node $Z \in \mathsf{W}$ that has a tail-to-tail meeting on ρ.*

3. *There is a node Z, such that Z and all Z's descendents are not in W, that has a head-to-head meeting on ρ.*

Example 6.1 *For the DAG in Figure 6.3, the following are some examples of chains that are blocked and that are not blocked.*

1. *The chain $[X, B, C, D, Y]$ is blocked by $\mathsf{W} = \{C\}$ because there is a head-to-tail meeting at C.*

2. *The chain $[X, B, C, D, Y]$ is blocked by $\mathsf{W} = \{C, H\}$ because there is a head-to-tail meeting at C.*

3. *The chain $[X, F, H, I, Y]$ is blocked by $\mathsf{W} = \{C, H\}$ because there is a tail-to-tail meeting at H.*

4. *The chain $[X, J, K, L, Y]$ is blocked by $\mathsf{W} = \{C, H\}$ because there is a head-to-head meeting at K, and K and M are both not in W.*

5. *The chain $[X, J, K, L, Y]$ is not blocked by $\mathsf{W} = \{C, H, K\}$ because there is a head-to-head meeting at K, and K is not in W.*

6. *The chain $[X, J, K, L, Y]$ is blocked by $\mathsf{W} = \{C, H, K, L\}$ because there is a head-to-tail meeting at L.*

We can now define d-separation.

Definition 6.2 *Suppose we have a DAG $\mathbb{G} = (\mathsf{V}, \mathsf{E})$ and a subset of nodes $\mathsf{W} \subseteq \mathsf{V}$. Then X and Y are **d-separated** by W in $\mathbb{G}$ if every chain between X and Y is blocked by W. We write*

$$I_{\mathbb{G}}(X, Y | \mathsf{W}).$$

Definition 6.3 *Suppose we have a DAG $\mathbb{G} = (\mathsf{V}, \mathsf{E})$ and three subsets of nodes X, $\mathsf{Y} \subseteq \mathsf{V}$, and W. We say X and Y are d-separated by W in $\mathbb{G}$ if for every $X \in \mathsf{X}$ and $Y \in \mathsf{Y}$, X and Y are d-separated by W. We write*

$$I_{\mathbb{G}}(\mathsf{X}, \mathsf{Y} | \mathsf{W}).$$

As you might have already suspected, d-separation recognizes all the conditional independencies entailed by the Markov condition. Specifically, we have the following theorem.

Theorem 6.1 *Suppose we have a DAG* $\mathbb{G} = (\mathsf{V}, \mathsf{E})$ *and three subsets of nodes* X, Y, *and* $\mathsf{W} \subseteq \mathsf{V}$. *Then* $\mathbb{G}$ *entails the conditional independency* $I_P(\mathsf{X}, \mathsf{Y}|\mathsf{W})$ *if and only if* $I_{\mathbb{G}}(\mathsf{X}, \mathsf{Y}|\mathsf{W})$.

Proof. *The proof can be found in [Neapolitan, 1990].* ∎

We stated the theorem for sets of variables, but clearly it holds for single variables. That is, if X contains a single variable X and Y contains a single variable Y, then $I_P(\mathsf{X}, \mathsf{Y}|\mathsf{W})$ is the same as $I_P(X, Y|\mathsf{W})$. We show examples of this simpler case next and investigate more complex sets in the exercises.

Example 6.2 *Owing to Theorem 6.1, the following are some conditional independencies the Markov condition entails for the DAG in Figure 6.3.*

Conditional Independency	Reason Conditional Independency Is Entailed
$I_P(X, Y|\{H, C\})$	$[X, F, H, I, Y]$ is blocked at H
	$[X, B, C, D, Y]$ is blocked at C
	$[X, J, K, L, Y]$ is blocked at K
$I_P(X, Y|\{F, D\})$	$[X, F, H, I, Y]$ is blocked at F
	$[X, B, C, D, Y]$ is blocked at D
	$[X, J, K, L, Y]$ is blocked at K
$I_P(X, Y|\{H, C, K, L\})$	$[X, F, H, I, Y]$ is blocked at H
	$[X, B, C, D, Y]$ is blocked at C
	$[X, J, K, L, Y]$ is blocked at L
$I_P(X, Y|\{H, C, M, L\})$	$[X, F, H, I, Y]$ is blocked at H
	$[X, B, C, D, Y]$ is blocked at C
	$[X, J, K, L, Y]$ is blocked at L

In the third row it is necessary to include L to obtain the independency because there is a head-to-head meeting at K on the chain $[X, J, K, L, Y]$, and $K \in \{H, C, K, L\}$. Similarly, in the fourth row, it is necessary to include L to obtain the independency because there is a head-to-head meeting at K on the chain $[X, J, K, L, Y]$, M is a descendent of K, and $M \in \{H, C, M, L\}$.

Example 6.3 *Owing to Theorem 6.1, the following are some conditional independencies the Markov condition does not entail for the DAG in Figure 6.3.*

Conditional Independency	Reason Conditional Independency Is Not Entailed
$I_P(X, Y|H)$	$[X, B, C, D, Y]$ is not blocked
$I_P(X, Y|D)$	$[X, F, H, I, Y]$ is not blocked
$I_P(X, Y|\{H, C, K\})$	$[X, J, K, L, Y]$ is not blocked
$I_P(X, Y|\{H, C, M\})$	$[X, J, K, L, Y]$ is not blocked

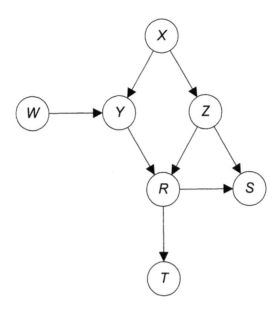

Figure 6.4: The Markov condition entails $I_P(W, X)$ for this DAG.

Example 6.4 *Owing to Theorem 6.1, the Markov condition entails the following conditional independency for the DAG in Figure 6.4.*

Conditional Independency	Reason Conditional Independency Is Entailed
$I_P(W, X)$	$[W, Y, X]$ is blocked at Y
	$[W, Y, R, Z, X]$ is blocked at R
	$[W, Y, R, S, Z, X]$ is blocked at S

Note that $I_P(W, X)$ is the same as $I_P(W, X | \varnothing)$, where $\varnothing$ is the empty set, and Y, R, S, and T are all not in $\varnothing$.

6.2 Faithfulness

Recall that a DAG entails a conditional independency if every probability distribution, which satisfies the Markov condition with the DAG, must have the conditional independency. Theorem 6.1 states that all and only d-separations are entailed conditional independencies. Do not misinterpret this result. It does not say that if some particular probability distribution P satisfies the Markov condition with a DAG $\mathbb{G}$, then P cannot have conditional independencies that $\mathbb{G}$ does not entail. Rather, it only says that P must have all the conditional independencies that are entailed. We illustrate the difference next.

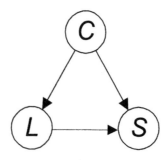

Figure 6.5: A complete DAG.

6.2.1   Unfaithful Probability Distributions

We show two examples of probability distributions that satisfy the Markov condition with a DAG and contain a conditional independency that is not entailed by the DAG.

Example 6.5 *A **complete DAG** is a DAG in which there is an edge between every pair of nodes. Figure 6.5 shows a complete DAG $\mathbb{G}$ containing three nodes, C, L, and S. A complete DAG entails no conditional independencies. So, every probability distribution P of C, L, and S satisfies the Markov condition with the DAG in Figure 6.5.*

Another way to look at this is to notice that the chain rule says that for every probability distribution P of C, L, and S, for all values c, l, and s,

$$P(c, l, s) = P(s|l, c)P(l|c)P(c).$$

So, P is equal to the product of its conditional distributions for the DAG in Figure 6.5, which means that, owing to Theorem 5.1, P satisfies the Markov condition with that DAG.

However, any probability distribution that has a conditional independency, will have a conditional independency that is not entailed by the complete DAG $\mathbb{G}$. For example, consider the joint probability distribution P of C, L, and S discussed in Example 5.1. We showed that

$$I_P(L, S|C).$$

Therefore, this distribution has a conditional independency that is not entailed by $\mathbb{G}$.

Example 6.6 *Consider the Bayesian network in Figure 6.6. The only conditional independency entailed by the Markov condition for the DAG $\mathbb{G}$ in that figure is $I_P(E, F|D)$. So, Theorem 6.1 says that all probability distributions that satisfy the Markov condition with $\mathbb{G}$ must have $I_P(E, F|D)$, which means*

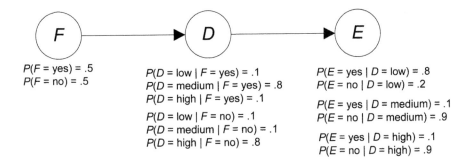

Figure 6.6: For the distribution P in this Bayesian network we have $I_P(E, F)$, but the Markov condition does not entail this conditional independency.

that the probability distribution P in the Bayesian network in Figure 6.6 must have $I_P(E, F|D)$. However, the theorem does not say that P cannot have other independencies. Indeed, it is left as an exercise to show that we have $I_P(E, F)$ for the distribution P in that Bayesian network.

We purposefully assigned values to the conditional distributions in the network in Figure 6.6 to achieve $I_P(E, F)$. If we randomly assigned values, we would be almost certain to obtain a probability distribution that does not have this independency. That is, Meek [1995] proved that almost all assignments of values to the conditional distributions in a Bayesian network will result in a probability distribution that only has conditional independencies entailed by the Markov condition.

Could actual phenomena in nature result in a distribution like that in Figure 6.6? Although we made up the numbers in the network in that figure, we patterned them after something that actually occurred in nature. Let the variables in the figure represent the following.

Variable	What the Variable Represents
F	Whether the subject takes finasteride
D	Subject's dihydro-testosterone level
E	Whether the subject suffers from erectile dysfunction

As shown in Example 5.5, dihydro-testosterone seems to be the hormone necessary for erectile function. Recall from Section 5.3.1 that Merck performed a study indicating that finasteride has a positive causal effect on hair growth. Finasteride accomplishes this by inhibiting the conversion of testosterone to dihydro-testosterone, the hormone responsible for hair loss. Given this, Merck feared that dihydro-testosterone would cause erectile dysfunction. That is, ordinarily if X has a causal influence on Y and Y has a causal influence on Z, then X has a causal influence on Z through Y. However, in a manipulation study Merck found that F does not appear to have a causal influence on E. That is, they learned $I_P(E, F)$. The explanation for this is that finasteride does not

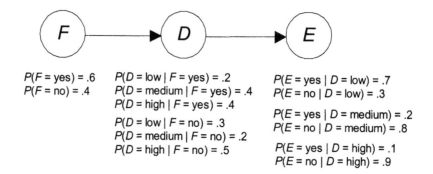

$P(F = \text{yes}) = .6$
$P(F = \text{no}) = .4$

$P(D = \text{low} \mid F = \text{yes}) = .2$
$P(D = \text{medium} \mid F = \text{yes}) = .4$
$P(D = \text{high} \mid F = \text{yes}) = .4$

$P(D = \text{low} \mid F = \text{no}) = .3$
$P(D = \text{medium} \mid F = \text{no}) = .2$
$P(D = \text{high} \mid F = \text{no}) = .5$

$P(E = \text{yes} \mid D = \text{low}) = .7$
$P(E = \text{no} \mid D = \text{low}) = .3$

$P(E = \text{yes} \mid D = \text{medium}) = .2$
$P(E = \text{no} \mid D = \text{medium}) = .8$

$P(E = \text{yes} \mid D = \text{high}) = .1$
$P(E = \text{no} \mid D = \text{high}) = .9$

Figure 6.7: The probability distribution in this Bayesian network is faithful to the DAG in the network.

lower dihydro-testosterone levels beneath some threshold level, and that threshold level is all that is necessary for erectile function. The numbers we assigned in the Bayesian network in Figure 6.6 reflect these causal relationships. The value of F has no effect on whether D is *low*, and D must be *low* to make the probability that E is *yes* high.

6.2.2 The Faithfulness Condition

The probability distribution in the Bayesian network in Figure 6.6 is said to be unfaithful to the DAG in that figure because it contains a conditional independency that is not entailed by the Markov condition. We have the following definition:

Definition 6.4 *Suppose we have a joint probability distribution P of the random variables in some set V and a DAG $\mathbb{G} = (V, E)$. We say that $(\mathbb{G}, P)$ satisfies the **faithfulness condition** if all and only the conditional independencies in P are entailed by $\mathbb{G}$. Furthermore, we say that P and $\mathbb{G}$ are faithful to each other.*

Notice that the faithfulness condition includes the Markov condition because *only* conditional independencies in P can be entailed by $\mathbb{G}$. However, it requires more; that is, it requires that *all* conditional independencies in P must be entailed by $\mathbb{G}$. As noted previously, almost all assignments of values to the conditional distributions will result in a faithful distribution. For example, it is left as an exercise to show that the probability distribution P in the Bayesian network in Figure 6.7 is faithful to the DAG in that figure. We arbitrarily assigned values to the conditional distributions in the figure. However, owing to the result in [Meek, 1995] that almost all assignments of values to the conditional distributions will lead to a faithful distribution, we were willing to bet the farm that this assignment would, too.

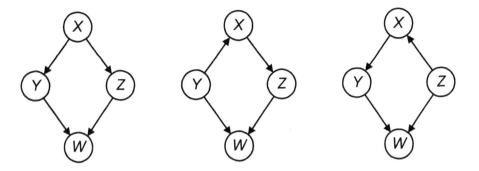

Figure 6.8: These DAGs are Markov equivalent, and there are no other DAGs that are Markov equivalent to them.

Is there some DAG faithful to the probability distribution in the Bayesian network in Figure 6.6? The answer is no, but it is beyond the scope of this book to show this. See [Neapolitan, 2004] for a proof of this fact. Intuitively, the DAG in Figure 6.6 represents the causal relationships among the variables, which means we should not be able to find a DAG that better represents the probability distribution.

6.3 Markov Equivalence

Many DAGs are equivalent in the sense that they have the same d-separations, which means they entail the same conditional independencies. For example, each of the DAGs in Figure 6.8 contains the d-separations $I_{\mathbb{G}}(Y, Z \mid X)$ and $I_{\mathbb{G}}(X, W \mid \{Y, Z\})$, and these are the only d-separations each has. After stating a formal definition of this equivalence, we give a theorem showing how it relates to probability distributions. Finally, we establish a criterion for recognizing this equivalence.

Definition 6.5 *Let* $\mathbb{G}_1 = (\mathsf{V}, \mathsf{E}_1)$ *and* $\mathbb{G}_2 = (\mathsf{V}, \mathsf{E}_2)$ *be two DAGs containing the same set of variables* V. *Then* $\mathbb{G}_1$ *and* $\mathbb{G}_2$ *are called **Markov equivalent** if for every three mutually disjoint subsets* $\mathsf{A}, \mathsf{B}, \mathsf{C} \subseteq \mathsf{V}$, A *and* B *are d-separated by* C *in* $\mathbb{G}_1$ *if and only if* A *and* B *are d-separated by* C *in* $\mathbb{G}_2$. *That is*

$$I_{\mathbb{G}_1}(\mathsf{A}, \mathsf{B}|\mathsf{C}) \Longleftrightarrow I_{\mathbb{G}_2}(\mathsf{A}, \mathsf{B}|\mathsf{C}).$$

Although the previous definition has only to do with graph properties, its application is in probability due to the following theorem:

Theorem 6.2 *Two DAGs are Markov equivalent if and only if they entail the same conditional independencies.*

Proof. *The proof follows immediately from Theorem 6.1.* ∎

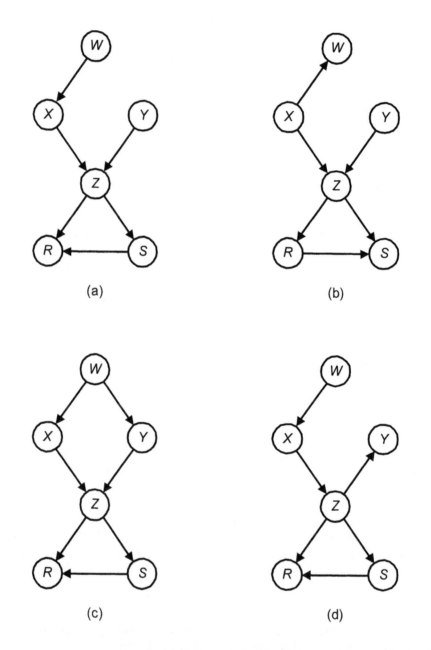

Figure 6.9: The DAGs in (a) and (b) are Markov equivalent. The DAGs in (c) and (d) are not Markov equivalent to the first two DAGs or to each other.

Corollary 6.1 *Let* $\mathbb{G}_1 = (V, E_1)$ *and* $\mathbb{G}_2 = (V, E_2)$ *be two DAGs containing the same set of variables* V. *Then* $\mathbb{G}_1$ *and* $\mathbb{G}_2$ *are Markov equivalent if and only if for every probability distribution* P *of* V, $(\mathbb{G}_1, P)$ *satisfies the Markov condition if and only if* $(\mathbb{G}_2, P)$ *satisfies the Markov condition.*

Proof. *The proof is left as an exercise.* ∎

The theorem that follows enables us to identify Markov equivalence. This theorem was first stated in [Pearl et al., 1989]. First we need this definition. We say there is an **uncoupled head-to-head meeting** at X on a chain in a DAG $\mathbb{G}$ if there is a head-to-head meeting $Y \to X \leftarrow Z$ at X and there is no edge between Y and Z.

Theorem 6.3 *Two DAGs* $\mathbb{G}_1$ *and* $\mathbb{G}_2$ *are Markov equivalent if and only if they have the same links (edges without regard for direction) and the same set of uncoupled head-to-head meetings.*

Proof. *The proof can be found in [Neapolitan, 2004].* ∎

Example 6.7 *The DAGs in Figures 6.9 (a) and 6.9 (b) are Markov equivalent because they have the same links and the only uncoupled head-to-head meeting in both is* $X \to Z \leftarrow Y$. *The DAG in Figure 6.9 (c) is not Markov equivalent to the first two because it has the link* $W - Y$. *The DAG in Figure 6.9 (d) is not Markov equivalent to the first two because, although it has the same links, it does not have the uncoupled head-to-head meeting* $X \to Z \leftarrow Y$. *Clearly, the DAGs in Figures 6.9 (c) and 6.9 (d) are not Markov equivalent to each other, either.*

Theorem 6.3 easily enables us to develop a polynomial-time algorithm for determining whether two DAGs are Markov equivalent. The algorithm would simply check whether the two DAGs have the same links and the same set of uncoupled head-to-head meetings. It is left as an exercise to write such an algorithm.

Furthermore, Theorem 6.3 gives us a simple way to represent a Markov equivalence class with a single graph. That is, we can represent a Markov equivalence class with a graph that has the same links and the same uncoupled head-to-head meeting as the DAGs in the class. Any assignment of directions to the undirected edges in this graph that does not create a new uncoupled head-to-head meeting or a directed cycle yields a member of the equivalence class.

Often there are edges other than uncoupled head-to-head meetings that must be oriented the same in Markov equivalent DAGs. For example, if all DAGs in a given Markov equivalence class have the edge $X \to Y$, and the uncoupled meeting $X \to Y - Z$ is not head-to-head, then all the DAGs in the equivalence class must have $Y - Z$ oriented as $Y \to Z$. So, we define a **DAG pattern** for a Markov equivalence class to be the graph that has the same links as the DAGs in the equivalence class and has oriented all and only the edges common to all the DAGs in the equivalence class. The directed links in a DAG pattern are called

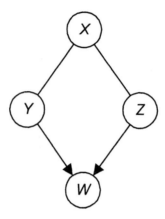

Figure 6.10: This DAG pattern represents the Markov equivalence class in Figure 6.8.

compelled edges. The DAG pattern in Figure 6.10 represents the Markov equivalence class in Figure 6.8. The DAG pattern in Figure 6.11 (b) represents the Markov equivalence class in Figure 6.11 (a). Notice that no DAG, which is Markov equivalent to each of the DAGs in Figure 6.11 (a), can have $W - U$ oriented as $W \leftarrow U$, because this would create another uncoupled head-to-head meeting.

6.4   Markov Blankets and Boundaries

A Bayesian network can have a large number of nodes, and the conditional probability of a given node can be affected by instantiating a distant node. However, it turns out that the instantiation of a set of close nodes can shield a node from the effect of all the other nodes. The next definition and theorem show this.

Definition 6.6 *Let* V *be a set of random variables, P be their joint probability distribution, and* $X \in \mathsf{V}$. *Then a **Markov blanket** M of X is any set of variables such that X is conditionally independent of all the other variables given* M. *That is,*

$$I_P(X, \mathsf{V} - (\mathsf{M} \cup \{X\})|\mathsf{M}).$$

Theorem 6.4 *Suppose $(\mathbb{G}, P)$ satisfies the Markov condition. Then for each variable X, the set of all parents of X, children of X, and parents of children of X is a Markov blanket of X.*

Proof. *Clearly the set of all parents of X, children of X, and parents of children of X d-separates X from the set of all other nodes in* V. *The proof, therefore, follows from Theorem 6.1.* ∎

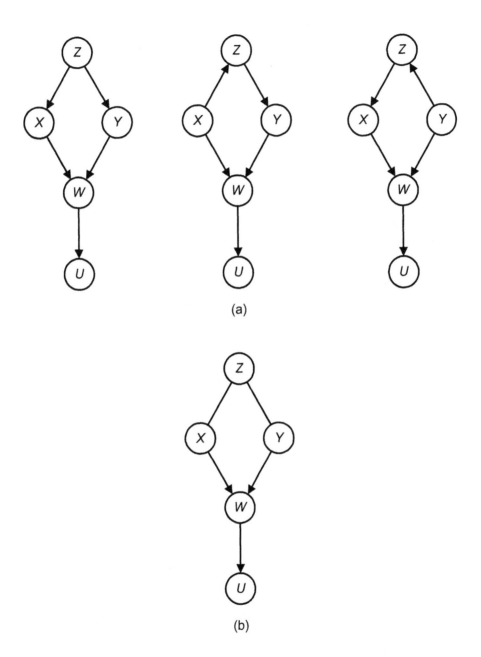

Figure 6.11: The DAG pattern in (b) represents the Markov equivalence class in (a).

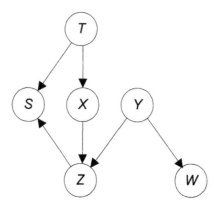

Figure 6.12: If P satisfies the Markov condition with this DAG, then $\{T, Y, Z\}$ is a Markov blanket of X.

Example 6.8 Suppose $(\mathbb{G}, P)$ satisfies the Markov condition where $\mathbb{G}$ is the DAG in Figure 6.12. Then, due to Theorem 6.4, $\{T, Y, Z\}$ is a Markov blanket of X. So, we have

$$I_P(X, \{S, W\} | \{T, Y, Z\}).$$

Example 6.9 Suppose $(\mathbb{G}, P)$ satisfies the Markov condition where $\mathbb{G}$ is the DAG in Figure 6.12, and P has the following conditional independency:

$$I_P(X, \{S, T, W\} | \{Y, Z\}).$$

Then the Markov blanket $\{T, Y, Z\}$ is not minimal in the sense that its subset $\{Y, Z\}$ is also a Markov blanket of X.

The last example motivates the definition that follows.

Definition 6.7 Let V be a set of random variables, P be their joint probability distribution, and $X \in \mathsf{V}$. Then a **Markov boundary** of X is any Markov blanket such that none of its proper subsets is a Markov blanket of X.

We have the following theorem.

Theorem 6.5 Suppose $(\mathbb{G}, P)$ satisfies the faithfulness condition. Then, for each variable X, the set of all parents of X, children of X, and parents of children of X is the unique Markov boundary of X.

Proof. The proof can be found in [Neapolitan, 2004]. ∎

Example 6.10 Suppose $(\mathbb{G}, P)$ satisfies the faithfulness condition where $\mathbb{G}$ is the DAG in Figure 6.12. Then, due to Theorem 6.5, $\{T, Y, Z\}$ is the unique Markov boundary of X.

EXERCISES

Section 6.1

Exercise 6.1 Consider the DAG $\mathbb{G}$ in Figure 6.2 (a). Prove that the Markov condition entails $I_P(C, G | \{A, F\})$ for $\mathbb{G}$.

Exercise 6.2 Suppose we add another variable R, an edge from F to R, and an edge from R to C to the DAG $\mathbb{G}$ in Figure 6.2 (a). The variable R might represent the professor's initial reputation. State which of the following conditional independencies you would feel are entailed by the Markov condition for $\mathbb{G}$. For each that you feel is entailed, try to prove it actually is.

1. $I_P(R, A)$

2. $I_P(R, A | F)$

3. $I_P(R, A | \{F, C\})$

Exercise 6.3 Show that the Markov condition entails the following conditional independencies for the DAG in Figure 6.4.

1. $I_P(X, R | \{Y, Z\})$

2. $I_P(X, T | \{Y, Z\})$

3. $I_P(W, T | R)$

4. $I_P(Y, Z | X)$

5. $I_P(W, S | \{R, Z\})$

6. $I_P(W, S | \{Y, Z\})$

7. $I_P(W, S | \{Y, X\})$

Exercise 6.4 Show that the Markov condition does not entail the following conditional independencies for the DAG in Figure 6.4.

1. $I_P(W, X | Y)$

2. $I_P(W, T | Y)$

Exercise 6.5 State which of the following conditional independencies are entailed by the Markov condition for the DAG in Figure 6.4.

1. $I_P(W, S | \{R, X\})$

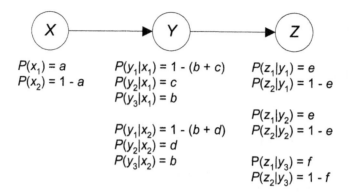

Figure 6.13: Any probability distribution P obtained by assigning values to the parameters in this network is not faithful to the DAG in the network because we have $I_P(X, Z)$.

2. $I_P(\{W, X\}, \{S, T\}|\{R, Z\})$

3. $I_P(\{Y, Z\}, T|\{R, S\})$

4. $I_P(\{X, S\}, \{W, T\}|\{R, Z\})$

5. $I_P(\{X, S, Z\}, \{W, T\}|R)$

6. $I_P(\{X, Z\}, W)$

7. $I_P(\{X, S\}, W)$

Exercise 6.6 Does the Markov condition entail $I_P(\{X, S\}, W|\mathsf{U})$ for any subset of variables U in the DAG in Figure 6.4?

Section 6.2

Exercise 6.7 Show $I_P(F, E)$ for the distribution P in the Bayesian network in Figure 6.6.

Exercise 6.8 Consider the Bayesian network in Figure 6.13. Show that for all assignments of values to a, b, c, d, e, and f that yield a probability distribution P, we have $I_P(X, Z)$. Such probability distributions are not faithful to the DAG in that figure, because X and Z are not d-separated by the empty set. Note that the probability distribution in Figure 6.6 is a member of this family of distributions.

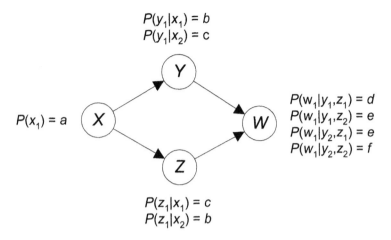

$P(y_1|x_1) = b$
$P(y_1|x_2) = c$

$P(x_1) = a$

$P(w_1|y_1,z_1) = d$
$P(w_1|y_1,z_2) = e$
$P(w_1|y_2,z_1) = e$
$P(w_1|y_2,z_2) = f$

$P(z_1|x_1) = c$
$P(z_1|x_2) = b$

Figure 6.14: Any probability distribution P obtained by assigning values to the parameters in this network is not faithful to the DAG in the network, because we have $I_P(X, W)$.

Exercise 6.9 Assign arbitrary values to the conditional distributions for the DAG in Figure 6.13, and see whether the resultant distribution is faithful to the DAG. Try to find an unfaithful distribution besides ones in the family shown in that figure.

Exercise 6.10 Consider the Bayesian network in Figure 6.14. Show that for all assignments of values to a, b, c, d, e, and f that yield a probability distribution P, we have $I_P(X, W)$. Such probability distributions are not faithful to the DAG in that figure, because X and W are not d-separated by the empty set.

If the edges in the DAG in Figure 6.14 represent causal influences, X and W would be independent if the causal effect of X on W through Y negated the causal effect of X on W through Z. If X represents an individual's age, W represents the individual's typing ability, Y represents the individual's experience, and Z represents the individual's manual dexterity, do you feel X and W might be independent for this reason?

Exercise 6.11 Consider the Bayesian network in Figure 6.15. Show that for all assignments of values to a, b, c, d, e, f, and g that yield a probability distribution P, we have $I_P(X, Y|Z)$. Such probability distributions are not faithful to the DAG in that figure because X and Y are not d-separated by Z.

If the edges in the DAG in Figure 6.15 represent causal influences, X and Y would be independent given Z, if no discounting occurred. Try to find some causal influences that might behave like this.

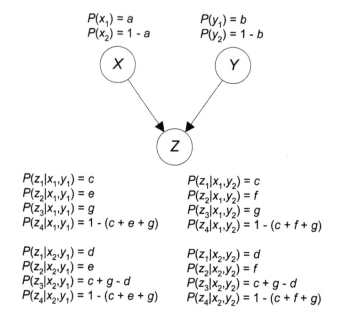

$P(x_1) = a$
$P(x_2) = 1 - a$

$P(y_1) = b$
$P(y_2) = 1 - b$

$P(z_1|x_1,y_1) = c$
$P(z_2|x_1,y_1) = e$
$P(z_3|x_1,y_1) = g$
$P(z_4|x_1,y_1) = 1 - (c + e + g)$

$P(z_1|x_1,y_2) = c$
$P(z_2|x_1,y_2) = f$
$P(z_3|x_1,y_2) = g$
$P(z_4|x_1,y_2) = 1 - (c + f + g)$

$P(z_1|x_2,y_1) = d$
$P(z_2|x_2,y_1) = e$
$P(z_3|x_2,y_1) = c + g - d$
$P(z_4|x_2,y_1) = 1 - (c + e + g)$

$P(z_1|x_2,y_2) = d$
$P(z_2|x_2,y_2) = f$
$P(z_3|x_2,y_2) = c + g - d$
$P(z_4|x_2,y_2) = 1 - (c + f + g)$

Figure 6.15: Any probability distribution P obtained by assigning values to the parameters in this network is not faithful to the DAG in the network, because we have $I_P(X, Y|Z)$.

Section 6.3

Exercise 6.12 Prove Corollary 6.1.

Exercise 6.13 Write an algorithm that determines whether two DAGs are Markov equivalent.

Section 6.4

Exercise 6.14 Apply Theorem 6.4 to find a Markov blanket for node Z in the DAG in Figure 6.12.

Exercise 6.15 Apply Theorem 6.4 to find a Markov blanket for node Y in Figure 6.4.

Chapter 7

Learning Bayesian Network Parameters

Until the early 1990s the DAG in a Bayesian network was ordinarily hand-constructed by a domain expert. Then the conditional probabilities were assessed by the expert, learned from data, or obtained using a combination of both techniques. Eliciting Bayesian networks from experts can be a laborious and difficult process in the case of large networks. As a result, researchers developed methods that could learn the DAG from data. Furthermore, they formalized methods for learning the conditional probabilities from data. We discuss these latter methods in this chapter; the next chapter concerns learning the DAG. In a Bayesian network the conditional probability distributions are called the **parameters**. In Section 7.1 we address the problem of learning a single parameter; in Section 7.2 we discuss learning all the parameters in a Bayesian network. We only discuss learning discrete parameters. Neapolitan [2004] shows a method for learning the parameters in a Gaussian Bayesian network.

7.1   Learning a Single Parameter

We can only learn parameters from data when the probabilities are relative frequencies, which were discussed in Section 2.3.1. So, this discussion pertains only to such probabilities. Although the method is based on rigorous mathematical results obtained by modeling an individual's subjective belief concerning a relative frequency, the method itself is quite simple. Here, we merely present the method. See [Neapolitan, 2004] for the mathematical development. After presenting a method for learning the probability of a binomial random variable, we extend the method to multinomial random variables. Finally, we provide guidelines for articulating our prior beliefs concerning probabilities.

7.1.1   Binomial Random Variables

We illustrate learning with a sequence of examples.

Example 7.1 *Recall the discussion concerning a thumbtack at the beginning of Section 2.3.1. We noted that a thumbtack could land on its flat end, which we call "heads," or it could land with the edge of the flat end and the point touching the ground, which we call "tails." Because the thumbtack is not symmetrical, we have no reason to apply the Principle of Indifference and assign probabilities of .5 to both outcomes. So, we need data to estimate the probability of heads. Suppose we toss the thumbtack 100 times, and it lands heads 65 of those times. Then the maximum likelihood estimate (MLE) is*

$$P(heads) \approx \frac{65}{100} = .65.$$

In general, if there are s heads in n trials, the MLE of the probability is

$$P(heads) \approx \frac{s}{n}.$$

Using the MLE seems reasonable when we have no prior belief concerning the probability. However, it is not so reasonable when we do have prior belief. Consider the next example.

Example 7.2 *Suppose you take a coin from your pocket, toss it 10 times, and it lands heads all those times. Then using the MLE we estimate*

$$P(heads) \approx \frac{10}{10} = 1.$$

After the coin landed heads 10 times, we would not bet as though we were certain that the outcome of the 11th toss will be heads. So, our belief concerning the $P(heads)$ is not the MLE value of 1. Assuming we believe the coins in our pockets are fair, should we instead maintain $P(heads) = .5$ after all 10 tosses landed heads? This might seem reasonable for 10 tosses, but it does not seem so reasonable if 1000 straight tosses landed heads. At some point we would start suspecting the coin was weighted to land heads. We need a method that incorporates one's prior belief with the data. The standard way to do this is for the probability assessor to ascertain integers a and b such that the assessor's experience is equivalent to having seen the first outcome (heads, in this case) occur a times and the second outcome occur b times in $m = a + b$ trials. Then the assessor's prior probabilities are

$$P(heads) = \frac{a}{m} \qquad P(tails) = \frac{b}{m}. \qquad (7.1)$$

After observing s heads and t tails in $n = s + t$ trials, the assessor's posterior probabilities are

$$P(heads|s,t) = \frac{a + s}{m + n} \qquad P(tails|s,t) = \frac{a + t}{m + n}. \qquad (7.2)$$

This posterior probability is called the **maximum a posterior probability (MAP)**. Note that we have used the symbol $=$ rather than $\approx$, and we have written the probability as a conditional probability rather than as an estimate. The reason is that this is a Bayesian technique, and Bayesians say that the value is their probability (belief) based on the data rather than saying it is an estimate of a probability (relative frequency).

We developed Equalities 7.1 and 7.2 based on intuitive grounds. The following theorem is a rigorous derivation of them.

Theorem 7.1 *Suppose we are about to repeatedly toss a thumbtack (or perform any repeatable experiment with two outcomes). Suppose further we assume exchangeability, and we represent our prior belief concerning the probability of heads using a **Dirichlet distribution** with parameters a and b. Then our prior probabilities are given by Equality 7.1, and after observing s heads and t tails in $n = s + t$ trials, our posterior probabilities are given by Equality 7.2.*

Proof. *The proof can be found in [Neapolitan, 2004]* . ∎

See [Neapolitan, 2004] for a discussion of the Dirichlet distribution and a derivation of Equality 7.2. Neapolitan [2004] also discusses how, when we use the Dirichlet distribution with parameters a and b to represent our prior belief concerning the probability of heads, then intuitively our experience is equivalent to having seen the heads occur a times and tails occur b times in $m = a + b$ trials. Briefly, the assumption of **exchangeability**, which was first developed by de Finetti in 1937, is that an individual assigns the same probability to all sequences of the same length containing the same number of each outcome. For

example, the individual assigns the same probability to these two sequences of
heads (H) and tails (T):

$$H, T, H, T, H, T, H, T, T, T \qquad \text{and} \qquad H, T, T, T, T, H, H, T, H, T.$$

Furthermore, the individual assigns the same probability to any other sequence
of ten tosses that has four heads and six tails.

Next, we show more examples. In these examples we only compute the
probability of the first outcome because the probability of the second outcome
is uniquely determined by it.

Example 7.3 *Suppose you are going to repeatedly toss a coin from your
pocket. Since you would feel it highly probable that the relative frequency
is around .5, you might feel your prior experience is equivalent to having seen
50 heads in 100 tosses. Therefore, you could represent your belief with $a = 50$
and $b = 50$. Then $m = 50 + 50 = 100$, and your prior probability of heads is*

$$P(heads) = \frac{a}{m} = \frac{50}{100} = .5.$$

After seeing 48 heads in 100 tosses, your posterior probability is

$$P(heads|48, 52) = \frac{a + s}{m + n} = \frac{50 + 48}{100 + 100} = .49.$$

*The notation $48, 52$ on the right of the conditioning bar in $P(heads|48, 52)$
represents the event that 48 heads and 52 tails have occurred.*

Example 7.4 *Suppose you are going to repeatedly toss a thumbtack. Based
on its structure, you might feel it should land heads about half the time, but
you are not nearly so confident as you were with the coin from your pocket. So,
you might feel your prior experience is equivalent to having seen 3 heads in 6
tosses. Then your prior probability of heads is*

$$P(heads) = \frac{a}{m} = \frac{3}{6} = .5.$$

After seeing 65 heads in 100 tosses, your posterior probability is

$$P(heads|65,35) = \frac{a + s}{m + n} = \frac{3 + 65}{6 + 100} = .64.$$

Example 7.5 *Suppose you are going to sample individuals in the United States
and determine whether they brush their teeth. In this case, you might feel your
prior experience is equivalent to having seen 18 individuals brush their teeth
out of 20 sampled. Then your prior probability of brushing is*

$$P(brushes) = \frac{a}{m} = \frac{18}{20} = .9.$$

*After sampling 100 individuals and learning that 80 brush their teeth, your
posterior probability is*

$$P(brushes|80, 20) = \frac{a + s}{m + n} = \frac{18 + 80}{20 + 100} = .82.$$

You could feel that if we have complete prior ignorance as to the probability, we should take $a = b = 0$. However, consider the next example.

Example 7.6 *Suppose we are going to sample dogs and determine whether or not they eat the potato chips we offer them. Since we have no idea whether a particular dog would eat potato chips, we assign $a = b = 0$, which means $m = 0 + 0 = 0$. Since we cannot divide a by m, we have no prior probability. Suppose next that we sample one dog, and that dog eats the potato chips. Our probability of the next dog eating potato chips is now*

$$P(eats|1,0) = \frac{a+s}{m+n} = \frac{0+1}{0+1} = 1.$$

This belief is not very reasonable, since it means that we are know certain that all dogs eat potato chips. Owing to difficulties such as this and more rigorous mathematical results, prior ignorance to a probability is usually modeled by taking $a = b = 1$, which means $m = 1 + 1 = 2$. If we use these values instead, our posterior probability when the first sampled dog was found to eat potato chips is given by

$$P(eats|1,0) = \frac{a+s}{m+n} = \frac{1+1}{2+1} = \frac{2}{3}.$$

Sometimes we want fractional values for a and b. Consider this example.

Example 7.7 *This example is taken from [Berry, 1996]. Glass panels in highrise buildings sometimes break and fall to the ground. A particular case involved 39 broken panels. In their quest for determining why the panels broke, the owners wanted to analyze the broken panels, but they could only recover three of them. These three were found to contain nickel sulfide (NiS), a manufacturing flaw that can cause panels to break. To determine whether they should hold the manufacturer responsible, the owners then wanted to determine how probable it was that all 39 panels contained NiS. So, they contacted a glass expert.*

The glass expert testified that among glass panels that break, only 5% contain NiS. However, NiS tends to be pervasive in production lots. So, given that the first panel sampled, from a particular production lot of broken panels, contains NiS, the expert felt the probability was .95 that the second panel sampled also contains NiS. It was known that all 39 panels came from the same production lot. So, if we model the expert's prior belief using values of a, b, and $m = a + b$, as discussed previously, we must have that the prior probability is given by

$$P(NiS) = \frac{a}{m} = .05.$$

Furthermore, the expert's posterior probability after finding that the first panel contains NiS must be given by

$$P(NiS|1,0) = \frac{a+1}{m+1} = .95.$$

Solving these last two equations for a and m yields

$$a = \frac{1}{360} \qquad m = \frac{20}{360}.$$

This is an alternative technique for assessing a and b. Namely, we assess the probability for the first trial. Then we assess the conditional probability for the second trial, given the first one is a "success." Once we have values of a and b, we can determine how likely it is that any one of the other 36 panels (the next one sampled) contains NiS after the first three sampled were found to contain it. We have that this probability is given by

$$P(NiS|3,0) = \frac{a+s}{m+n} = \frac{1/360 + 3}{20/360 + 3} = .983.$$

Notice how the expert's probability of NiS quickly changed from being very small to being very large. This is because the values of a and m are so small. We are really most interested in whether all 36 remaining panels contain NiS. It is left as an exercise to show that this probability is given by

$$\prod_{i=0}^{35} \frac{1/360 + 3 + i}{20/360 + 3 + i} = .866.$$

7.1.2 Multinomial Random Variables

The method just discussed readily extends to multinomial random variables. We have the following theorem.

Theorem 7.2 *Suppose we are about to repeatedly perform an experiment with k outcomes $x_1, x_2, \ldots, x_k$. Suppose further we assume exchangeability, and we represent our prior belief concerning the probability of heads using a **Dirichlet distribution** with parameters $a_1, a_2, \ldots, a_k$. Then our prior probabilities are*

$$P(x_1) = \frac{a_1}{m} \qquad P(x_2) = \frac{a_2}{m} \qquad \cdots \qquad P(x_k) = \frac{a_k}{m},$$

where $m = a_1 + a_2 + \cdots + a_k$. After seeing x_1 occur s_1 times, x_2 occur s_2 times, $\ldots$, and x_n occur s_n times in $n = s_1 + s_2 + \cdots + s_k$ trials, our posterior probabilities are as follows:

$$P(x_1|s_1, s_2, \ldots, s_k) = \frac{a_1 + s_1}{m + n}$$

$$P(x_2|s_1, s_2, \ldots, s_k) = \frac{a_2 + s_2}{m + n}$$

$$\vdots$$

$$P(x_k|s_1, s_2, \ldots, s_k) = \frac{a_k + s_k}{m + n}.$$

Proof. *The proof can be found in [Neapolitan, 2004] .* ∎

We ascertain the numbers $a_1, a_2, \ldots, a_k$ by equating our experience to having seen the first outcome occur a_1 times, the second outcome occur a_2 times, $\ldots$, and the last outcome occur a_k times.

Example 7.8 *Suppose we have an asymmetrical-, six-sided die, and we have little idea of the probability of each side coming up. However, it seems that all sides are equally likely. So, we assign*

$$a_1 = a_2 = \cdots = a_6 = 3.$$

Then our prior probabilities are as follows:

$$P(1) = P(2) = \cdots = P(6) = \frac{a_i}{n} = \frac{3}{18} = .16667.$$

Suppose next we throw the die 100 times, with the following results:

Outcome	Number of Occurrences
1	10
2	15
3	5
4	30
5	13
6	27

We then have

$$P(1|10, 15, 5, 30, 13, 27) = \frac{a_1 + s_1}{m + n} = \frac{3 + 10}{18 + 100} = .110$$

$$P(2|10, 15, 5, 30, 13, 27) = \frac{a_2 + s_2}{m + n} = \frac{3 + 15}{18 + 100} = .153$$

$$P(3|10, 15, 5, 30, 13, 27) = \frac{a_3 + s_3}{m + n} = \frac{3 + 5}{18 + 100} = .067$$

$$P(4|10, 15, 5, 30, 13, 27) = \frac{a_4 + s_4}{m + n} = \frac{3 + 30}{18 + 100} = .280$$

$$P(5|10, 15, 5, 30, 13, 27) = \frac{a_5 + s_5}{m + n} = \frac{3 + 13}{18 + 100} = .136$$

$$P(6|10, 15, 5, 30, 13, 27) = \frac{a_6 + s_6}{m + n} = \frac{3 + 27}{18 + 100} = .254.$$

7.1.3 Guidelines for Articulating Prior Belief

Next we give some guidelines for choosing the values that represent our prior beliefs.

Binary Random Variables

The guidelines for binary random variables are as follows:

1. $a = b = 1$: We use these values when we feel we have no knowledge at all concerning the value of the probability. We might also use these values to try to achieve objectivity in the sense that we impose none of our beliefs concerning the probability on the learning algorithm. We only impose the fact that we know, at most, two things can happen. An example might be when we are learning the probability of lung cancer given smoking from data, and we want to communicate our result to the scientific community. The scientific community would not be interested in our prior belief; rather, it would be interested only in what the data had to say. Essentially, when we use these values, the posterior probability represents the belief of an individual who has no prior belief concerning the probability.

2. $a, b > 1$: These values mean that we feel it is likely that the probability of the first outcome is a/m. The larger the values of a and b, the more we believe this. We would use such values when we want to impose our beliefs concerning the relative frequency on the learning algorithm. For example, if we were going to toss a coin taken from a pocket, we might take $a = b = 50$.

3. $a, b < 1$: These values mean that we feel it is likely that the probability of one of the outcomes is high, although we are not committed to which one it is. If we take $a = b \approx 0$ (almost 0), then we are almost certain that the probability of one of the outcomes is very close to 1. We would also use values like these when we want to impose our beliefs concerning the probability on the learning algorithm. Example 7.7 shows a case in which we would choose values less than 1. Notice that such prior beliefs are quickly overwhelmed by data. For example, if $a = b = .1$, and we saw the first outcome x_1 occur in a single trial, we have

$$P(x_1 | 1, 0) = \frac{.1 + 1}{.2 + 1} = .917. \tag{7.3}$$

Intuitively, we thought *a priori* that the probability of one of the outcomes was high. The fact that it took the value x_1 once makes us believe that it is probably that outcome.

Multinomial Random Variables

The guidelines are essentially the same as those for binomial random variables, but we restate them for the sake of clarity:

1. $a_1 = a_2 = \cdots = a_k = 1$: We use these values when we feel we have no knowledge at all concerning the probabilities. We might also use these values to try to achieve objectivity in the sense that we impose none of our beliefs concerning the probability on the learning algorithm. We only impose the fact that we know, at most, k things can happen. An example might be learning the probability of low, medium, and high blood pressure from data, which we want to communicate to the scientific community.

2. $a_1 = a_2 = \cdots = a_k > 1$: These values mean that we feel it is likely that the probability of the kth value is around a_k/m. The larger the values of a_k are, the more we believe this. We would use such values when we want to impose our beliefs concerning the probability on the learning algorithm. For example, if we were going to toss an ordinary die, we might take $a_1 = a_2 = \cdots = a_6 = 50$.

3. $a_1 = a_2 = \cdots = a_k < 1$: These values mean that we feel it is likely that only a few outcomes are probable. We would use such values when we want to impose our beliefs concerning the probabilities on the learning algorithm. For example, suppose we know there are 1,000,000 different species in the world, and we are about to land on an uncharted island. We might feel it probable that not very many of the species are present. So, if we considered the probabilities with which we encountered different species, we would not consider likely probabilities that resulted in a lot of different species. Therefore, we might take $a_i = 1/1{,}000{,}000$ for all i.

7.2 Learning Parameters in a Bayesian Network

The method for learning parameters in a Bayesian network follows readily from the method for learning a single parameter. We illustrate the method with binomial variables. It extends readily to the case of multinomial variables (see [Neapolitan, 2004]). After showing the method, we discuss equivalent sample sizes.

7.2.1 Procedure for Learning Parameters

Consider the two-node network in Figure 7.1 (a). We call such a network a **Bayesian network for parameter learning**. For each probability in the network there is a pair (a_{ij}, b_{ij}). The i indexes the variable; the j indexes the value of the parent(s) of the variable. For example, the pair (a_{11}, b_{11}) is for the first variable (X) and the first value of its parent (in this case there is a default of one parent value since X has no parent). The pair (a_{21}, b_{21}) is for the second variable (Y) and the first value of its parent, namely x_1. The pair (a_{22}, b_{22}) is for the second variable (Y) and the second value of its parent, namely x_2. We have attempted to represent prior ignorance as to the value of all probabilities by taking $a_{ij} = b_{ij} = 1$. We compute the prior probabilities using these pairs, just as we did when we were considering a single parameter. We have the following:

$$P(x_1) = \frac{a_{11}}{a_{11} + b_{11}} = \frac{1}{1+1} = \frac{1}{2}$$

$$P(y_1|x_1) = \frac{a_{21}}{a_{21} + b_{21}} = \frac{1}{1+1} = \frac{1}{2}$$

$$P(y_1|x_2) = \frac{a_{22}}{a_{22} + b_{22}} = \frac{1}{1+1} = \frac{1}{2}.$$

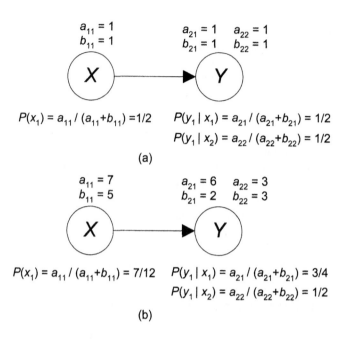

Figure 7.1: A Bayesian network for parameter learning appears in (a); the updated network based on the data in Figure 7.2 appears in (b).

When we obtain data, we use an (s_{ij}, t_{ij}) pair to represent the counts for the ith variable when the variable's parents have their jth value. Suppose we obtain the data in Figure 7.2. The values of the (s_{ij}, t_{ij}) pairs are shown in that figure. We have that $s_{11} = 6$ because x_1 occurs six times, and $t_{11} = 4$ because x_2 occurs four times. Of the six times that x_1 occurs, y_1 occurs five times and y_2 occurs one time. So, $s_{21} = 5$ and $t_{21} = 1$. Of the four times that x_2 occurs, y_1 occurs two times and y_2 occurs two times. So, $s_{22} = 2$ and $t_{22} = 2$. To determine the posterior probability distribution based on the data, we update each conditional probability with the counts relative to that conditional probability. Since we want an updated Bayesian network, we recompute the values of the (a_{ij}, b_{ij}) pairs. We therefore have the following:

$$
\begin{aligned}
a_{11} &= a_{11} + s_{11} = 1 + 6 = 7 \\
b_{11} &= b_{11} + t_{11} = 1 + 4 = 5
\end{aligned}
$$

$$
\begin{aligned}
a_{21} &= a_{21} + s_{21} = 1 + 5 = 6 \\
b_{21} &= b_{21} + t_{21} = 1 + 1 = 2
\end{aligned}
$$

$$
\begin{aligned}
a_{22} &= a_{22} + s_{22} = 1 + 2 = 3 \\
b_{22} &= b_{22} + t_{22} = 1 + 2 = 3.
\end{aligned}
$$

$$
s_{11} = 6 \quad
\begin{array}{ccc}
\text{Case} & X & Y \\
1 & x_1 & y_1 \\
2 & x_1 & y_1 \\
3 & x_1 & y_1 \\
4 & x_1 & y_1 \\
5 & x_1 & y_1 \\
6 & x_1 & y_2 \\
7 & x_2 & y_1 \\
8 & x_2 & y_1 \\
9 & x_2 & y_2 \\
10 & x_2 & y_2
\end{array}
\quad
\begin{array}{l}
s_{21} = 5 \\
\\
t_{21} = 1 \\
s_{22} = 2 \\
t_{22} = 2
\end{array}
$$

$t_{11} = 4$

Figure 7.2: Data on 10 cases.

We then compute the new values of the parameters:

$$
P(x_1) = \frac{a_{11}}{a_{11} + b_{11}} = \frac{7}{7 + 5} = \frac{7}{12}
$$

$$
P(y_1|x_1) = \frac{a_{21}}{a_{21} + b_{21}} = \frac{6}{6 + 2} = \frac{3}{4}
$$

$$
P(y_1|x_2) = \frac{a_{22}}{a_{22} + b_{22}} = \frac{3}{3 + 3} = \frac{1}{2}.
$$

The updated network is shown in Figure 7.1 (b).

7.2.2 Equivalent Sample Size

There is a problem with the way we represented prior ignorance in the preceding subsection. Although it seems natural to set $a_{ij} = b_{ij} = 1$ to represent prior ignorance of all the conditional probabilities, such assignments are not consistent with the metaphor we used for articulating these values. Recall that we said the probability assessor is to choose values of a and b such that the assessor's experience is equivalent to having seen the first outcome occur a times in $a + b$ trials. Therefore, if we set $a_{11} = b_{11} = 1$, the assessor's experience is equivalent to having seen x_1 occur one time in two trials. However, if we set $a_{21} = b_{21} = 1$, the assessor's experience is equivalent to having seen y_1 occur one time out of the two times x_1 occurred. This is not consistent. First, we are saying x_1 occurred once; then we are saying it occurred twice. Aside from this inconsistency, we obtain odd results if we use these priors.

Consider the Bayesian network for parameter learning in Figure 7.3 (a). If we update that network with the data in Figure 7.2, we obtain the network in Figure 7.3 (b). The DAG in Figure 7.3 (a) is Markov equivalent to the one

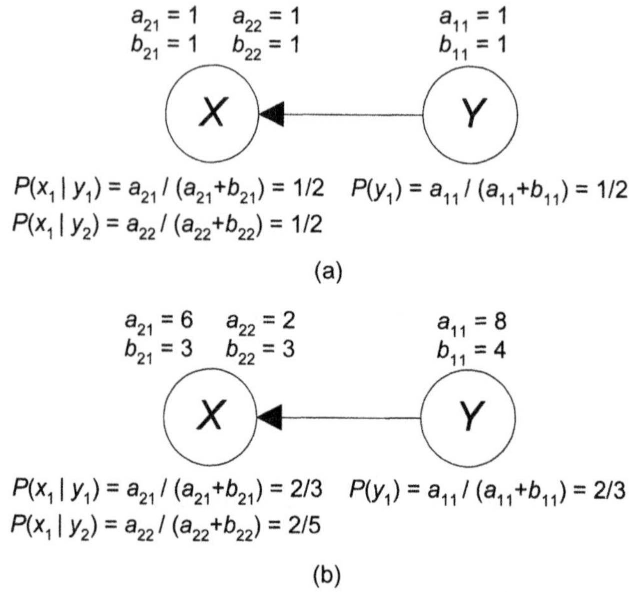

$a_{21} = 1$ $a_{22} = 1$ $a_{11} = 1$
$b_{21} = 1$ $b_{22} = 1$ $b_{11} = 1$

$$\boxed{X} \longleftarrow \boxed{Y}$$

$P(x_1 \mid y_1) = a_{21} / (a_{21}+b_{21}) = 1/2$ $P(y_1) = a_{11} / (a_{11}+b_{11}) = 1/2$
$P(x_1 \mid y_2) = a_{22} / (a_{22}+b_{22}) = 1/2$

(a)

$a_{21} = 6$ $a_{22} = 2$ $a_{11} = 8$
$b_{21} = 3$ $b_{22} = 3$ $b_{11} = 4$

$$\boxed{X} \longleftarrow \boxed{Y}$$

$P(x_1 \mid y_1) = a_{21} / (a_{21}+b_{21}) = 2/3$ $P(y_1) = a_{11} / (a_{11}+b_{11}) = 2/3$
$P(x_1 \mid y_2) = a_{22} / (a_{22}+b_{22}) = 2/5$

(b)

Figure 7.3: A Bayesian network initialized for parameter learning appears in (a); the updated network based on the data in Figure 7.2 appears in (b).

in Figure 7.1 (b). It seems that if we represent the same prior beliefs with equivalent DAGs, then the posterior distributions based on data should be the same. In this case we have attempted to represent prior ignorance as to all probabilities with the networks in Figure 7.1 (a) and Figure 7.3 (a). So, the posterior distributions based on the data in Figure 7.2 should be the same. However, from the Bayesian network in Figure 7.1 (b) we have

$$P(x_1) = \frac{7}{12} = .583,$$

whereas from the Bayesian network in Figure 7.3 (b) we have

$$\begin{aligned}
P(x_1) &= P(x_1|y_1)P(y_1) + P(x_1|y_2)P(y_2) \\
&= \frac{2}{3} \times \frac{2}{3} + \frac{2}{5} \times \frac{1}{3} = .578.
\end{aligned}$$

We see that we obtain different posterior probabilities. Such results are not only odd, but unacceptable since we have attempted to model the same prior belief with the Bayesian networks in Figures 7.1 (a) and 7.3 (a), but we end up with different posterior beliefs.

We can eliminate this difficulty by using a prior equivalent sample size. That is, we specify values of a_{ij} and b_{ij} that could actually occur in a prior sample that exhibit the conditional independencies entailed by the DAG. For example, given the network $X \rightarrow Y$, if we specify that $a_{21} = b_{21} = 1$, this

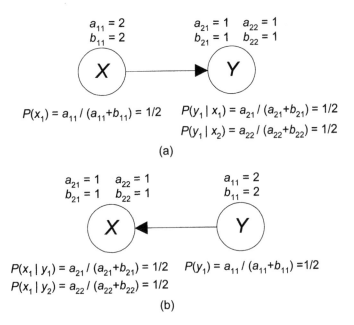

Figure 7.4: Bayesian networks for parameter learning containing prior equivalent sample sizes.

means our prior sample must have x_1 occurring two times. So, we need to specify $a_{11} = 2$. Similarly, if we specify that $a_{22} = b_{22} = 1$, this means that our prior sample must have x_2 occurring two times. So, we need to specify $b_{11} = 2$. Note that we are not saying we actually have a prior sample. We are saying that the probability assessor's beliefs are represented by a prior sample. Figure 7.4 shows prior Bayesian networks using equivalent sample sizes. Notice that the values of a_{ij} and b_{ij} in these networks represent the following prior sample:

Case	X	Y
1	x_1	y_1
2	x_1	y_2
3	x_2	y_1
4	x_2	y_2

It is left as an exercise to show that if we update both the Bayesian networks in Figure 7.4 using the data in Figure 7.2, we obtain the same posterior probability distribution. This result is true, in general. We state it as a theorem, but first we give a formal definition of a prior equivalent sample size.

Definition 7.1 *Suppose we specify a Bayesian network for parameter learning in the case of binomial variables. If there is a number N such that for all i and j*

$$a_{ij} + b_{ij} = P(\mathsf{pa}_{ij}) \times N,$$

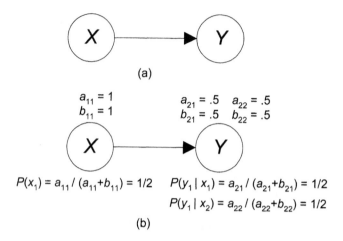

Figure 7.5: Given the DAG in (a) and that X and Y are binomial variables, the Bayesian network for parameter learning in (b) represents prior ignorance.

where pa_{ij} denotes the jth instantiation of the parents of the ith variable, then we say the network has **prior equivalent sample size** N.

This definition is a bit hard to grasp by itself. The following theorem, whose proof can be found in [Neapolitan, 2004], yields a way to represent uniform prior distributions, which is often what we want to do.

Theorem 7.3 *Suppose we specify a Bayesian network for parameter learning in the case of binomial variables and assign for all i and j*

$$a_{ij} = b_{ij} = \frac{N}{2q_i}$$

where N is a positive integer and q_i is the number of instantiations of the parents of the ith variable. Then the resultant Bayesian network has equivalent sample size N, and the joint probability distribution in the Bayesian network is uniform.

We can represent prior ignorance by applying the preceding theorem with $N = 2$. The next example illustrates the technique.

Example 7.9 *Suppose we start with the DAG in Figure 7.5 (a) and X and Y are binary variables. Set $N = 2$. Then, using the method in Theorem 7.3, we have*

$$a_{11} = b_{11} = \frac{N}{2q_1} = \frac{2}{2 \times 1} = 1$$

$$a_{21} = b_{21} = a_{22} = b_{22} = \frac{N}{2q_2} = \frac{2}{2 \times 2} = .5.$$

We obtain the Bayesian network for parameter learning in Figure 7.5 (b).

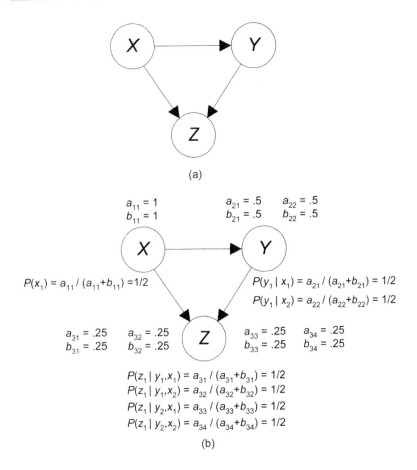

Figure 7.6: Given the DAG in (a) and that X, Y, and Z are binomial variables, the Bayesian network for parameter learning in (b) represents prior ignorance.

Note that we obtained fractional values for a_{21}, b_{21}, a_{22}, and b_{22} in the preceding example, which might seem odd. However, the sum of these values is

$$a_{21} + b_{21} + a_{22} + b_{22} = .5 + .5 + .5 + .5 = 2.$$

So, these fractional values are consistent with the metaphor that says we represent prior ignorance of the $P(Y)$ by assuming that the assessor's experience is equivalent to having seen two trials (see Section 7.1.3). The following is an intuitive justification for the reason that these values should be fractional. Recall from Section 7.1.3 that we said we use fractional values when we feel it is likely that the probability of one of the outcomes is high, although we are not committed to which one it is. The smaller the values, the more likely we feel that this is the case. Now the more parents a variable has, the smaller are the values of a_{ij} and b_{ij} when we set $N = 2$. Intuitively, this seems reasonable because when a variable has many parents and we know the values of the parents, we know a lot about the state of the variable, and therefore, it is more likely that the probability of one of the outcomes is high.

Example 7.10 *Suppose we start with the DAG in Figure 7.6 (a), and X and Y are binary variables. If we set $N = 2$ and use the method in Theorem 7.3, then*

$$a_{11} = b_{11} = \frac{N}{2q_1} = \frac{2}{2 \times 1} = 1$$

$$a_{21} = b_{21} = a_{22} = b_{22} = \frac{N}{2q_2} = \frac{2}{2 \times 2} = .5$$

$$a_{31} = b_{31} = a_{32} = b_{32} = a_{33} = b_{33} = a_{34} = b_{34} = \frac{N}{2q_3} = \frac{2}{2 \times 4} = .25.$$

We obtain the Bayesian network for parameter learning in Figure 7.6 (b).

EXERCISES

Section 7.1

Exercise 7.1 For some two-outcome experiment that you can repeat indefinitely (such as the tossing of a thumbtack), determine the number of occurrences a and b of each outcome that you feel your prior experience is equivalent to having seen. Determine the probability of the first outcome.

Exercise 7.2 Assume that you feel your prior experience concerning the relative frequency of smokers in a particular bar is equivalent to having seen 14 smokers and 6 nonsmokers.

1. You then decide to poll individuals in the bar and ask them if they smoke. What is your probability of the first individual you poll being a smoker?

2. Suppose that after polling 10 individuals, you obtain these data (the value 1 means the individual smokes and 2 means the individual does not smoke):

$$\{1, 2, 2, 2, 2, 1, 2, 2, 2, 1\}.$$

What is your probability that the next individual you poll is a smoker?

3. Suppose that after polling 1000 individuals (it is a big bar), you learn that 312 are smokers. What is your probability that the next individual you poll is a smoker? How does this probability compare to your prior probability?

Exercise 7.3 In Example 7.7 it was left as an exercise to show that the probability that all 36 remaining window panels contain NiS is given by

$$\prod_{i=0}^{35} \frac{1/360 + 3 + i}{20/360 + 3 + i} = .866.$$

Show this.

Exercise 7.4 Suppose that you are about to watch Sam and Dave race several times, and Sam looks substantially athletically inferior to Dave. So, you give Sam a probability of .1 of winning the first race. However, you feel that if Sam wins once, he should usually win. So, given that Sam wins the first race, you give him a .8 probability of winning the next one.

1. Using the method shown in Example 7.7, compute your prior values of a and b.

2. Determine your probability that Sam will win the first five races.

3. Suppose next that Sam wins four of the first five races. Determine your probability that Sam will win the sixth race.

Exercise 7.5 Find a rectangular block (not a cube) and label the sides. Determine values of $a_1, a_2, \ldots, a_6$ that represent your prior probability concerning each side coming up when you throw the block.

1. What is your probability of each side coming up on the first throw?

2. Throw the block 20 times. Compute your probability of each side coming up on the next throw.

Exercise 7.6 Suppose that you are going to sample individuals who have smoked two packs of cigarettes or more daily for the past 10 years. You will determine whether each individual's systolic blood pressure is ≤ 100, 101-120, 121-140, 141-160, or ≥ 161. Determine values of $a_1, a_2, \ldots, a_5$ that represent your prior probability of each blood pressure range.

1. Next you sample such smokers. What is your probability of each blood pressure range for the first individual sampled?

2. Suppose that after sampling 100 individuals, you obtain the following results:

Blood Pressure Range	# of Individuals in This Range
≤ 100	2
101-120	15
121-140	23
141-160	25
≥ 161	35

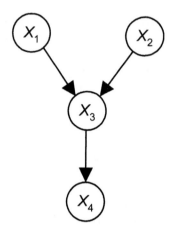

Figure 7.7: A DAG.

Compute your probability of each range for the next individual sampled.

Section 7.2

Exercise 7.7 Suppose that we have the Bayesian network for parameter learning in Figure 7.6 (b), and we have the following data:

Case	X	Y	Z
1	x_1	y_2	z_1
2	x_1	y_1	z_2
3	x_2	y_1	z_1
4	x_2	y_2	z_1
5	x_1	y_2	z_1
6	x_2	y_2	z_2
7	x_1	y_2	z_1
8	x_2	y_1	z_2
9	x_1	y_2	z_1
10	x_1	y_1	z_1
11	x_1	y_2	z_1
12	x_2	y_1	z_2
13	x_1	y_2	z_1
14	x_2	y_2	z_2
15	x_1	y_2	z_1

Determine the updated Bayesian network for parameter learning.

Exercise 7.8 Use the method in Theorem 7.3 to develop Bayesian networks for parameter learning with equivalent sample sizes 1, 2, 4, and 8 for the DAG in Figure 7.7.

Chapter 8

Learning Bayesian Network Structure

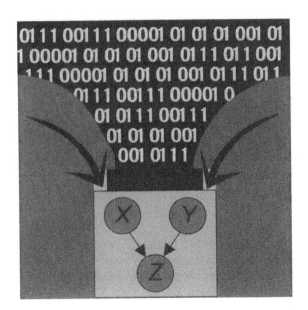

In a Bayesian network, the DAG is called the **structure** and, as mentioned at the beginning of the previous chapter, the conditional probability distributions are called the *parameters*. In the previous chapter we discussed learning the parameters. In this chapter we address learning the structure. In Section 8.1, we formalize the notion of structure learning. Section 8.2 concerns score-based structure learning. Constraint-based structure learning is the focus of Section 8.3. In Section 8.4 we apply structure learning to inferring causal influences from data. When we do not have a large amount of data, ordinarily a unique DAG cannot be learned. In such cases, we can often still learn something of interest concerning the variables in the network by model averaging. We cover model averaging in Section 8.5. When there is a large number of variables, it

Case	Sex	Height (inches)	Wage ($)
1	female	64	30,000
2	male	64	30,000
3	female	64	40,000
4	female	64	40,000
5	female	68	30,000
6	female	68	40,000
7	male	64	40,000
8	male	64	50,000
9	male	68	40,000
10	male	68	50,000
11	male	70	40,000
12	male	70	50,000

Table 8.1: Data on 12 workers.

is necessary to do approximate structure learning. This idea is discussed in Section 8.6. Finally, Section 8.7 presents learning packages that implement the methods discussed in earlier sections.

8.1 Model Selection

Structure learning consists of learning the DAG in a Bayesian network from data. To accomplish this, we need to learn a DAG that satisfies the Markov condition with the probability distribution P that is generating the data. Note that we do not know P; all we know are the data. The process of learning such a DAG is called **model selection**.

Example 8.1 *Suppose we want to model the probability distribution P of sex, height, and wage for American workers. We may obtain the data on 12 workers appearing in Table 8.1. We don't know the probability distribution P. However, from these data we want to learn a DAG that is likely to satisfy the Markov condition with P.*

If our only goal was simply learning some DAG that satisfied the Markov condition with P, our task would be trivial because, as discussed in Section 6.2.1, a probability distribution P satisfies the Markov condition with every complete DAG containing the variables over which P is defined. Recall that our goal with a Bayesian network is to represent a probability distribution succinctly. A complete DAG does not accomplish this because, if there are n binomial variables, the last variable in a complete DAG would require 2^{n-1} conditional distributions. To represent a distribution P succinctly, we need to find a sparse DAG (one containing few edges) that satisfies the Markov condition with P. Next we present methods for doing this.

8.2   Score-Based Structure Learning

In **score-based structure learning**, we assign a score to a DAG based on how well the DAG fits the data. Here we discuss two such scores.

8.2.1   The Bayesian Score

The most straightforward score, called the **Bayesian score**, is the probability of the data D given the DAG. We present this score shortly. However, first we need to discuss the probability of data.

Probability of Data

Suppose that we are going to toss the same coin two times in a row. Let X_1 be a random variable whose value is the result of the first toss, and let X_2 be a random variable whose value is the result of the second toss. If we know that the probability of heads for this coin is .5 and make the usual assumption that the outcomes of the two tosses are independent, we have

$$P(X_1 = heads, X_2 = heads) = \frac{1}{2} \times \frac{1}{2} = \frac{1}{4}.$$

This is a standard result. Suppose now that we are going to toss a thumbtack two times in a row. Suppose further we represent our prior belief concerning the probability of heads using a Dirichlet distribution with parameters a and b (as discussed in Section 7.1.1), and we represent prior ignorance of the probability of heads by taking $a = b = 1$. If the outcome of the first toss is heads, using the notation developed in Section 7.1.1, our updated probability of heads is

$$P(heads|1,0) = \frac{a+1}{a+b+1} = \frac{1+1}{1+1+1} = \frac{2}{3}.$$

Heads is more probable for the second toss because our belief has changed owing to heads occurring on the first toss. So, using our current notation in which we have articulated two random variables, we have that

$$P(X_2 = heads|X_1 = heads) = P(heads|1,0) = \frac{2}{3},$$

and

$$
\begin{aligned}
P(X_1 = heads, X_2 = heads) &= P(X_2 = heads|X_1 = heads)P(X_1 = heads) \\
&= \frac{2}{3} \times \frac{1}{2} = \frac{1}{3}
\end{aligned}
$$

$$
\begin{aligned}
P(X_1 = heads, X_2 = tails) &= P(X_2 = tails|X_1 = heads)P(X_1 = heads) \\
&= \frac{1}{3} \times \frac{1}{2} = \frac{1}{6}
\end{aligned}
$$

$$
\begin{aligned}
P(X_1 = tails, X_2 = heads) &= P(X_2 = heads|X_1 = tails)P(X_1 = tails) \\
&= \frac{1}{3} \times \frac{1}{2} = \frac{1}{6}
\end{aligned}
$$

$$
\begin{aligned}
P(X_1 = tails, X_2 = tails) &= P(X_2 = tails|X_1 = tails)P(X_1 = tails) \\
&= \frac{2}{3} \times \frac{1}{2} = \frac{1}{3}.
\end{aligned}
$$

It might seem odd to you that the four outcomes do not have the same probability. However, recall that we do not know the probability of heads. Therefore, we learn something about the probability of heads from the result of the first toss. If heads occurs on the first toss, that probability goes up; if tails occurs, it goes down. So, two consecutive heads or two consecutive tails are more probable *a priori* than a head followed by a tail or a tail followed by a head.

This result extends readily to a sequence of tosses. For example, suppose we toss the thumbtack three times. Then, owing to the chain rule,

$$P(X_1 = heads, X_2 = tails, X_3 = tails)$$

$$
\begin{aligned}
&= P(X_3 = tails|X_2 = tails, X_1 = heads)P(X_2 = tails|X_1 = heads) \\
&\quad P(X_1 = heads) \\
&= \frac{b+1}{a+b+2} \times \frac{b}{a+b+1} \times \frac{a}{a+b} \\
&= \frac{1+1}{1+1+2} \times \frac{1}{1+1+1} \times \frac{1}{1+1} = .0833.
\end{aligned}
$$

We get the same probability regardless of the order of the outcomes as long as the number of heads and tails is the same. For example,

$$P(X_1 = tails, X_2 = tails, X_3 = heads)$$

$$
\begin{aligned}
&= P(X_3 = heads|X_2 = tails, X_1 = tails)P(X_2 = tails|X_1 = tails) \\
&\quad P(X_1 = tails) \\
&= \frac{a}{a+b+2} \times \frac{b+1}{a+b+1} \times \frac{b}{a+b} \\
&= \frac{1}{1+1+2} \times \frac{2}{1+1+1} \times \frac{1}{1+1} = .0833.
\end{aligned}
$$

We now have the following theorem.

Theorem 8.1 *Suppose that we are about to repeatedly toss a thumbtack (or perform any repeatable experiment with two outcomes). Suppose further that we assume exchangeability, and we represent our prior belief concerning the probability of heads using a Dirichlet distribution with parameters a and b, where a and b are positive integers and $m = a + b$. Let D be data that consist of s heads and t tails in n trials. Then*

$$
P(\mathsf{D}) = \frac{(m-1)!}{(m+n-1)!} \times \frac{(a+s-1)!(b+t-1)!}{(a-1)!(b-1)!}.
$$

Proof. *Since the probability is the same regardless of the order in which the heads and tails occur, we can assume all the heads occur first. We therefore have (as before, the notation s, t on the right side of the conditioning bar means that we have seen s heads and t tails) the following:*

$P(\mathsf{D})$

$$
= P(X_{s+t} = tails | s, t-1) \cdots P(X_{s+2} = tails | s, 1) P(X_{s+1} = tails | s, 0)
$$
$$
P(X_s = heads | s-1, 0) \cdots P(X_2 = heads | 1, 0) P(X_1 = heads)
$$

$$
= \frac{b+t-1}{a+b+s+t-1} \times \cdots \frac{b+1}{a+b+s+1} \times \frac{b}{a+b+s} \times
$$
$$
\frac{a+s-1}{a+b+s-1} \times \cdots \frac{a+1}{a+b+1} \times \frac{a}{a+b}
$$

$$
= \frac{(a+b-1)!}{(a+b+s+t-1)!} \times \frac{(a+s-1)!}{(a-1)!} \times \frac{(b+t-1)!}{(b-1)!}
$$

$$
= \frac{(m-1)!}{(m+n-1)!} \times \frac{(a+s-1)!(b+t-1)!}{(a-1)!(b-1)!}.
$$

The first equality is due to Theorem 7.1. This completes the proof. ■

Example 8.2 *Suppose that, before tossing a thumbtack, we assign $a = 3$ and $b = 5$ to model the slight belief that tails is more probable than heads. We then toss the coin ten times and obtain four heads and six tails. Owing to the preceding theorem, the probability of obtaining these data D is given by*

$$
P(\mathsf{D}) = \frac{(m-1)!}{(m+n-1)!} \times \frac{(a+s-1)!(b+t-1)!}{(a-1)!(b-1)!}
$$

$$
= \frac{(8-1)!}{(8+10-1)!} \times \frac{(3+4-1)!(5+6-1)!}{(3-1)!(5-1)!} = .00077.
$$

Note that the probability of these data is very small. This is because there are many possible outcomes (namely 2^{10}) of tossing a thumbtack ten times.

Theorem 8.1 holds, even if a and b are not integers. We merely state the result here.

Theorem 8.2 *Suppose we are about to repeatedly toss a thumbtack (or perform any repeatable experiment with two outcomes), we assume exchangeability, and we represent our prior belief concerning the probability of heads using a Dirichlet distribution with parameters a and b, where a and b are positive real numbers, and $m = a + b$. Let D be data that consist of s heads and t tails in n trials. Then*

$$
P(\mathsf{D}) = \frac{\Gamma(m)}{\Gamma(m+n)} \times \frac{\Gamma(a+s)\Gamma(b+t)}{\Gamma(a)\Gamma(b)}. \tag{8.1}
$$

Proof. *The proof can be found in [Neapolitan, 2004].* ■

In the preceding theorem Γ denotes the gamma function. When n is an integer ≥ 1, we have that

$$\Gamma(n) = (n-1)! \,.$$

So, the preceding theorem generalizes Theorem 8.1.

Example 8.3 *Recall that in Example 7.7 we set $a = 1/360$ and $b = 19/360$. Then after seeing three windows containing NiS, our updated values of a and b became as follows:*

$$a = \frac{1}{360} + 3 = \frac{1081}{360}$$

$$b = \frac{19}{360} + 0 = \frac{19}{360}.$$

We then wanted the probability that the next 36 windows would contain NiS. This is the probability of obtaining data with $s = 36$ and $t = 0$. Owing to the previous theorem, the probability of these data D is given by

$$
\begin{aligned}
P(\mathsf{D}) &= \frac{\Gamma(m)}{\Gamma(m+n)} \times \frac{\Gamma(a+s)\Gamma(b+t)}{\Gamma(a)\Gamma(b)} \\[2mm]
&= \frac{\Gamma(\frac{1100}{360})}{\Gamma(\frac{1100}{360}+36)} \times \frac{\Gamma(\frac{1081}{360}+36)\Gamma(\frac{19}{360}+0)}{\Gamma(\frac{1081}{360})\Gamma(\frac{19}{360})} \\[2mm]
&= .866.
\end{aligned}
$$

Recall that we obtained this same result by direct computation at the end of Example 7.7.

We developed the method for computing the probability of data for the case of binomial variables. It extends readily to multinomial variables. See [Neapolitan, 2004] for that extension.

Learning DAG Models Using the Bayesian Score

Next we show how we score a DAG model by computing the probability of the data given the model and how we use that score to learn a DAG model.

Suppose we have a Bayesian network for learning, as discussed in Section 7.2. For example, we might have the network in Figure 8.1 (a). Here we call such a network a **DAG model**. We can score a DAG model $\mathbb{G}$ based on data D by determining how probable the data are given the DAG model. That is, we compute $P(\mathsf{D}|\mathbb{G})$. The formula for this probability is the same as that developed in Theorem 8.2, except there is a term of the form in Equality 12.2 for each probability in the network. So, the probability of data D given the DAG model

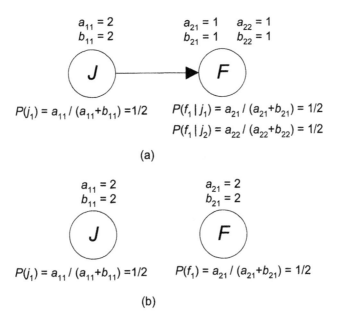

Figure 8.1: The network in (a) models that J has a causal effect on F, whereas the network in (b) models that neither variable causes the other.

$\mathbb{G}_1$ in Figure 8.1 (a) is given by

$$P(\mathsf{D}|\mathbb{G}_1) = \frac{\Gamma(m_{11})}{\Gamma(m_{11}+n_{11})} \times \frac{\Gamma(a_{11}+s_{11})\Gamma(b_{11}+t_{11})}{\Gamma(a_{11})\Gamma(b_{11})} \times \quad (8.2)$$

$$\frac{\Gamma(m_{21})}{\Gamma(m_{21}+n_{21})} \times \frac{\Gamma(a_{21}+s_{21})\Gamma(b_{21}+t_{21})}{\Gamma(a_{21})\Gamma(b_{21})} \times$$

$$\frac{\Gamma(m_{22})}{\Gamma(m_{22}+n_{22})} \times \frac{\Gamma(a_{22}+s_{22})\Gamma(b_{22}+t_{22})}{\Gamma(a_{22})\Gamma(b_{22})}.$$

The data used in each term include only the data relevant to the conditional probability the term represents. This is exactly the same scheme that was used to learn parameters in Section 7.2. For example, the value of s_{21} is the number of cases that have J equal to j_2 and F equal to f_1.

Similarly, the probability of data D given the DAG model $\mathbb{G}_2$ in Figure 8.1 (b) is given by

$$P(\mathsf{D}|\mathbb{G}_2) = \frac{\Gamma(m_{11})}{\Gamma(m_{11}+n_{11})} \times \frac{\Gamma(a_{11}+s_{11})\Gamma(b_{11}+t_{11})}{\Gamma(a_{11})\Gamma(b_{11})} \times \quad (8.3)$$

$$\frac{\Gamma(m_{21})}{\Gamma(m_{21}+n_{21})} \times \frac{\Gamma(a_{21}+s_{21})\Gamma(b_{21}+t_{21})}{\Gamma(a_{21})\Gamma(b_{21})}.$$

Note that the values of a_{11}, s_{11}, and so on in Equality 8.3 are the ones relevant to $\mathbb{G}_2$ and are not the same values as those in Equality 8.2. We have not explicitly shown their dependence on the DAG model, for the sake of notational simplicity.

Example 8.4 *Suppose we want to determine whether job status (J) has a causal effect on whether someone defaults on a loan (F). Furthermore, we articulate just two values for each variable as follows:*

Variable	Value	When the Variable Takes This Value
J	j_1	Individual is a white collar worker
	j_2	Individual is a blue collar worker
F	f_1	Individual has defaulted on a loan at least once
	f_2	Individual has never defaulted on a loan

We represent the assumption that J has a causal effect on F with the DAG model $\mathbb{G}_1$ in Figure 8.1 (a) and the assumption that neither variable has a causal effect on the other with the DAG model $\mathbb{G}_2$ in Figure 8.1 (b). We assume that F does not have a causal effect on J, so we do not model this situation. Note that in both models we used a prior equivalent sample size of four and we represented the prior belief that all probabilities are .5. In general, we can use whatever prior sample size and prior belief that best model our prior knowledge. The only requirement is that both DAG models have the same prior equivalent sample size.

Suppose that next we obtain the data D in the following table:

Case	J	F
1	j_1	f_1
2	j_1	f_1
3	j_1	f_1
4	j_1	f_1
5	j_1	f_2
6	j_2	f_1
7	j_2	f_2
8	j_2	f_2

Then, owing to Equality 8.2,

$$
\begin{aligned}
P(\mathsf{D}|\mathbb{G}_1) &= \frac{\Gamma(m_{11})}{\Gamma(m_{11}+n_{11})} \times \frac{\Gamma(a_{11}+s_{11})\Gamma(b_{11}+t_{11})}{\Gamma(a_{11})\Gamma(b_{11})} \times \\
&\quad \frac{\Gamma(m_{21})}{\Gamma(m_{21}+n_{21})} \times \frac{\Gamma(a_{21}+s_{21})\Gamma(b_{21}+t_{21})}{\Gamma(a_{21})\Gamma(b_{21})} \times \\
&\quad \frac{\Gamma(m_{22})}{\Gamma(m_{22}+n_{22})} \times \frac{\Gamma(a_{22}+s_{22})\Gamma(b_{22}+t_{22})}{\Gamma(a_{22})\Gamma(b_{22})} \\
&= \frac{\Gamma(4)}{\Gamma(4+8)} \times \frac{\Gamma(2+5)\,\Gamma(2+3)}{\Gamma(2)\Gamma(2)} \times \\
&\quad \frac{\Gamma(2)}{\Gamma(2+5)} \times \frac{\Gamma(1+4)\,\Gamma(1+1)}{\Gamma(1)\Gamma(1)} \times \\
&\quad \frac{\Gamma(2)}{\Gamma(2+3)} \times \frac{\Gamma(1+1)\,\Gamma(1+2)}{\Gamma(1)\Gamma(1)} \\
&= 7.2150 \times 10^{-6}.
\end{aligned}
$$

Furthermore,

$$P(D|\mathbb{G}_2) = \frac{\Gamma(m_{11})}{\Gamma(m_{11} + n_{11})} \times \frac{\Gamma(a_{11} + s_{11})\Gamma(b_{11} + t_{11})}{\Gamma(a_{11})\Gamma(b_{11})} \times$$

$$\frac{\Gamma(m_{21})}{\Gamma(m_{21} + n_{21})} \times \frac{\Gamma(a_{21} + s_{21})\Gamma(b_{21} + t_{21})}{\Gamma(a_{21})\Gamma(b_{21})} \times$$

$$= \frac{\Gamma(4)}{\Gamma(4 + 8)} \times \frac{\Gamma(2 + 5)\Gamma(2 + 3)}{\Gamma(2)\Gamma(2)} \times$$

$$\frac{\Gamma(4)}{\Gamma(4 + 8)} \times \frac{\Gamma(2 + 5)\Gamma(2 + 3)}{\Gamma(2)\Gamma(2)}$$

$$= 6.7465 \times 10^{-6}.$$

If our prior belief is that neither model is more probable than the other, we assign

$$P(\mathbb{G}_1) = P(\mathbb{G}_2) = .5.$$

Then, owing to Bayes' Theorem,

$$P(\mathbb{G}_1|D) = \frac{P(D|\mathbb{G}_1)P(\mathbb{G}_1)}{P(D|\mathbb{G}_1)P(\mathbb{G}_1) + P(D|\mathbb{G}_2)P(\mathbb{G}_2)}$$

$$= \frac{7.2150 \times 10^{-6} \times .5}{7.2150 \times 10^{-6} \times .5 + 6.7465 \times 10^{-6} \times .5}$$

$$= .517$$

and

$$P(\mathbb{G}_2|D) = \frac{P(D|\mathbb{G}_2)P(\mathbb{G}_2)}{P(D)}$$

$$= \frac{6.7465 \times 10^{-6}(.5)}{7.2150 \times 10^{-6} \times .5 + 6.7465 \times 10^{-6} \times .5}$$

$$= .483.$$

We select (learn) DAG $\mathbb{G}_1$ and conclude that it is more probable that job status does have a causal effect on whether someone defaults on a loan.

Example 8.5 Suppose we are doing the same study as in Example 8.4, and we obtain the data in the following table:

Case	J	F
1	j_1	f_1
2	j_1	f_1
3	j_1	f_1
4	j_1	f_1
5	j_2	f_2
6	j_2	f_2
7	j_2	f_2
8	j_2	f_2

Then it is left as an exercise to show

$$P(\mathsf{D}|\mathbb{G}_1) = 8.6580 \times 10^{-5}$$

$$P(\mathsf{D}|\mathbb{G}_2) = 4.6851 \times 10^{-6},$$

and if we assign the same prior probability to both DAG models,

$$P(\mathbb{G}_1|\mathsf{D}) \quad = \quad .949$$
$$P(\mathbb{G}_2|\mathsf{D}) \quad = \quad .051.$$

We select (learn) DAG $\mathbb{G}_1$. Notice that the causal model is substantially more probable. This makes sense because even though there are not many data, it exhibits complete dependence.

Example 8.6 Suppose we are doing the same study as in Example 8.4, and we obtain the data D in the following table:

Case	J	F
1	j_1	f_1
2	j_1	f_1
3	j_1	f_2
4	j_1	f_2
5	j_2	f_1
6	j_2	f_1
7	j_2	f_2
8	j_2	f_2

Then it is left as an exercise to show

$$P(\mathsf{D}|\mathbb{G}_1) = 2.4050 \times 10^{-6}$$

$$P(\mathsf{D}|\mathbb{G}_2) = 4.6851 \times 10^{-6},$$

and if we assign the same prior probability to both DAG models,

$$P(\mathbb{G}_1|\mathsf{D}) \quad = \quad .339$$
$$P(\mathbb{G}_2|\mathsf{D}) \quad = \quad .661.$$

We select (learn) DAG $\mathbb{G}_2$. Notice that it is somewhat more probable that the two variables are independent. This makes sense since the data exhibit complete independence.

Learning DAG Patterns

The DAG $F \rightarrow J$ is Markov equivalent to the DAG in Figure 8.1 (a). Intuitively, we would expect it to have the same score. As long as we use a prior equivalent sample size (see Section 8.2.1), they will have the same score. This is discussed in [Neapolitan, 2004]. In general, we cannot distinguish Markov-equivalent DAGs based on data. So, we are actually learning Markov equivalence classes (DAG patterns) when we learn a DAG model from data.

Scoring Larger DAG Models

We illustrated scoring using only two variables. The general formula for the score when there are n variables and the variables are binomial is as follows:

Theorem 8.3 *Suppose we have a DAG* $\mathbb{G} = (V, E)$ *where* V *is a set of binomial random variables, we assume exchangeability, and we use a Dirichlet distribution to represent our prior belief for each conditional probability distribution of every variable in* V. *Suppose further we have data* D *consisting of a set of data items such that each data item is a vector of values of all the variables in* V. *Then*

$$P(D|\mathbb{G}) = \prod_{i=1}^{n} \prod_{j=1}^{q_i} \frac{\Gamma(N_{ij})}{\Gamma(N_{ij} + M_{ij})} \frac{\Gamma(a_{ij} + s_{ij})\Gamma(b_{ij} + t_{ij})}{\Gamma(a_{ij})\Gamma(b_{ij})} \qquad (8.4)$$

where

1. n *is the number of variables.*

2. q_i *is the number of different instantiations of the parents of* X_i.

3. a_{ij} *is our ascertained prior belief concerning the number of times* X_i *took its first value when the parents of* X_i *had their* jth *instantiation.*

4. b_{ij} *is our ascertained value prior belief concerning the number of times* X_i *took its second value when the parents of* X_i *had their* jth *instantiation.*

5. s_{ij} *is the number of times in the data that* X_i *took its first value when the parents of* X_i *had their* jth *instantiation.*

6. t_{ij} *is the number of times in the data that* X_i *took its second value when the parents of* X_i *had their* jth *instantiation.*

7. N_{ij} *and* M_{ij} *are as follows:*

$$N_{ij} = a_{ij} + b_{ij}$$

$$M_{ij} = s_{ij} + t_{ij}.$$

Proof. *The proof can be found in [Neapolitan, 2004].* ∎

Note that, other than n, all the variables defined in the previous theorem depend on $\mathbb{G}$, but we do not show that dependency for the sake of simplicity.

The corresponding theorem when there are n variables and the variables are multinomial is as follows:

Theorem 8.4 *Suppose we have a DAG* $\mathbb{G} = (V, E)$ *where* V *is a set of multinomial random variables, we assume exchangeability, and we use a Dirichlet distribution to represent our prior belief for each conditional probability distribution of every variable in* V. *Suppose further we have data* D *consisting of a set of* *data items such that each data item is a vector of values of all the variables in* V. *Then*

$$P(\mathsf{D}|\mathbb{G}) = \prod_{i=1}^{n} \prod_{j=1}^{q_i} \frac{\Gamma(N_{ij})}{\Gamma(N_{ij} + M_{ij})} \prod_{k=1}^{r_i} \frac{\Gamma(a_{ijk} + s_{ijk})}{\Gamma(a_{ijk})} \qquad (8.5)$$

where

1. *n is the number of variables.*

2. q_i *is the number of different instantiations of the parents of* X_i.

3. a_{ijk} *is our ascertained prior belief concerning the number of times* X_i *took its kth value when the parents of* X_i *had their jth instantiation.*

4. s_{ijk} *is the number of times in the data that* X_i *took its kth value when the parents of* X_i *had their jth instantiation.*

5. N_{ij} *and* M_{ij} *are as follows:*

$$N_{ij} = \sum_k a_{ijk}$$

$$M_{ij} = \sum_k s_{ijk}.$$

Proof. *The proof can be found in [Neapolitan, 2004].* ∎

Note that Equality 8.4 is a special case of Equality 8.5. We call the $P(\mathsf{D}|\mathbb{G})$, obtained using the assumptions in the previous theorem, the **Bayesian score assuming Dirichlet priors**, but ordinarily we only say **Bayesian score**. We denote it as follows:

$$score_{Bayesian}(\mathbb{G} : \mathsf{D}) = P(\mathsf{D}|\mathbb{G}).$$

Neapolitan [2004] develops a Bayesian score for scoring Gaussian Bayesian networks. That is, each variable is assumed to be a function of its parents as shown in Equality 5.1 in Section 5.3.2.

8.2.2   The BIC Score

The **Bayesian information criterion (BIC) score** is as follows:

$$BIC(\mathbb{G} : \mathsf{D}) = \ln\left(P(\mathsf{D}|\hat{\mathsf{P}}, \mathbb{G})\right) - \frac{d}{2}\ln m,$$

where m is the number of data items, d is the dimension of the DAG model, and $\hat{\mathsf{P}}$ is the set of maximum likelihood values of the parameters. The dimension is the number of parameters in the model.

The BIC score is intuitively appealing because it contains (1) a term that shows how well the model predicts the data when the parameter set is equal to its ML value, and (2) a term that punishes for model complexity. Another nice feature of the BIC is that it does not depend on the prior distribution of the parameters, which means there is no need to assess one.

Example 8.7 *Suppose we have the DAG models in Figure 8.1 and the data in Example 8.4. That is, we have the data* D *in the following table:*

Case	J	F
1	j_1	f_1
2	j_1	f_1
3	j_1	f_1
4	j_1	f_1
5	j_1	f_2
6	j_2	f_1
7	j_2	f_2
8	j_2	f_2

For the DAG model in Figure 8.1 (a) we have that

$$\hat{P}(j_1) = \frac{5}{8}$$

$$\hat{P}(f_1|j_1) = \frac{4}{5}$$

$$\hat{P}(f_1|j_2) = \frac{1}{3}.$$

Since there are three parameters in the model, $d = 3$. We then have that

$$P(\mathsf{D}|\hat{P}, \mathbb{G}_1)$$

$$= \left[\hat{P}(f_1|j_1)\hat{P}(j_1)\right]^4 \left[\hat{P}(f_2|j_1)\hat{P}(j_1)\right] \left[\hat{P}(f_1|j_2)\hat{P}(j_2)\right] \left[\hat{P}(f_2|j_2)\hat{P}(j_2)\right]^2$$

$$= \left(\frac{4}{5}\frac{5}{8}\right)^4 \left(\frac{1}{5}\frac{5}{8}\right) \left(\frac{1}{3}\frac{3}{8}\right) \left(\frac{2}{3}\frac{3}{8}\right)^2$$

$$= 6.1035 \times 10^{-5},$$

and therefore

$$BIC\,(\mathbb{G}_1 : \mathsf{D}) = \ln\left(P(\mathsf{D}|\hat{P}, \mathbb{G}_1)\right) - \frac{d}{2}\ln m$$

$$= \ln\left(6.1035 \times 10^{-5}\right) - \frac{3}{2}\ln 8$$

$$= -12.823.$$

For the DAG model in Figure 8.1 (b) we have that

$$\hat{P}(j_1) = \frac{5}{8}$$

$$\hat{P}(f_1) = \frac{5}{8}.$$

Since there are three parameters in the model, $d = 2$. We then have that

$$P(\mathsf{D}|\hat{\mathsf{P}}, \mathbb{G}_2)$$

$$
\begin{aligned}
&= \left[\hat{P}(f_1)\hat{P}(j_1)\right]^4 \left[\hat{P}(f_2|)\hat{P}(j_1)\right] \left[\hat{P}(f_1)\hat{P}(j_2)\right] \left[\hat{P}(f_2)\hat{P}(j_2)\right]^2 \\
&= \left(\frac{5}{8}\frac{5}{8}\right)^4 \left(\frac{3}{8}\frac{5}{8}\right) \left(\frac{5}{8}\frac{3}{8}\right) \left(\frac{3}{8}\frac{3}{8}\right)^2 \\
&= 2.529\,2 \times 10^{-5},
\end{aligned}
$$

and therefore

$$
\begin{aligned}
BIC\,(\mathbb{G}_2 : \mathsf{D}) &= \ln\left(P(\mathsf{D}|\hat{\mathsf{P}}, \mathbb{G}_2)\right) - \frac{d}{2}\ln m \\
&= \ln\left(2.529\,2 \times 10^{-5}\right) - \frac{2}{2}\ln 8 \\
&= -12.644.
\end{aligned}
$$

Note that although the data were more probable given $\mathbb{G}_1$, $\mathbb{G}_2$ won because it is less complex.

Looking at Examples 8.4 and 8.7, we see that the Bayesian score and the BIC can choose different DAG models. The reason is that the data set is small. In the limit they will both choose the same DAG model because the BIC is asymptotically correct. A scoring criterion for DAG models is called **asymptotically correct** if for a sufficiently large data set it chooses the DAG that maximizes $P(\mathsf{D}|\mathbb{G})$.

8.2.3   Consistent Scoring Criteria

We've presented two methods for scoring DAG models. There are others, several of which are discussed in [Neapolitan, 2004]. The question remains as to the quality of these scores. The probability distribution generating the data is called the **generative distribution**. Our goal with a Bayesian network is to represent the generative distribution succinctly. A consistent scoring criterion will almost certainly do this if the data set is large. Specifically, we say a DAG **includes** a probability distribution P if every conditional independency entailed by the DAG is in P. A **consistent scoring criterion** for DAG models has the following two properties:

1. As the size of the data set approaches infinity, the probability approaches one that a DAG that includes P will score higher than a DAG that does not include P.

2. As the size of the data set approaches infinity, the probability approaches one that a smaller DAG that includes P will score higher than a larger DAG that includes P.

Both the Bayesian score and the BIC are consistent scoring criteria.

8.2.4 How Many DAGs Must We Score?

When there are not many variables, we can exhaustively score all possible
DAGs. We then select the DAG(s) with the highest score. However, when
the number of variables is not small, it is computationally unfeasible to find
the maximizing DAGs by exhaustively considering all DAG patterns. That is,
Robinson [1977] showed that the number of DAGs containing n nodes is given
by the following recurrence:

$$f(n) = \sum_{i=1}^{n}(-1)^{i+1}\binom{n}{i}2^{i(n-i)}f(n-i) \qquad n > 2$$
$$f(0) = 1$$
$$f(1) = 1.$$

It is left as an exercise to show $f(2) = 3$, $f(3) = 25$, $f(5) = 29,000$, and
$f(10) = 4.2 \times 10^{18}$. Furthermore, Chickering [1996] proved that for certain
classes of prior distributions, the problem of finding a highest-scoring DAG
is NP-hard. So, researchers developed heuristic DAG search algorithms. We
discuss such algorithms in Section 8.6.1.

8.2.5 Using the Learned Network to Do Inference

Once we learn a DAG from data, we can then learn the parameters. The result
will be a Bayesian network that we can use to do inference for the next case.
The next example illustrates the technique.

Example 8.8 *Suppose we have the situation and data in Example 8.4. That
is, we have the data D in the following table:*

Case	J	F
1	j_1	f_1
2	j_1	f_1
3	j_1	f_1
4	j_1	f_1
5	j_1	f_2
6	j_2	f_1
7	j_2	f_2
8	j_2	f_2

*Then, as shown in Example 8.4, we would learn the DAG in Figure 8.1 (a).
Next we can update the conditional probabilities in the Bayesian network for
learning in Figure 8.1 (a) using the preceding data and the parameter learning
technique discussed in Section 7.2. The result is the Bayesian network in Figure
8.2.*

*Suppose now that we find out that Sam has $F = f_2$. That is, Sam has
never defaulted on a loan. We can use the Bayesian network to compute the*

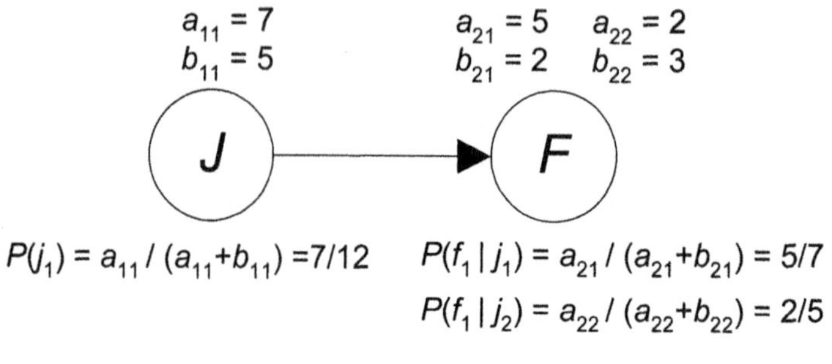

$$a_{11} = 7 \qquad\qquad a_{21} = 5 \qquad a_{22} = 2$$
$$b_{11} = 5 \qquad\qquad b_{21} = 2 \qquad b_{22} = 3$$

$$\boxed{J} \longrightarrow \boxed{F}$$

$$P(j_1) = a_{11} / (a_{11}+b_{11}) = 7/12 \qquad P(f_1 \mid j_1) = a_{21} / (a_{21}+b_{21}) = 5/7$$
$$P(f_1 \mid j_2) = a_{22} / (a_{22}+b_{22}) = 2/5$$

Figure 8.2: An updated Bayesian network for learning based on the data in Example 8.8.

probability that Sam is a white-collar worker. For this simple network we can just use Bayes' Theorem as follows:

$$P(j_1|f_1) = \frac{P(f_1|j_1)P(j_1)}{P(f_1|j_1)P(j_1) + Pf_1|j_2)P(j_2)}$$

$$= \frac{(5/7)\,(7/12)}{(5/7)\,(7/12) + (2/5)\,(5/12)} = .714.$$

The probabilities in the previous calculation are all conditional on the data D *and the DAG that we select. However, once we select a DAG and learn the parameters, we do not bother to show this dependence.*

8.3 Constraint-Based Structure Learning

Next we discuss a quite different structure learning technique called **constraint-based learning**. In this approach, we try to learn a DAG from the conditional independencies in the generative probability distribution P. First, we illustrate the constraint-based approach by showing how to learn a DAG faithful to a probability distribution. This is followed by a discussion of embedded faithfulness. Finally, we present causal learning.

8.3.1 Learning a DAG Faithful to P

Recall that $(\mathbb{G}, P)$ satisfies the faithfulness condition if all and only the conditional independencies in P are entailed by $\mathbb{G}$. After discussing why we would want to learn a DAG faithful to a probability distribution P, we illustrate learning such a DAG.

Why We Want a Faithful DAG

Consider again the objects in Figure 2.1. In Example 2.23, we let P assign $1/13$ to each object in the figure, and we defined these random variables on the set containing the objects:

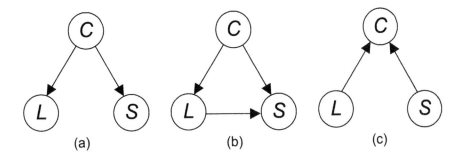

Figure 8.3: If the only conditional independency in P is $I_P(L, S|C)$, then P satisfies the Markov condition with the DAGs in (a) and (b), and P satisfies the faithfulness condition only with the DAG in (a).

Variable	Value	Outcomes Mapped to This Value
L	l_1	All objects containing an A
	l_2	All objects containing a B
S	s_1	All square objects
	s_2	All diamond-shaped objects
C	c_1	All black objects
	c_2	All white objects

We then showed that L and S are conditionally independent given C. That is, using the notation established in Section 2.2.2, we showed that

$$I_P(L, S|C).$$

In Example 5.1, we showed that this implies that P satisfies the Markov condition with the DAG in Figure 8.3 (a). However, P also satisfies the Markov condition with the complete DAG in Figure 8.3 (b). P does not satisfy the Markov condition with the DAG in Figure 8.3 (c) because that DAG entails $I_P(L, S)$ and this independency is not in P. The DAG in Figure 8.3 (b) does not represent P very well because it does not entail a conditional independency that is in P, namely $I_P(L, S|C)$. This is a violation of the faithfulness condition. Of the DAGs in Figure 8.3, only the one in Figure 8.3 (a) is faithful to P.

If we can find a DAG that is faithful to a probability distribution P, we have achieved our goal of representing P succinctly. That is, if there are DAGs faithful to P, then those DAGs are the smallest DAGs that include P (see [Neapolitan, 2004]). We say *DAGs* because if a DAG is faithful to P, then clearly any Markov-equivalent DAG is also faithful to P. For example, the DAGs $L \rightarrow C \rightarrow S$ and $S \leftarrow C \leftarrow L$, which are Markov equivalent to the DAG $L \leftarrow C \rightarrow S$, are also faithful to the probability distribution P concerning the objects in Figure 2.1. As we shall see, not every probability distribution has a DAG that is faithful to it. However, if there are DAGs faithful to a probability distribution, it is relatively easy to discover them. We discuss learning a faithful DAG next.

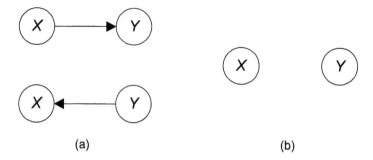

(a)   (b)

Figure 8.4: If the set of conditional independencies is $\{I_P(X,Y)\}$, we must have the DAG in (b), whereas if it is $\varnothing$, we must have one of the DAGs in (a).

Learning a Faithful DAG

We assume that we have a sample of entities from the population over which the random variables are defined, and we know the values of the variables of interest for the entities in the sample. The sample could be a random sample, or it could be obtained from passive data. From this sample, we have deduced the conditional independencies among the variables. A method for deducing conditional independencies and obtaining a measure of our confidence in them is described in [Spirtes et al., 1993; 2000] and [Neapolitan, 2004]. Our confidence in the DAG we learn is no greater than our confidence in these conditional independencies.

Example 8.9 *It is left as an exercise to show that the data shown in Example 8.1 exhibit this conditional independency:*

$$I_P(Height, Wage|Sex).$$

Therefore, from these data we can conclude, with a certain measure of confidence, that this conditional independency exists in the population at large.

Next we give a sequence of examples in which we learn a DAG that is faithful to the probability distribution of interest. These examples illustrate how a faithful DAG can be learned from the conditional independencies if one exists. We stress again that the DAG is faithful to the conditional independencies we have learned from the data. We are not certain that these are the conditional independencies in the probability distribution for the entire population.

Example 8.10 *Suppose* V *is our set of observed variables,* V $= \{X,Y\}$*, and the set of conditional independencies in* P *is*

$$\{I_P(X,Y)\}.$$

We want to find a DAG faithful to P*. We cannot have either of the DAGs in Figure 8.4 (a). The reason is that these DAGs do not entail that* X *and* Y

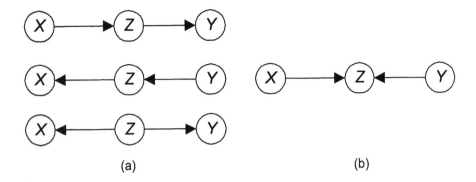

Figure 8.5: If the set of conditional independencies is $\{I_P(X,Y)\}$, we must have the DAG in (b); if it is $\{I_P(X,Y|Z)\}$, we must have one of the DAGs in (a).

are independent, which means the faithfulness condition is not satisfied. So, we must have the DAG in Figure 8.4 (b). We conclude that P is faithful to the DAG in Figure 8.4 (b).

Example 8.11 Suppose $\mathsf{V} = \{X,Y\}$ and the set of conditional independencies in P is the empty set

$$\varnothing.$$

That is, there are no independencies. We want to find a DAG faithful to P. We cannot have the DAG in Figure 8.4 (b). The reason is that this DAG entails that X and Y are independent, which means that the Markov condition is not satisfied. So, we must have one of the DAGs in Figure 8.4 (a). We conclude that P is faithful to both the DAGs in Figure 8.4 (a). Note that these DAGs are Markov equivalent.

Example 8.12 Suppose $\mathsf{V} = \{X,Y,Z\}$, and the set of conditional independencies in P is

$$\{I_P(X,Y)\}.$$

We want to find a DAG faithful to P. There can be no edge between X and Y in the DAG owing to the reason given in Example 8.10. Furthermore, there must be edges between X and Z and between Y and Z owing to the reason given in Example 8.11. We cannot have any of the DAGs in Figure 8.5 (a). The reason is that these DAGs entail $I_P(X,Y|Z)$, and this conditional independency is not present. So, the Markov condition is not satisfied. Furthermore, the DAGs do not entail $I_P(X,Y)$. So, the DAG must be the one in Figure 8.5 (b). We conclude that P is faithful to the DAG in Figure 8.5 (b).

Example 8.13 Suppose $\mathsf{V} = \{X,Y,Z\}$ and the set of conditional independencies in P is

$$\{I_P(X,Y|Z)\}.$$

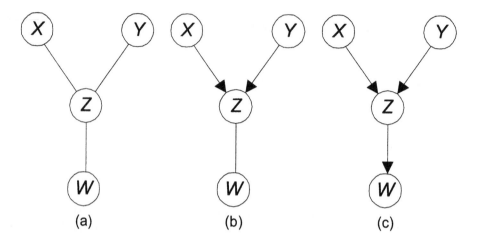

Figure 8.6: If the set of conditional independencies is $\{I_p(X, \{Y, W\}),$ $I_P(Y, \{X, Z\})\}$, we must have the DAG in (c).

We want to find a DAG faithful to P. Owing to reasons similar to those given before, the only edges in the DAG must be between X and Z and between Y and Z. We cannot have the DAG in Figure 8.5 (b). The reason is that this DAG entails $I(X, Y)$, and this conditional independency is not present. So, the Markov condition is not satisfied. So, we must have one of the DAGs in Figure 8.5 (a). We conclude that P is faithful to all the DAGs in Figure 8.5 (a).

We now state a theorem whose proof can be found in [Neapolitan, 2004]. At this point your intuition should suspect that it is true.

Theorem 8.5 *If $(\mathbb{G}, P)$ satisfies the faithfulness condition, then there is an edge between X and Y if and only if X and Y are not conditionally independent given any set of variables.*

Example 8.14 *Suppose $\mathsf{V} = \{X, Y, Z, W\}$ and the set of conditional independencies in P is*

$$\{I_p(X, Y), \quad I_P(W, \{X, Y\}|Z)\}.$$

We want to find a DAG faithful to P. Owing to Theorem 8.5, the links (edges without regard for direction) must be as shown in Figure 8.6 (a). We must have the directed edges shown in Figure 8.6 (b) because we have $I_p(X, Y)$. Therefore, we must also have the directed edge shown in Figure 8.6 (c) because we do not have $I_P(W, X)$. We conclude that P is faithful to the DAG in Figure 8.6 (c).

Example 8.15 *Suppose $\mathsf{V} = \{X, Y, Z, W\}$ and the set of conditional independencies in P is*

$$\{I_P(X, \{Y, W\}), \quad I_P(Y, \{X, Z\})\}.$$

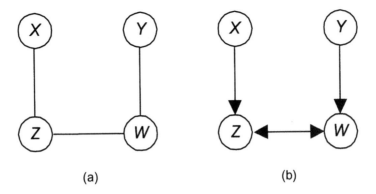

Figure 8.7: If the set of conditional independencies is $\{I_P(X, \{Y, W\}),$ $I_P(Y, \{X, Z\})\}$ and we try to find a DAG faithful to P, we obtain the graph in (b), which is not a DAG.

We want to find a DAG faithful to P. Owing to Theorem 8.5, we must have the links shown in Figure 8.7 (a). Now, if we have the chain $X \rightarrow Z \rightarrow W$, $X \leftarrow Z \leftarrow W$, or $X \leftarrow Z \rightarrow W$, then we do not have $I_P(X, W)$. So, we must have the chain $X \rightarrow Z \leftarrow W$. Similarly, we must have the chain $Y \rightarrow W \leftarrow Z$. So, our graph must be the one in Figure 8.7 (b). However, this graph is not a DAG. The problem here is that there is no DAG faithful to P.

Example 8.16 *Again suppose* $\mathsf{V} = \{X, Y, Z, W\}$ *and the set of conditional independencies in* P *is*

$$\{I_P(X, \{Y, W\}), \quad I_P(Y, \{X, Z\})\}.$$

As shown in the previous example, there is no DAG faithful to P*. However, this does not mean we cannot find a more succinct way to represent* P *than using a complete DAG.* P *satisfies the Markov condition with each of the DAGs in Figure 8.8. That is, the DAG in Figure 8.8 (a) entails*

$$\{I_P(X, Y), \quad I_P(Y, Z)\}$$

and these conditional independencies are both in P*, whereas the DAG in Figure 8.8 (b) entails*

$$\{I_P(X, Y), \quad I_P(X, W)\}$$

and these conditional independencies are both in P*. However,* P *does not satisfy the faithfulness condition with either of these DAGs because the DAG in Figure 8.8 (a) does not entail* $I_P(X, W)$*, whereas the DAG in Figure 8.8 (b) does not entail* $I_P(Y, Z)$*.*

Each of these DAGs is as succinct as we can represent the probability distribution. So, when there is no DAG faithful to a probability distribution P*, we can still represent* P *much more succinctly than we would by using the complete DAG. A structure learning algorithm tries to find the most succinct*

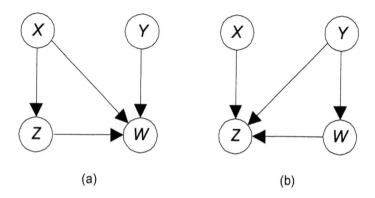

(a)   (b)

Figure 8.8: If the set of conditional independencies is $\{I_P(X, \{Y, W\}),$ $I_P(Y, \{X, Z\})\}$, P satisfies the Markov condition with both these DAGs.

representation. *Depending on the number of alternatives of each variable, one of the DAGs in Figure 8.8 may actually be a more succinct representation than the other because it contains fewer parameters. A constraint-based learning algorithm could not distinguish between the two, but a score-based one could. See [Neapolitan, 2004] for a complete discussion of this matter.*

8.3.2   Learning a DAG in Which P Is Embedded Faithfully

In a sense we compromised in Example 8.16 because the DAG we learned did not entail all the conditional independencies in P. This is fine if our goal is to learn a Bayesian network that will later be used to do inference. However, another application of structure learning is *causal learning*, which is discussed in the next subsection. When we're learning causes it would be better to find a DAG in which P is embedded faithfully. We discuss embedded faithfulness next.

Definition 8.1 *Suppose we have a joint probability distribution P of the random variables in some set V and a DAG $\mathbb{G} = (W, E)$ such that $V \subseteq W$. We say that $(\mathbb{G}, P)$ satisfy the **embedded faithfulness condition** if all and only the conditional independencies in P are entailed by $\mathbb{G}$, restricted to variables in V. Furthermore, we say that P is embedded faithfully in $\mathbb{G}$.*

Example 8.17 *Again suppose $V = \{X, Y, Z, W\}$ and the set of conditional independencies in P is*

$$\{I_P(X, \{Y, W\}), \quad I_P(Y, \{X, Z\})\}.$$

Then P is embedded faithfully in the DAG in Figure 8.9. It is left as an exercise to show this. By including the variable H in the DAG, we are able to entail all and only the conditional independencies in P restricted to variables in V.

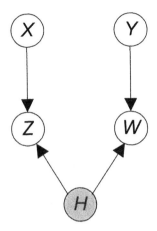

Figure 8.9: If the set of conditional independencies in P is $\{I_P(X, \{Y, W\}), I_P(Y, \{X, Z\})\}$, then P is embedded faithfully in this DAG.

Variables such as H are called **hidden variables** because they are not among the observed variables.

8.4 Causal Learning

In many, if not most, applications the variables of interest have causal relationships to each other. For example, the variables in Example 8.1 are causally related in that sex has a causal effect on height and may have a causal effect on wage. If the variables are causally related, we can learn something about their causal relationships when we learn the structure of the DAG from data. However, we must make certain assumptions to do this. We discuss these assumptions and causal learning next.

8.4.1 Causal Faithfulness Assumption

Recall from Section 5.3.2 that if we assume the observed probability distribution P of a set of random variables V satisfies the Markov condition with the causal DAG $\mathbb{G}$ containing the variables, we say we are making the **causal Markov assumption**, and we call $(\mathbb{G}, P)$ a **causal network**. Furthermore, we concluded that the causal Markov assumption is justified for a causal graph if the following conditions are satisfied:

1. There are no hidden common causes. That is, all common causes are represented in the graph.

2. There are no causal feedback loops. That is, our graph is a DAG.

3. Selection bias is not present.

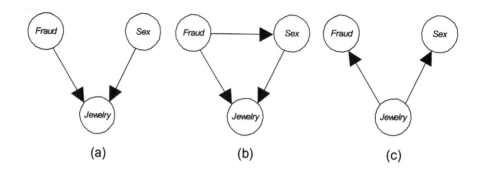

Figure 8.10: If the only causal relationships are that *Fraud* and *Sex* have causal influences on *Jewelry*, then the causal DAG is the one in (a). If we make the causal Markov assumption, only the DAG in (c) is ruled out if we observe $I_P(Fraud, Sex)$.

Recall the discussion concerning credit card fraud in Section 5.1. Suppose that both fraud and sex do indeed have a causal effect on whether jewelry is purchased, and there are no other causal relationships among the variables. Then the causal DAG containing these variables is the one in Figure 8.10 (a). If we make the causal Markov assumption, we must have $I_P(Fraud, Sex)$.

Suppose now that we do not know the causal relationships among the variables, we make the causal Markov assumption, and we learn only the conditional independency $I_P(Fraud, Sex)$ from data. Can we conclude that the causal DAG must be the one in Figure 8.10 (a)? No, we cannot because P also satisfies the Markov condition with the DAG in Figure 8.10 (b). This concept is a bit tricky to understand. However, recall that we are assuming that we do not know the causal relationships among the variables. As far as we know, they could be the ones in Figure 8.10 (b). If the DAG in Figure 8.10 (b) were the causal DAG, the causal Markov assumption would still be satisfied when the only conditional independency is $I_P(Fraud, Sex)$, because that DAG satisfies the Markov condition with P. So, if we make only the causal Markov assumption, we cannot distinguish the causal DAGs in Figures 8.10 (a) and 8.10 (b) based on the conditional independency $I_P(Fraud, Sex)$. The causal Markov assumption only enables us to rule out causal DAGs that contain conditional independencies that are not in P.

One such DAG is the one in Figure 8.10 (c). We need to make the causal faithfulness assumption to conclude that the causal DAG is the one in Figure 8.10 (a). That assumption is as follows: If we assume that the observed probability distribution P of a set of random variables V satisfies the faithfulness condition with the causal G containing the variables, we say we are making the **causal faithfulness assumption**. If we make the causal faithfulness assumption, then if we find a unique DAG that is faithful to P, the edges in that DAG must represent causal influences. This is illustrated by the following examples.

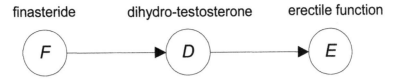

Figure 8.11: Finasteride and erectile function are independent.

Example 8.18 *Recall that in Example 8.12 we showed that if* $V = \{X, Y, Z\}$ *and the set of conditional independencies in P is*

$$\{I_P(X, Y)\},$$

the only DAG faithful to P is the one in Figure 8.5 (b). If we make the causal faithfulness assumption, this DAG must be the causal DAG, which means we can conclude that X and Y each cause Z. This is the exact same situation as that illustrated earlier concerning fraud, sex, and jewelry. Therefore, if we make the causal faithfulness assumption, we can conclude that the causal DAG is the one in Figure 8.10 (a) based on the conditional independency $I_P(Fraud, Sex)$.

Example 8.19 *In Example 8.13 we showed that if* $V = \{X, Y, Z\}$ *and the set of conditional independencies in P is*

$$\{I_P(X, Y|Z)\},$$

all the DAGs in Figure 8.5 (a) are faithful to P. So, if we make the causal faithfulness assumption, we can conclude that one of these DAGs is the causal DAG, but we do not know which one.

Example 8.20 *In Example 8.14 we showed that if* $V = \{X, Y, Z, W\}$ *and the set of conditional independencies in P is*

$$\{I_p(X, Y), \quad I_P(W, \{X, Y\}|Z)\},$$

the only DAG faithful to P is the one in Figure 8.6 (c). So, we can conclude that X and Y each cause Z and Z causes W.

When is the causal faithfulness assumption justified? It requires the three conditions mentioned previously for the causal Markov assumption plus one more, which we discuss next. Recall from Section 6.2.1 that the causal relationships among finasteride (F), dihydro-testosterone (D), and erectile dysfunction (E) have clearly been found to be those depicted in Figure 8.11. However, as discussed in that section, we have $I_P(F, E|D)$. We would expect a causal mediary to transmit an effect from its antecedent to its consequence, but in this case it does not. As also discussed in Section 6.2.1, the explanation is that finasteride cannot lower dihydro-testosterone levels beyond a certain threshold level, and that level is all that is needed for erectile function. So, we have $I_P(F, E)$.

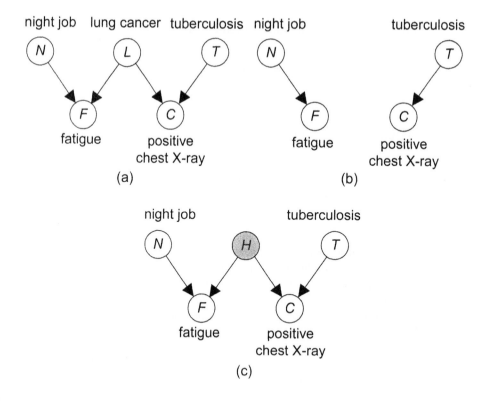

Figure 8.12: If the causal relationships are those shown in (a), P is not faithful to the DAG in (b), but P is embedded faithfully in the DAG in (c).

The Markov condition does not entail $I_P(F, E)$ for the causal DAG in Figure 8.11. It only entails $I_P(F, E|D)$. So, the causal faithfulness assumption is not justified. If we learned the conditional independencies in the probability distribution of these variables from data, we would learn the following set of independencies:

$$\{I_P(F, E),\quad I_P(F, E|D)\}.$$

There is no DAG that entails both these conditional independencies, so no DAG could be learned from such data.

The causal faithfulness assumption is usually justified when the three conditions listed previously for the causal Markov assumption are satisfied and when we do not have unusual causal relationships, as in the finasteride example. So, the causal faithfulness assumption is ordinarily justified for a causal graph if the following conditions are satisfied:

1. There are no hidden common causes. That is, all common causes are represented in the graph.

2. There are no causal feedback loops. That is, our graph is a DAG.

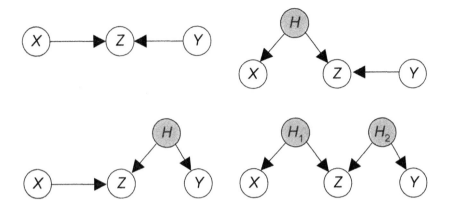

Figure 8.13: If we make the causal embedded faithfulness assumption and our set of conditional independencies is $\{I_P(X, Y)\}$, the causal relationships could be the ones in any of these DAGs.

3. Selection bias is not present.

4. All intermediate causes transmit influences from their antecedents to their consequences.

8.4.2 Causal Embedded Faithfulness Assumption

It seems that the main exception to the causal faithfulness assumption (and the causal Markov assumption) is the presence of hidden common causes. Even in the example concerning sex, height, and wage (Example 8.1), perhaps there is a genetic trait that makes people grow taller and also gives them some personality trait that helps them compete better in the job market. Our next assumption eliminates the requirement that there are no hidden common causes. If we assume that the observed probability distribution P of a set of random variables V is embedded faithfully in a causal DAG containing the variables, we say that we are making the **causal embedded faithfulness assumption**. The causal embedded faithfulness assumption is usually justified when the conditions for the causal faithfulness assumption are satisfied, except that hidden common causes may be present.

Next we illustrate the causal embedded faithfulness assumption. Suppose that the causal DAG in Figure 8.12 (a) satisfies the causal faithfulness assumption. However, we only observe $\mathsf{V} = \{N, F, C, T\}$. Then the causal DAG containing the observed variables is the one in Figure 8.12 (b). The DAG in Figure 8.12 (b) entails $I_P(F, C)$, and this conditional independency is not entailed by the DAG in Figure 8.12 (a). Therefore, the observed distribution $P(\mathsf{V})$ does not satisfy the Markov condition with the causal DAG in Figure 8.12 (b), which means the causal faithfulness assumption is not warranted. However, $P(\mathsf{V})$ is embedded faithfully in the DAG in Figure 8.12 (c). So, the causal embedded

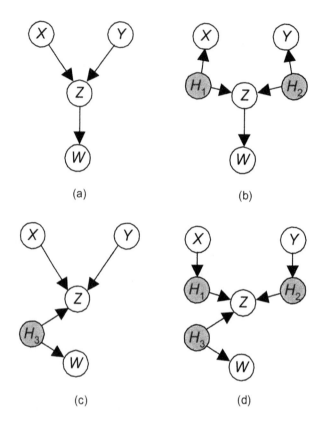

Figure 8.14: If our set of conditional independencies is $\{I_P(X, Y),$ $I_P(W, \{X, Y\}|Z)\}$, then P is embedded faithfully in the DAGs in (a) and (b) but not in the DAGs in (c) and (d).

faithfulness assumption is warranted. Note that this example illustrates a situation in which we identify four variables and two of them have a hidden common cause. That is, we have not identified lung cancer as a feature of humans.

Let's see how much we can learn about causal influences when we make only the causal embedded faithfulness assumption.

Example 8.21 *Recall that in Example 8.18, $\mathsf{V} = \{X, Y, Z\}$, our set of conditional independencies was*

$$\{I_P(X, Y)\},$$

and we concluded that X and Y each caused Z while making the causal faithfulness assumption. However, the probability distribution is embedded faithfully in all the DAGs in Figure 8.13. So, if we make only the causal embedded faithfulness assumption, it could be that X causes Z, or it could be that X and Z have a hidden common cause. The same holds for Y and Z.

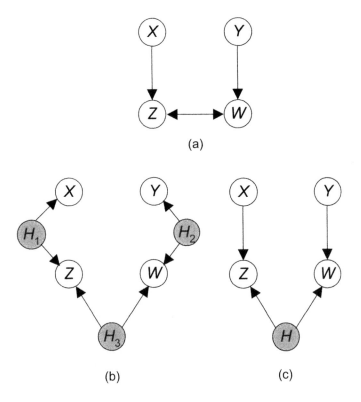

Figure 8.15: If our set of conditional independencies is $\{I_P(X, \{Y, W\}),$ $I_P(Y, \{X, Z\})$, we can conclude that Z and W have a hidden common cause.

While making only the more reasonable causal embedded faithfulness assumption, we were not able to learn any causal influences in the previous example. Can we ever learn a causal influence while making only this assumption? The next example shows that we can.

Example 8.22 *Recall that in Example 8.20,* $\mathsf{V} = \{X, Y, Z, W\}$ *and our set of conditional independencies was*

$$\{I_P(X, Y), \quad I_P(W, \{X, Y\}|Z)\}.$$

In this case the probability distribution P *is embedded faithfully in the DAGs in Figures 8.14 (a) and 8.14 (b). However, it is not embedded faithfully in the DAGs in Figure 8.14 (c) or 8.14 (d). The reason is that these latter DAGs entail* $I_P(X, W)$, *and we do not have this conditional independency. That is, the Markov condition says* X *must be independent of its nondescendents conditional on its parents. Since* X *has no parents, this means that* X *must simply be independent of its nondescendents, and* W *is one of its nondescendents. If we make the causal embedded faithfulness assumption, we conclude that* Z *causes* W.

Example 8.23 *Recall that in Example 8.16,* $V = \{X, Y, Z, W\}$, *our set of conditional independencies was*

$$\{I_P(X, \{Y, W\}), I_P(Y, \{X, Z\}),$$

and we obtained the graph in Figure 8.15 (a) when we tried to learn a DAG faithful to P. We concluded that there is no DAG faithful to P. Then in Example 8.17 we showed that P is embedded faithfully in the DAG in Figure 8.15 (c). P is also embedded faithfully in the DAG in Figure 8.15 (b). If we make the causal embedded faithfulness assumption, we conclude that Z and W have a hidden common cause.

8.5   Model Averaging

Heckerman et al. [1999] illustrate that when the number of variables is small and the amount of data is large, one structure can be orders of magnitude more likely than any other. In such cases, model selection yields good results. However, recall that in Example 8.4 we had few data, we obtained $P(\mathbb{G}_1|D) = .517$ and $P(\mathbb{G}_2|D) = .483$, and we chose (learned) DAG $\mathbb{G}_1$ because it was most probable. Then in Example 8.8 we used a Bayesian network containing DAG $\mathbb{G}_1$ to do inference for Sam. Since the probabilities of the two models are so close, it seems somewhat arbitrary to choose $\mathbb{G}_1$. So, model selection does not seem appropriate. Next, we describe another approach.

Instead of choosing a single DAG and then using it to do inference, we could use the Law of Total Probability to do the inference as follows: We perform the inference using each DAG and multiply the result (a probability value) by the posterior probability of the DAG. This is called **model averaging**.

Example 8.24 *Recall that based on the data in Example 8.4 we learned that*

$$P(\mathbb{G}_1|D) = .517$$

and

$$P(\mathbb{G}_2|D) = .483.$$

In Example 8.8 we updated a Bayesian network containing $\mathbb{G}_1$ based on the data to obtain the Bayesian network in Figure 8.16 (a). If in the same way we update a Bayesian network containing $\mathbb{G}_2$, we obtain the Bayesian network in Figure 8.16 (b). Given that Sam has never defaulted on a loan ($F = f_2$), we can then use model averaging to compute the probability that Sam is a white-collar worker, as follows:[1]

$$
\begin{aligned}
P(j_1|f_1, D) &= P(j_1|f_1, \mathbb{G}_1)P(\mathbb{G}_1|D) + P(j_1|f_1, \mathbb{G}_2)P(\mathbb{G}_2|D) \\
&= (.714)(.517) + (7/12)(.483) = .651.
\end{aligned}
$$

---

[1] Note that we substitute $P(\mathbb{G}_1|D)$ for $P(\mathbb{G}_1|f_1, D)$. They are not exactly equal, but we are assuming that the data set is sufficiently large that the dependence of the DAG models on the current case can be ignored.

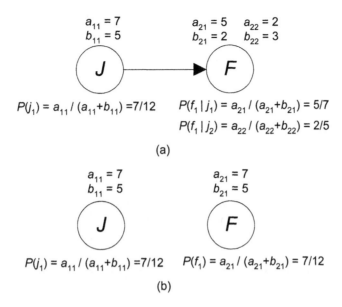

$$P(j_1) = a_{11} / (a_{11}+b_{11}) = 7/12 \qquad P(f_1 | j_1) = a_{21} / (a_{21}+b_{21}) = 5/7$$

$$P(f_1 | j_2) = a_{22} / (a_{22}+b_{22}) = 2/5$$

(a)

$$P(j_1) = a_{11} / (a_{11}+b_{11}) = 7/12 \qquad P(f_1) = a_{21} / (a_{21}+b_{21}) = 7/12$$

(b)

Figure 8.16: Updated Bayesian network for learning based on the data in Examples 8.4 and 8.8.

The result that $P(j_1|f_1, \mathbb{G}_1) = .714$ was obtained in Example 8.8, although in that example we did not show the dependence on $\mathbb{G}_1$ because that DAG was the only DAG considered. The result that $P(j_1|f_1, \mathbb{G}_2) = 7/12$ is obtained directly from the Bayesian network in Figure 8.16 (b) because J and F are independent in that network.

Example 8.24 illustrated using model averaging to do inference. The following example illustrates using it to learn partial structure.

Example 8.25 *Suppose we have three random variables X_1, X_2, and X_3. Then the possible DAG patterns are the ones in Figure 8.17. We might be interested in the probability that a feature of the DAG pattern is present. For example, we might be interested in the probability that there is an edge between X_1 and X_2. Given the five DAG patterns in which there is an edge, this probability is 1, and given the six DAG pattern in which there is no edge, this probability is 0. Let gp denote a DAG pattern. If we let F be a random variable whose value is present if a feature is present,*

$$P(F = present|\mathrm{D}) = \sum_{gp} P(F = present|gp, \mathrm{D})P(gp|\mathrm{D})$$

$$= \sum_{gp} P(F = present|gp)P(gp|\mathrm{D}),$$

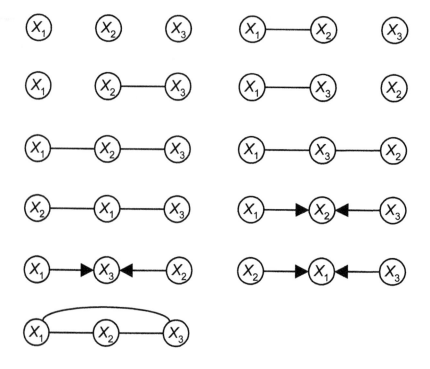

Figure 8.17: The 11 DAG patterns when there are three nodes.

where

$$P(F = present|gp) = \begin{cases} 1 & \textit{if the feature is present in gp} \\ 0 & \textit{if the feature is not present in gp.} \end{cases}$$

You may wonder what event a feature represents. For example, what event does an edge between X_1 and X_2 represent? This event is the event that X_1 and X_2 are not independent and are not conditionally independent given X_3 in the actual relative frequency distribution of the variables. Another possible feature is that there is a directed edge from X_1 to X_2. This feature is the event that, assuming that the relative frequency distribution admits a faithful DAG representation, there is a directed edge from X_1 to X_2 in the DAG pattern faithful to that distribution. Similarly, the feature that there is a directed path from X_1 to X_2 represents the event that there is a directed path from X_1 to X_2 in the DAG pattern faithful to that distribution.

Given that we are only discussing the relative frequency distribution, these events are ordinarily not of great interest. However, if we are discussing causality, they tell us something about the causal relationships among the variables. For example, recall that in Example 5.16 we mentioned that Chapter 12 discusses learning from data how the proteins produced by one gene have a causal effect on the level of mRNA (called the *gene expression level*) of another gene. Ordinarily there are thousands of genes (variables), but typically we have at most only a few thousand data items. In such cases, there are often many structures that are equally likely. So, choosing one particular structure is somewhat arbitrary. However, in these cases we are not always interested in learning the entire structure. That is, rather than needing the structure for inference and decision making, we are only interested in learning relationships among some of the variables. In particular, in the gene expression example, we are interested in the dependence and causal relationships between the expression levels of certain genes (see [Lander and Shenoy, 1999]).

As is the case for model selection, when the number of possible DAGs is large, we cannot average over all DAGs. In these situations we heuristically search for high-probability DAGs, and then we average over them. In particular, in the gene expression example, since there are thousands of variables, we could not average over all of them. We discuss approximate model averaging in Section 8.6.2.

8.6 Approximate Structure Learning

Recall from Section 8.2.4 that when the number of variables is not small, it is computationally unfeasible to find the maximizing DAGs by exhaustively considering all DAGs. Therefore, researchers have developed heuristic search algorithms. We discuss such algorithms next.

8.6.1 Approximate Model Selection

Here we discuss heuristic search algorithms for model selection. First we present algorithms that search over DAGs.

Algorithms That Search Over DAGs

We present two algorithms in which the search space consists of DAGs. Specifically, the search space is the set of all DAGs containing n nodes, where n is our number of random variables. In these algorithms, our goal is find a DAG with maximum score, where our scoring criterion could be the Bayesian score, the BIC score, or some other score. Therefore, we will simply refer to the score as $score\,(\mathbb{G}:\mathsf{D})$, where D is our data.

The K2 Algorithm If we use either the Bayesian score or the BIC score, the score for the entire DAG is a product of local scores for each node. For example, Theorem 8.4 obtains the result that the Bayesian score, in the case of multinomial variables, is given by

$$score_{Bayesian}(\mathbb{G}:\mathsf{D}) = P(\mathsf{D}|\mathbb{G}) = \prod_{i=1}^{n}\prod_{j=1}^{q_i^{\mathbb{G}}}\frac{\Gamma(N_{ij}^{\mathbb{G}})}{\Gamma(N_{ij}^{\mathbb{G}} + M_{ij}^{\mathbb{G}})}\prod_{k=1}^{r_i}\frac{\Gamma(a_{ijk}^{\mathbb{G}} + s_{ijk}^{\mathbb{G}})}{\Gamma(a_{ijk}^{\mathbb{G}})}.$$

See Theorem 8.4 for the definition of the variables in this formula. Note that we have now explicitly shown their dependence on $\mathbb{G}$. Let $\mathsf{PA}_i^{\mathbb{G}}$ denote the parents of X_i in $\mathbb{G}$. For each node X_i, the value of

$$\prod_{j=1}^{q_i^{\mathbb{G}}} \frac{\Gamma(N_{ij}^{\mathbb{G}})}{\Gamma(N_{ij}^{\mathbb{G}} + M_{ij}^{\mathbb{G}})} \prod_{k=1}^{r_i} \frac{\Gamma(a_{ijk}^{\mathbb{G}} + s_{ijk}^{\mathbb{G}})}{\Gamma(a_{ijk}^{\mathbb{G}})}$$

depends only on parameter values stored locally at X_i, and data values of X_i and nodes in $\mathsf{PA}_i^{\mathbb{G}}$. Define

$$score(X_i, \mathsf{PA}_i^{\mathbb{G}} : \mathsf{D}) = \prod_{j=1}^{q_i^{\mathbb{G}}} \frac{\Gamma(N_{ij}^{\mathbb{G}})}{\Gamma(N_{ij}^{\mathbb{G}} + M_{ij}^{\mathbb{G}})} \prod_{k=1}^{r_i} \frac{\Gamma(a_{ijk}^{\mathbb{G}} + s_{ijk}^{\mathbb{G}})}{\Gamma(a_{ijk}^{\mathbb{G}})}.$$

Cooper and Herskovits [1992] developed a greedy search algorithm that tries to maximize the score of the DAG by maximizing these local scores. That is, for each variable X_i they locally find a value PA_i that approximately maximizes $score(X_i, \mathsf{PA}_i^{\mathbb{G}} : \mathsf{D})$. The single operation in this search algorithm is the addition of a parent to a node. The algorithm proceeds as follows: We assume an ordering of the nodes such that if X_i precedes X_j in the order, an arc from X_j to X_i is not allowed. Let $Pred(X_i)$ be the set of nodes that precede X_i in the ordering, We initially set the parents $\mathsf{PA}_i^{\mathbb{G}}$ of X_i to empty and compute $score(X_i, \mathsf{PA}_i^{\mathbb{G}} : \mathsf{D})$. Next we visit the nodes in sequence according to the ordering. When we visit X_i, we determine the node in $Pred(X_i)$ that most increases $score(X_i, \mathsf{PA}_i^{\mathbb{G}} : \mathsf{D})$. We "greedily" add this node to PA_i. We continue doing this until the addition of no node increases $score(X_i, \mathsf{PA}_i^{\mathbb{G}} : \mathsf{D})$. Pseudocode for this algorithm follows. The algorithm is called K2 because it evolved from a system named Kutató [Herskovits and Cooper, 1990].

Algorithm 8.1: K2

Problem: Find a DAG that approximates maximizing $score(\mathbb{G} : \mathsf{D})$.

Inputs: A set V of n random variables; an upper bound u on the number of parents a node may have; data D.

Outputs: n sets of parent nodes PA_i, where $1 \leq i \leq n$, in a DAG that approximates maximizing $score(\mathbb{G} : \mathsf{D})$.

```
void K2   (set_of_variables V, int u,
          data D, for 1 ≤ i ≤ n parent_set& PAᵢ)
{
   for (i = 1; i <= n; i + +) {      // n is the number of variables.
      PAᵢᴳ = ∅;
      Pₒₗd = score(Xᵢ, PAᵢᴳ : D);
      findmore = true;
```

```
while (findmore && |PA_i^G| < u) {
    Z = node in Pred(X_i) − PA_i that maximizes
        score(X_i, PA_i^G ∪ {Z} : D);
    P_new = score(X_i, PA_i^G ∪ {Z} : D) ;
    if (P_new > P_old) {
        P_old = P_new;
        PA_i^G = PA_i^G ∪ {Z};
    }
    else
        findmore = false;
}
}
}
```

Neapolitan [2004] analyzes the algorithm. Furthermore, he shows an example in which the algorithm was provided with a prior order and learned a DAG from 10,000 cases sampled at random from the ALARM Bayesian network [Beinlich et al., 1989]. The DAG learned was identical to the one in the ALARM network except that one edge was missing.

You might wonder where we could obtain the ordering required by Algorithm 8.1. Such an ordering could possibly be obtained from domain knowledge such as a time ordering of the variables. For example, we might know that in patients, smoking precedes bronchitis and lung cancer and that each of these conditions precedes fatigue and a positive chest X-ray.

When a model searching algorithm need only locally recompute a few scores to determine the score for the next model under consideration, we say the algorithm has **local scoring updating**. A model with local scoring updating is considerably more efficient than one without it. Clearly, the K2 algorithm has local scoring updating.

An Algorithm without a Prior Ordering We present a straightforward greedy search algorithm that does not require a time ordering. The search space is again the set of all DAGs containing the n variables, and the set **DAGOPS** of operations is as follows:

1. If two nodes are not adjacent, add an edge between them in either direction.

2. If two nodes are adjacent, remove the edge between them.

3. If two nodes are adjacent, reverse the edge between them.

All operations are subject to the constraint that the resultant graph does not contain a cycle. The set of all DAGs that can be obtained from $\mathbb{G}$ by applying one of the operations is called $\mathsf{Nbhd}(\mathbb{G})$. If $\mathbb{G}' \in \mathsf{Nbhd}(\mathbb{G})$, we say $\mathbb{G}'$ is in the **neighborhood** of $\mathbb{G}$. Clearly, this set of operations is **complete** for the search space. That is, for any two DAGs $\mathbb{G}$ and $\mathbb{G}'$ there exists a sequence of operations that transforms $\mathbb{G}$ to $\mathbb{G}'$. The reverse edge operation is not needed for the operations to be complete, but it increases the connectivity of the space without adding too much complexity, which typically leads to a better search. Furthermore, when we use a greedy search algorithm, including edge reversals often seems to lead to a better local maximum.

The algorithm proceeds as follows: We start with a DAG with no edges. At each step of the search, of all those DAGs in the neighborhood of our current DAG, we "greedily" choose the one that maximizes $score\,(\mathbb{G} : \mathsf{D})$. We halt when no operation increases this score.

Note that in each step, if an edge to X_i is added or deleted, we need only re-evaluate $score(X_i, \mathsf{PA}_i : \mathsf{D})$. If an edge between X_i and X_j is reversed, we need only reevaluate $score(X_i, \mathsf{PA}_i : \mathsf{D})$ and $score(X_j, \mathsf{PA}_j : \mathsf{D})$. Therefore, this algorithm has local scoring updating. The algorithm follows:

Algorithm 8.2: DAG Search

Problem: Find a DAG that approximates maximizing $score\,(\mathbb{G} : \mathsf{D})$.

Inputs: A set V of n random variables; data D.

Outputs: A set of edges E in a DAG that approximates maximizing $score\,(\mathbb{G} : \mathsf{D})$.

```
void DAG_search (set_of_variables V, data D,
                 set_of_edges& E)
{
  E = ∅; G = (V, E);
  do
    if (any DAG in the neighborhood of our current DAG
    increases score (G : D))
        modify E according to the one that increases score (G : D) the most;
  while (some operation increases score (G : D));
}
```

A problem with a greedy search algorithm is that it could halt at a candidate solution that locally maximizes the objective function rather than globally maximizes it (see [Xiang et al., 1996]). One way for dealing with this problem is iterated hill-climbing. In iterated hill-climbing, local search is done until a local maximum is obtained. Then the current structure is randomly perturbed, and the process is repeated. Finally, the maximum over local maxima is used. Other methods for attempting to avoid local maxima include simulated annealing [Metropolis et al., 1953], best-first search [Korf, 1993], and Gibb's sampling [Neapolitan, 2004].

Searching over DAG Patterns

We present an algorithm that searches over DAG patterns. First, we discuss why we might want to do this.

Why Search over DAG Patterns? Although Algorithms 8.1 and 8.2 find a DAG $\mathbb{G}$ rather than a DAG pattern, we can use them to find a DAG pattern by determining the DAG pattern gp representing the Markov equivalence class to which $\mathbb{G}$ belongs. Since $score(gp : \mathsf{D}) = score(\mathbb{G} : \mathsf{D})$, we have approximated maximizing $score(\mathsf{D}, gp)$. However, as discussed in [Anderson et al., 2007], there are a number of potential problems in searching for a DAG instead of a DAG pattern. Briefly, we discuss two of the problems. The first is efficiency. By searching over DAGs, the algorithm can waste time encountering and rescoring DAGs in the same Markov equivalence class. A second problem has to do with priors. If we search over DAGs, we are implicitly assigning equal priors to all DAGs, which means that DAG patterns containing more DAGs will have higher prior probability. For example, if there are n nodes, the complete DAG pattern (representing no conditional independencies) contains $n!$ DAGs, whereas the pattern with no edges (representing that all variables are mutually independent) contains just one DAG. On the other hand, Gillispie and Pearlman [2001] show that an asymptotic ratio of the number of DAGs to DAG patterns equal to about 3.7 is reached when the number of nodes is only 10. Therefore, on average the number of DAGs in a given equivalence class is small, and perhaps our concern about searching over DAGs is not necessary. Contrariwise, in simulations performed by Chickering [2001] the average number of DAGs in the equivalence classes over which his algorithm searched were always greater than 8.5 and in one case was 9.7×10^{19}.

When performing model selection, assigning equal priors to DAGs is not necessarily a serious problem, since we will finally select a high-scoring DAG that corresponds to a high-scoring DAG pattern. However, as discussed in Section 8.6.2, it can be a more serious problem in the case of model averaging.

The GES Algorithm In 1997 Meek developed an algorithm called **Greedy Equivalent Search (GES)**, which searches over DAG patterns and has the following property: If there is a DAG pattern faithful to P, as the size of the data set approaches infinity, the limit of the probability of finding a DAG pattern faithful to P is equal to 1. In 2002 Chickering proved this is the case. We describe the algorithm next.

In what follows we denote the equivalence class represented by DAG pattern gp by **gp**. GES is a two-phase algorithm that searches over DAG patterns. In the first phase, DAG pattern gp' is in the neighborhood of DAG pattern gp, denoted $\mathsf{Nbhd}^+(gp)$, if there is some DAG $\mathbb{G} \in \mathbf{gp}$ for which a single edge addition results in a DAG $\mathbb{G}' \in \mathbf{gp}'$. Starting with the DAG pattern containing no edges, we repeatedly replace the current DAG pattern gp by the DAG pattern in $\mathsf{Nbhd}^+(gp)$ that has the highest score of all DAG patterns in $\mathsf{Nbhd}^+(gp)$. We do this until there is no DAG pattern in $\mathsf{Nbhd}^+(gp)$ that increases the score.

The second phase is completely analogous to the first phase. In this phase, DAG pattern gp' is in the neighborhood of DAG pattern gp, denoted $\mathsf{Nbhd}^-(gp)$, if there is some DAG $\mathbb{G} \in gp$ for which a single edge deletion results in a DAG $\mathbb{G}' \in gp'$. Starting with the DAG pattern obtained in the first phase, we repeatedly replace the current DAG pattern gp by the DAG pattern in $\mathsf{Nbhd}^-(gp)$ that has the highest score of all DAG patterns in $\mathsf{Nbhd}^-(gp)$. We do this until there is no DAG pattern in $\mathsf{Nbhd}^-(gp)$ that increases the score.

It is left as an exercise to write this algorithm.

Neapolitan [2004] discusses other algorithms that search over DAG patterns.

8.6.2 Approximate Model Averaging

As mentioned previously, Heckerman et al. [1999] illustrate that when the number of variables is small and the amount of data is large, one structure can be orders of magnitude more likely than any other. In such cases, approximate model selection yields good results. However, if the amount of data is not large, it seems more appropriate to do inference by approximately averaging over models as illustrated in Example 8.24. Another application of approximate model averaging would be to learn partial structure when the amount of data is small relative to the number of variables. In Example 8.25, we discussed how we might do this to learn how the protein transcription factors produced by one gene have a causal effect on the level of mRNA of another gene.

Approximate Model Averaging Using MCMC

Next we discuss how we can heuristically search for high-probability structures and then average over them using the Markov Chain Monte Carlo (MCMC) method.

Recall our two examples of model averaging (Examples 8.24 and 8.25). The first involved computing a conditional probability over all possible DAGs as follows:

$$P(x|y, \mathsf{D}) = \sum_{\mathbb{G}} P(x|y, \mathbb{G}, \mathsf{D}) P(\mathbb{G}|\mathsf{a}, \mathsf{D}).$$

The second involved computing the probability a feature is present as follows:

$$P(F = present|\mathsf{D}) = \sum_{gp} P(F = present|gp) P(gp|\mathsf{D}).$$

In general, these problems involve some function of the DAG or DAG pattern and possibly the data and a probability distribution of the DAGs or DAG patterns conditional on the data. So, we can represent the general problem to be the determination of

$$\sum_{gp} f(gp, \mathsf{D}) P(gp|\mathsf{D}), \tag{8.6}$$

where f is some function of gp and possibly D, and P is some probability distribution of the DAG patterns. Although we represented the problem in terms of DAG patterns, we could sum over DAGs instead.

To approximate the value of Expression 8.6 using MCMC, our stationary distribution $\mathbf{r}$ is $P(gp|\mathsf{D})$. Ordinarily we can compute $P(\mathsf{D}|gp)$ but not $P(gp|\mathsf{D})$. However, if we assume that the prior probability $P(gp)$ is the same for all DAG patterns,

$$
\begin{aligned}
P(gp|\mathsf{D}) &= \frac{P(\mathsf{D}|gp)P(gp)}{P(\mathsf{D})} \\
&= kP(\mathsf{D}|gp)P(gp),
\end{aligned}
$$

where k does not depend on gp. If we use Equality 3.14 or 3.15 as our expression for $\boldsymbol{\alpha}$, k cancels out of the expression, which means that we can use $P(\mathsf{D}|gp)$ in the expression for $\boldsymbol{\alpha}$. Note that we do not have to assign equal prior probabilities to all DAG patterns. That is, we could use $P(\mathsf{D}|gp)P(gp)$ in the expression for $\boldsymbol{\alpha}$ also.

If we average over DAGs instead of DAG patterns, the problem is the determination of

$$
\sum_{\mathbb{G}} f(\mathbb{G}, \mathsf{D})P(\mathbb{G}|\mathsf{D}),
$$

where f is some function of $\mathbb{G}$ and possibly D, and P is some probability distribution of the DAGs. As is the case for DAG patterns, if we assume that the prior probability $P(\mathbb{G})$ is the same for all DAGs, then $P(\mathbb{G}|\mathsf{D}) = kP(\mathsf{D}|\mathbb{G})$, and we can use $P(\mathsf{D}|\mathbb{G})$ in the expression for $\boldsymbol{\alpha}$. However, we must realize what this assumption entails. If we assign equal prior probabilities to all DAGs, DAG patterns containing more DAGs will have higher prior probability.

As noted previously, when we perform model selection, assigning equal prior probabilities to DAGs is not necessarily a serious problem, since we will finally select a high-scoring DAG that corresponds to a high-scoring DAG pattern. However, in performing model averaging, a given DAG pattern will be included in the average according to the number of DAGs in the pattern. For example, there are three DAGs corresponding to the DAG pattern $X - Y - Z$ but only one corresponding to DAG pattern $X \rightarrow Y \leftarrow Z$. So, by assuming that all DAGs have the same prior probability, we are assuming that the prior probability that the actual relative frequency distribution has the set of conditional independencies $\{I_P(X, Z|Y)\}$ is three times the prior probability that it has the set of conditional independencies $\{I_P(X, Z)\}$. Even more dramatic, there are $n!$ DAGs corresponding to the complete DAG pattern and only one corresponding to the DAG pattern with no edges. So, we are assuming that the prior probability that there are no conditional independencies is far greater than the prior probability that the variables are mutually independent.

This assumption has consequences as follows: Suppose, for example, that the correct DAG pattern is $X : Y$, which denotes the DAG pattern with no edge, and the feature of interest is $I_P(X, Y)$. Since the feature is present, our results are better if we confirm it. Therefore, averaging over DAG patterns has a better result because, by averaging over DAG patterns, we are assigning a prior probability of $1/2$ to the feature, whereas by averaging over DAGs, we are only assigning a prior probability of $1/3$ to the feature. On the other hand, if the correct DAG pattern is $X - Y$, the feature is not present, which means that our results are better if we disconfirm. Therefore, averaging over DAGs is better.

We see then that we need look at the ensemble of all relative frequency distributions rather than any one to discuss which method might be "correct." If relative frequency distributions are distributed uniformly in nature and we assign equal prior probabilities to all DAG patterns, then $P(F = present|\mathsf{D})$, obtained by averaging over DAG patterns, is the relative frequency with which we are investigating a relative frequency distribution with this feature when we are observing these data. So, averaging over DAG patterns is "correct." On the other hand, if relative frequency distributions are distributed in nature according to the number of DAGs in DAG patterns and we assign equal prior probabilities to all DAGs, then $P(F = present|\mathsf{D})$, obtained by averaging over DAGs, is the relative frequency with which we are investigating a relative frequency distribution with this feature when we are observing these data. So, averaging over DAGs is "correct." Although it seems reasonable to assume that relative frequency distributions are distributed uniformly in nature, some feel that a relative frequency distribution, represented by a DAG pattern containing a larger number of DAGs, may occur more often because there are more causal relationships that can give rise to it.

Approximate Averaging over DAGs

After presenting a straightforward algorithm, we simplify it.

A Straightforward Algorithm   We show how to use MCMC to approximate averaging over DAGs. Our set of states is the set of all possible DAGs containing the variables in the application, and our stationary distribution is $P(\mathbb{G}|\mathsf{D})$, but as noted previously, we can use $P(\mathsf{D}|\mathbb{G})$ in our expression for $\boldsymbol{\alpha}$. Recall from Section 8.6.1 that $\mathsf{Nbhd}(\mathbb{G})$ is the set of all DAGs that differ from $\mathbb{G}$ by one edge addition, one edge deletion, or one edge reversal. Clearly $\mathbb{G}_j \in \mathsf{Nbhd}(\mathbb{G}_i)$ if and only if $\mathbb{G}_i \in \mathsf{Nbhd}(\mathbb{G}_j)$. However, since adding or reversing an edge can create a cycle, if $\mathbb{G}_j \in \mathsf{Nbhd}(\mathbb{G}_i)$ it is not necessarily true that $\mathsf{Nbhd}(\mathbb{G}_i)$ and $\mathsf{Nbhd}(\mathbb{G}_j)$ contain the same number of elements. For example, if $\mathbb{G}_i$ and $\mathbb{G}_j$ are the DAGs in Figures 8.18 (a) and (b) respectively, then $\mathbb{G}_j \in \mathsf{Nbhd}(\mathbb{G}_i)$. However, $\mathsf{Nbhd}(\mathbb{G}_i)$ contains five elements because adding the edge $X_3 \rightarrow X_1$ would create a cycle, whereas $\mathsf{Nbhd}(\mathbb{G}_j)$ contains six elements. We create our transition matrix $\mathbf{Q}$ as follows: For each pair of states $\mathbb{G}_i$ and $\mathbb{G}_j$ we set

$$q_{ij} = \begin{cases} \dfrac{1}{|\mathsf{Nbhd}(\mathbb{G}_i)|} & \mathbb{G}_j \in \mathsf{Nbhd}(\mathbb{G}_i) \\[2em] 0 & \mathbb{G}_j \notin \mathsf{Nbhd}(\mathbb{G}_i) \end{cases},$$

where $|\mathsf{Nbhd}(\mathbb{G}_i)|$ returns the number of elements in the set. Since $\mathbf{Q}$ is not symmetric, we use Equality 3.14 rather than Equality 3.15 to compute α_{ij}. Specifically, our steps are as follows:

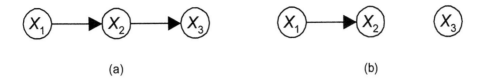

(a) (b)

Figure 8.18: These DAGs are in each other's neighborhoods, but their neighborhoods do not contain the same number of elements.

1. If the DAG at the trial k is $\mathbb{G}_i$ choose a DAG uniformly from $\mathsf{Nbhd}(\mathbb{G}_i)$. Suppose that DAG is $\mathbb{G}_j$.

2. Choose the DAG for trial $k+1$ to be $\mathbb{G}_j$ with probability

$$
\alpha_{ij} = \begin{cases} 1 & \dfrac{P(\mathsf{D}|\mathbb{G}_j) \times |\mathsf{Nbhd}(\mathbb{G}_i)|}{P(\mathsf{D}|\mathbb{G}_i) \times |\mathsf{Nbhd}(\mathbb{G}_j)|} \geq 1 \\[4mm] \dfrac{P(\mathsf{D}|\mathbb{G}_j)\,|\mathsf{Nbhd}(\mathbb{G}_i)|}{P(\mathsf{D}|\mathbb{G}_i)\,|\mathsf{Nbhd}(\mathbb{G}_j)|} & \dfrac{P(\mathsf{D}|\mathbb{G}_j) \times |\mathsf{Nbhd}(\mathbb{G}_i)|}{P(\mathsf{D}|\mathbb{G}_i) \times |\mathsf{Nbhd}(\mathbb{G}_j)|} \leq 1 \end{cases},
$$

and to be $\mathbb{G}_i$ with probability $1 - \alpha_{ij}$.

A Simplification It is burdensome to compute the sizes of the neighborhoods of the DAGs in each step. Alternatively, we could include DAGs with cycles in the neighborhoods. That is, $\mathsf{Nbhd}(\mathbb{G}_i)$ is the set of all graphs (including ones with cycles) that differ from $\mathbb{G}_i$ by one edge addition, one edge deletion, or one edge reversal. It is not hard to see that then the size of every neighborhood is equal $n(n-1)$. We therefore define

$$
q_{ij} = \begin{cases} \dfrac{1}{n(n-1)} & \mathbb{G}_j \in \mathsf{Nbhd}(\mathbb{G}_i) \\[4mm] 0 & \mathbb{G}_j \notin \mathsf{Nbhd}(\mathbb{G}_i) \end{cases}.
$$

If we are currently in state $\mathbb{G}_i$ and we obtain a graph $\mathbb{G}_j$ that is not a DAG, we set $P(\mathsf{D}|\mathbb{G}_j) = 0$ (effectively making r_j zero). In this way α_{ij} is zero, the graph is not chosen, and we stay at $\mathbb{G}_i$ in this step. Since $\mathbf{Q}$ is now symmetric, we can use Equality 3.15 to compute α_{ij}. Notice that our theory was developed by assuming that all values in the stationary distribution are positive, which is not currently the case. However, Tierney [1996] shows that convergence also follows if we allow 0 values as discussed here.

Neapolitan [2004] develops a similar method that averages over DAG patterns.

8.7 Software Packages for Learning

Based on considerations such as those illustrated in Section 8.3.1, Spirtes et al. [1993, 2000] developed an algorithm that finds the DAG faithful to P from the conditional independencies in P when there is a DAG faithful to P. Spirtes et al. [1993, 2000] further developed an algorithm that learns a DAG in which P is embedded faithfully from the conditional independencies in P when such a DAG exists. These algorithms have been implemented in the Tetrad software package [Scheines et al., 1994], which can be downloaded for free from www.phil.cmu.edu/projects/tetrad/.

The Tetrad software package also has a module that uses the GES algorithm along with the BIC score to learn a Bayesian network from data.

Other Bayesian network learning packages include the following:

- Belief Network Power Constructor (constraint-based approach),
  www.cs.ualberta.ca/~jcheng/bnpc.htm.

- Bayesware (structure and parameters), www.bayesware.com/.

- Bayes Net Toolbox, bnt.sourceforge.net/.

- Probabilistic Net Library, www.eng.itlab.unn.ru/?dir=139.

EXERCISES

Section 8.2

Exercise 8.1 Suppose we have the two models in Figure 8.1 and the following data:

Case	J	F
1	j_1	f_1
2	j_1	f_1
3	j_1	f_1
4	j_1	f_1
5	j_1	f_1
6	j_1	f_2
7	j_2	f_2
8	j_2	f_2
9	j_2	f_2
10	j_2	f_1

1. Score each DAG model using the Bayesian score and compute their posterior probabilities, assuming that the prior probability of each model is .5.

2. Create a data set containing 20 records by duplicating the data in the table one time, and score the models using this 20-record data set. How have the scores changed?

3. Create a data set containing 200 records by duplicating the data in the table 19 times, and score the models using this 200-record data set. How have the scores changed?

Exercise 8.2 Assume that we have the models and data set discussed in Exercise 8.1. Using model averaging, compute the following:

1. $P(j_1|f_1, D)$ when D consists of our original 10 records.

2. $P(j_1|f_1, D)$ when D consists of our original 20 records.

3. $P(j_1|f_1, D)$ when D consists of our original 200 records.

Section 8.3

Exercise 8.3 Suppose $V = \{X, Y, Z, U, W\}$ and the set of conditional independencies in P is

$$\{I_P(X, Y) \quad I_P(\{W, U\}, \{X, Y\}|Z) \quad I_P(U, \{X, Y, Z\}|W)\}.$$

Find all DAGs faithful to P.

Exercise 8.4 Suppose $V = \{X, Y, Z, U, W\}$ and the set of conditional independencies in P is

$$\{I_P(X, Y) \quad I_P(X, Z) \quad I_P(Y, Z) \quad I_P(U, \{X, Y, Z\}|W)\}.$$

Find all DAGs faithful to P.

Exercise 8.5 Suppose $V = \{X, Y, Z, U, W\}$ and the set of conditional independencies in P is

$$\{I_P(X, Y|U) \quad I_P(U, \{Z, W\}|\{X, Y\}) \quad I_P(\{X, Y, U\}, W|Z)\}.$$

Find all DAGs faithful to P.

Exercise 8.6 Suppose $V = \{X, Y, Z, W, T, V, R\}$ and the set of conditional independencies in P is

$$\{I_P(X, Y|Z) \quad I_P(T, \{X, Y, Z, V\}|W)$$

$$I_P(V, \{X, Z, W, T\}|Y) \quad I_P(R, \{X, Y, Z, W\}|\{T, V\})\}.$$

Find all DAGs faithful to P.

Exercise 8.7 Suppose $V = \{X, Y, Z, W, U\}$ and the set of conditional independencies in P is

$$\{I_P(X, \{Y, W\}|U) \quad I_P(Y, \{X, Z\}|U)\}.$$

1. Is there any DAG faithful to P?

2. Find DAGs in which P is embedded faithfully.

Section 8.4

Exercise 8.8 If we make the causal faithfulness assumption, determine what causal influences we can learn in each of the following cases:

1. Given the conditional independencies in Exercise 8.3

2. Given the conditional independencies in Exercise 8.4

3. Given the conditional independencies in Exercise 8.5

4. Given the conditional independencies in Exercise 8.6

Exercise 8.9 If we make only the causal embedded faithfulness assumption, determine what causal influences we can learn in each of the following cases:

1. Given the conditional independencies in Exercise 8.3

2. Given the conditional independencies in Exercise 8.4

3. Given the conditional independencies in Exercise 8.5

4. Given the conditional independencies in Exercise 8.6

5. Given the conditional independencies in Exercise 8.7

Section 8.5

Exercise 8.10 In Example 8.24, we computed $P(j_1|f_1, \mathsf{D})$ using model averaging. Use the same technique to compute $P(j_1|f_2, \mathsf{D})$.

Exercise 8.11 Assume that there are three variables X_1, X_2, and X_3, and that all DAG patterns have the same posterior probability $(1/11)$ given the data. Compute the probability of the following features being present given the data (assuming faithfulness):

1. $I_p(X_1, X_2)$

2. $\neg I_p(X_1, X_2)$

3. $I_p(X_1, X_2|X_3)$ and $\neg I_p(X_1, X_2)$

4. $\neg I_p(X_1, X_2|X_3)$ and $I_p(X_1, X_2)$

Section 8.7

Exercise 8.12 Using Tetrad (or some other Bayesian network learning algorithm), learn a DAG from the data in Table 8.1. Next learn the parameters for the DAG. Can you suspect any causal influences from the learned DAG?

Exercise 8.13 Create a data file containing 120 records from the data in Table 8.1 by duplicating the data nine times. Using Tetrad (or some other Bayesian network learning algorithm), learn a DAG from this larger data set. Next learn the parameters for the DAG. Compare these results to those obtained in Exercise 8.12.

Exercise 8.14 Suppose we have the following variables:

Variable	What the Variable Represents
H	Parents' smoking habits
I	Income
S	Smoking
L	Lung cancer

and the following data:

Case	H	I	S	L
1	Yes	30,000	Yes	Yes
2	Yes	30,000	Yes	Yes
3	Yes	30,000	Yes	No
4	Yes	50,000	Yes	Yes
5	Yes	50,000	Yes	Yes
6	Yes	50,000	Yes	No
7	Yes	50,000	No	No
8	No	30,000	Yes	Yes
9	No	30,000	Yes	Yes
10	No	30,000	Yes	No
11	No	30,000	No	No
12	No	30,000	No	No
14	No	50,000	Yes	Yes
15	No	50,000	Yes	Yes
16	No	50,000	Yes	No
17	No	50,000	No	No
18	No	50,000	No	No
19	No	50,000	No	No

Using Tetrad (or some other Bayesian network learning algorithm) learn a DAG from these data. Next learn the parameters for the DAG. Can you suspect any causal influences from the learned DAG?

Exercise 8.15 Create a data file containing 190 records from the data in Exercise 8.14 by duplicating the data nine times. Using Tetrad (or some other Bayesian network learning algorithm), learn a DAG from this larger data set. Next learn the parameters for the DAG. Compare these results to those obtained in Exercise 8.14.

Part III

Bioinformatics Applications

Chapter 9

Nonmolecular Evolutionary Genetics

Evolution is the process of change in the genetic makeup of populations, **Evolutionary genetics** is the study of this change. In nonmolecular evolutionary genetics, we look only at the level of the allele and study how alleles might replace other alleles over time. In molecular evolutionary genetics, we investigate how nucleotide sequences change over time. In this chapter we discuss nonmolecular evolutionary genetics; in the next chapter we discuss molecular evolutionary genetics.

It is believed that the changes in genetic makeup are due to mutations. We discuss two possible means by which mutations can result in changes in alleles and allele relative frequencies in a population: namely natural selection and genetic drift. By **allele relative frequency** we mean the relative frequency of an allele in the population. For example, if in a diploid population a particular

225

gene has two alleles A and a; if there are 100 individuals with genotype AA, 50 individuals with genotype Aa, and 150 individuals with genotype aa, the relative frequency of allele A is

$$\frac{2 \times 100 + 50}{2 \times 100 + 2 \times 50 + 2 \times 150} = .41667.$$

Natural selection is the process by which organisms that have traits that better enable them to adapt to environmental pressures such as predators, changes in climate, or competition for food or mates will tend to survive and reproduce in greater numbers than other similar organisms, thereby increasing the existence of those favorable traits in future generations. So, natural selection can result in an increase in the relative frequencies of alleles that impart to the individual these favorable traits. The process of the change in allele relative frequencies due only to chance is called **genetic drift**.

Before discussing natural selection and genetic drift, in Section 9.1 we investigate ways in which allele relative frequencies might change if neither of them occurred. Section 9.2 presents allele relative frequency changes due to natural selection; Section 9.3 shows how genetic drift can account for changes in the relative frequencies of alleles. In Section 9.4 we present results obtained when we assume that natural selection and genetic drift are acting simultaneously. Finally, in Section 9.5 we discuss the rate of substitution, and we show that new alleles due to advantageous mutations become fixed in the population far more often than new alleles due to neutral mutations.

9.1   No Mutations, Selection, or Genetic Drift

Here we discuss allele relative frequencies if there were no mutations, no selection, and no genetic drift. First we discuss a population consisting of haploid organisms.

9.1.1   Haploid Population

Suppose we have a haploid population consisting of two strains, A and a, which produce asexually. By a *strain* we mean a set of organisms that have the same genotype. That is, they have the exact same genetic makeup. We assume that there are no mutations and no selection. So, the offspring are genetic duplicates of the parents. Furthermore, each member of each strain is expected to give rise to the same number of offspring. Finally, we assume that the population undergoes synchronous reproduction with nonoverlapping generations.

Define the following variables:

N_t: The total number of individuals in generation t

p_t: The relative frequency of strain A in generation t

$q_t = 1 - p_t$: The relative frequency of strain a in generation t

w: The expected value of the number of offspring for each individual in each strain

Then the expected value of the number of individuals in strain A in generation $t + 1$ is equal to (E denotes expected value)

$$E(\# \text{ offspring}|strain = A)N_t p_t = w N_t p_t.$$

Similarly, the expected value of the number of individuals in strain a in generation $t + 1$ is equal to

$$E(\# \text{ offspring}|strain = a)N_t p_t = w N_t p_t.$$

Owing to the law of large numbers, if we have an essentially infinite population, we can be almost certain that these expected values will be the actual values in generation $t + 1$. We will assume that they are. Then the relative frequency of strain A in generation $t + 1$ is

$$\begin{aligned}
p_{t+1} &= \frac{w N_t p_t}{w N_t p_t + w N_t q_t} \\[2mm]
&= \frac{p_t}{p_t + q_t} = p_t.
\end{aligned}$$

So, the relative frequency does not change even though the number of individuals in strain A may increase or decrease depending on whether $w > 1$.

9.1.2 Diploid Populations

Next we investigate changes in allele relative frequencies in a diploid population in the absence of natural selection and genetic drift. Again, we assume that the population undergoes synchronous reproduction with nonoverlapping generations. Assume that the population has a gene with two alleles A and a and the relative frequencies of the genotypes in the population in generation t are as follows:

Genotype	AA	Aa	aa
Relative frequency	p_{AA}	p_{Aa}	p_{aa}

If we choose a gamete at random from the population, the probability that the gamete contains the A allele is as follows (the random variable *genotype* is the genotype of the individual from which we obtained the allele):

$$\begin{aligned}
P(allele = A) &= P(allele = A|genotype = AA)P(genotype = AA) + \\
&\quad P(allele = A|genotype = Aa)P(genotype = Aa) + \\
&\quad P(allele = A|genotype = aa)P(genotype = aa) \\
&= 1 \times p_{AA} + .5 \times p_{Aa} + 0 \times p_{aa} \\
&= p_{AA} + .5 p_{Aa}.
\end{aligned}$$

Similarly,

$$P(allele = a) = p_{aa} + .5p_{Aa}.$$

The probability of an individual in the generation $t + 1$ having the various genotypes is then as follows:

$$
\begin{aligned}
P(genotype = AA) &= P(allele_1 = A)P(allele_2 = A) \\
&= (p_{AA} + .5p_{Aa})(p_{AA} + .5p_{Aa}) \\
&= (p_{AA} + .5p_{Aa})^2
\end{aligned}
\tag{9.1}
$$

$$
\begin{aligned}
P(genotype = aa) &= P(allele_1 = a)P(allele_2 = a) \\
&= (p_{aa} + .5p_{Aa})(p_{aa} + .5p_{Aa}) \\
&= (p_{aa} + .5p_{Aa})^2
\end{aligned}
$$

$$
\begin{aligned}
P(genotype = Aa) &= P(allele_1 = A)P(allele_2 = a) + \\
&\quad\ \ P(allele_1 = a)P(allele_2 = A) \\
&= (p_{AA} + .5p_{Aa})\,(p_{aa} + .5p_{Aa}) + \\
&\quad\ \ (p_{aa} + .5p_{Aa})\,(p_{AA} + .5p_{Aa}) \\
&= 2(p_{AA} + .5p_{Aa})\,(p_{aa} + .5p_{Aa}).
\end{aligned}
$$

Owing to the law of large numbers, if we have an essentially infinite population, we can be almost certain that these probabilities will be the relative frequencies of the genotypes in generation $t + 1$. We will assume that they are.

Now if we choose a gamete at random from generation $t + 1$, we have that (for the sake of brevity, we drop the reference to the random variables in the expressions on the right)

$P(allele = A)$

$$
\begin{aligned}
&= P(A|AA)P(AA) + P(A|Aa)P(Aa) + P(A|aa)P(aa) \\
&= 1(p_{AA} + .5p_{Aa})^2 + .5 \times 2(p_{AA} + .5p_{Aa})\,(p_{aa} + .5p_{Aa}) + 0(p_{aa} + .5p_{Aa})^2 \\
&= (p_{AA} + .5p_{Aa})\,[(p_{AA} + .5p_{Aa}) + (p_{aa} + .5p_{Aa})] \\
&= (p_{AA} + .5p_{Aa})(p_{AA} + p_{Aa} + p_{aa}) \\
&= (p_{AA} + .5p_{Aa})(1).
\end{aligned}
\tag{9.2}
$$

Similarly,

$$P(allele = a) = p_{aa} + .5p_{Aa}.$$

These are the same probabilities as in generation t. Again assuming an essentially infinite population, this means that the relative frequencies of the genotypes in generation $t + 2$ will be the same as those in generation $t + 1$. By an inductive argument they will be the same in all future generations. That is, once we pass the first generation, the relative frequencies will remain the same (Note that they are not necessarily the same in generation 1 and generation 2.) This is called a **Hardy-Weinberg Equilibrium**.

The following assumptions are necessary to our derivation of the Hardy-Weinberg Equilibrium:

1. There are only two alleles.

2. We have nonoverlapping generations. That is, one generation gives rise to another, whereupon the parents do not reproduce again and are no longer counted as part of the population.

3. There are equal genotype relative frequencies in the two sexes. This enables us to assume the same genotype relative frequencies for both parents.

4. Individuals mate at random.

5. There is a near infinite population, so we can assume that the relative frequency of a genotype in the next generation is about equal to the probability of any one individual having that genotype.

6. If an individual has genotype Aa, the probability of one of its gametes having allele A is .5.

7. There is no immigration. In this way, all members of the next generation come from the present generation.

8. There is no differential emigration. In this way, any emigration that does occur does not change the relative frequency of a genotype.

9. There are no mutations. In this way, each individual passes on to the next generation an allele exactly like one of its own.

10. There is no differential fertility of the genotypes. In this way, the probability of a given genotype being the genotype of the parent of a given individual is equal to its relative frequency among all genotypes.

11. There is no differential viability of the genotypes. This means that any death that occurs between the zygote and the adult stages does not alter the genotype relative frequencies.

Felsenstein [2007] shows that a Hardy-Weinberg Equilibrium still holds if we have multiple alleles. Felsenstein [2007] further shows that in the case of nonoverlapping generations, a Hardy-Weinberg Equilibrium is approached as the members from the initial generation die out. We will be concerned with relaxing assumptions 9, 10, and 11 and then investigating how natural selection can change allele relative frequencies, next relaxing assumption 5 and then seeing how genetic drift can change allele relative frequencies.

9.2   Natural Selection

Natural selection is the process by which organisms that have traits (alleles) that better enable them to adapt to environmental pressures such as predators, changes in climate, or competition for food or mates will tend to survive and reproduce in greater numbers than other similar organisms, thereby ensuring the perpetuation of those favorable traits in succeeding generations. It is believed to be the primary force that causes evolution to be adaptive. Natural selection is a direct violation of two of the Hardy-Weinberg assumptions: namely, the assumptions of no differential fertility and no differential viability. Less directly, it violates the assumption of no mutations, for if there were no mutations, there would be no allele changes that could enable an organism to be more fertile or more viable.

Next, assuming that there are mutations, differential fertility, and differential viability, we investigate how allele relative frequencies change over time. The predominant allele in a population is called the **wild type**, and a new allele is called a **mutant**. We are interested in determining whether and when a mutant allele replaces a wild-type allele. First, we investigate this matter in a haploid population.

9.2.1   Natural Selection in a Haploid Population

As in Section 9.1.1, suppose we have two strains A and a. Define the following variables:

v_A: The probability of an individual from strain A surviving to adulthood (called **viability**).

v_a: The probability of an individual from strain a surviving to adulthood.

f_A: The expected value of the number of offspring for an adult from strain A (called **fertility**).

f_a: The expected value of the number of offspring for an adult from strain a.

The expected value of the number of offspring for a strain A individual is given by

$E(\# \text{ offspring}|strain = A)$

$$
\begin{aligned}
&= E(\# \text{ offspring} \mid \text{survives to adult})P(\text{survives to adult}) + \\
&\quad E(\# \text{ offspring} \mid \text{does not survive to adult})P(\text{does not survive to adult}) \\
&= f_A v_A + 0(1 - v_A) \\
&= f_A v_A.
\end{aligned}
$$

We call this expected value the **absolute fitness** W_A of strain A. Similarly, $W_a = f_a v_a$ is the absolute fitness of strain a.

If we choose strain a as the reference point, we define the **relative fitness** of the strains as follows:

$$w_a = \frac{W_a}{W_a} = 1$$

$$w_A = \frac{W_A}{W_a}.$$

Example 9.1 *The following table shows possible values of these variables for a particular population:*

Strain	A	a
Viability as larva	$v_A = .5$	$v_a = .4$
Fertility as adult	$f_A = 6$	$f_a = 5$
Absolute fitness	$W_A = f_A v_A = .5 \times 6 = 3$	$W_a = f_a v_a = .4 \times 5 = 2$
Relative fitness	$w_A = W_A/W_a = 3/2 = 1.5$	$w_a = W_a/W_a = 2/2 = 1$

As in Section 9.1.1, define these variables:

N_t: The total number of individuals in generation t.

p_t: The relative frequency of strain A in generation t.

$q_t = 1 - p_t$: The relative frequency of strain a in generation t.

Then the expected value of the number of individuals in strain A in generation $t + 1$ is as follows:

$$E(\# \text{ offspring}|strain = A)N_t p_t = W_A N_t p_t.$$

Similarly, the expected value of the number of individuals in strain a in generation $t + 1$ is as follows:

$$E(\# \text{ offspring}|strain = a)N_t p_t = W_a N_t q_t.$$

If we again assume a near infinite population, then the relative frequencies in generation $t + 1$ are about as follows:

$$
\begin{aligned}
p_{t+1} &= \frac{W_A N_t p_t}{W_A N_t p_t + W_a N_t q_t} \\
&= \frac{W_A p_t}{W_A p_t + W_a q_t} \\
&= \frac{(W_A/W_a)\, p_t}{(W_A/W_a)\, p_t + (W_a/W_a)\, q_t} \\
&= \frac{w_A p_t}{w_A p_t + q_t}.
\end{aligned}
$$

Similarly,

$$q_{t+1} = \frac{q_t}{w_A p + q_t}.$$

We therefore have

$$\frac{p_{t+1}}{q_{t+1}} = \frac{\frac{w_A p_t}{w_A p_t + q_t}}{\frac{q_t}{w_A p + q_t}} = w_A \frac{p_t}{q_t}. \tag{9.3}$$

Owing to this recurrence,

$$\frac{p_t}{q_t} = w_A^t \frac{p_0}{q_0}.$$

Solving for p_t we have

$$p_t = \frac{w_A^t}{\frac{q_0}{p_0} + w_A^t}. \tag{9.4}$$

If $w_A > 1$, then

$$\lim_{t \to \infty} p_t = \lim_{t \to \infty} \frac{w_A^t}{\frac{q_0}{p_0} + w_A^t} = \frac{1}{\lim_{t \to \infty} \left(\frac{q_0}{p_0 w_A^t} \right) + 1} = \frac{1}{0 + 1} = 1.$$

So, the relative frequency of the more fit strain will approach 1 as the number of generations approaches infinity. Unless the total number of individuals is increasing, the less fit strain will be driven to extinction.

Often the relative fitness is denoted as $w_A = 1 + s$, where $-1 \le s < \infty$. Equality 9.4, then, is as follows:

$$p_t = \frac{(1 + s)^t}{\left(\frac{q_0}{p_0} + (1 + s)^t \right)}.$$

The next example illustrates how long it will take for a more fit mutant to overtake the wild type.

Example 9.2 *Suppose $s = .01$ and $p_0 = .001$. Then our function for p_t is*

$$p_t = \frac{(1 + .01)^t}{\left(\frac{.999}{.001} + (1 + .01)^t \right)}.$$

Figure 9.1 plots this function. We see that by the 1000th generation, most of the individuals are from strain A, even though initially only .1% were from strain A.

9.2.2   Natural Selection in a Diploid Population

In what follows, we assume there is one gene with two alleles A and a which determine the fitness of the individuals. It is possible to generalize to more alleles (see [Felsenstein, 2007]). It might seem odd that we assume one gene determines the fitness. However, if we assume that the fitness of the entire

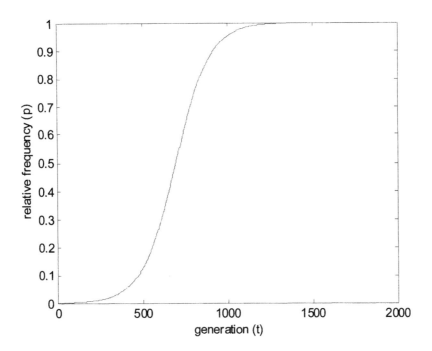

Figure 9.1: The relative frequency of strain A as a function of the generation when $p_0 = .001$ and $s = .01$.

organism is determined by the product of the fitness contribution of each allele, then the contribution to the fitness of an individual gene can be computed by assuming that gene determines the overall fitness. We show this at the end of this section. Note that in the previous discussion concerning a haploid population, we were discussing strains and not alleles, so it was not necessary to make any assumption about the fitness being determined by a product. Note further that when we derived the Hardy-Weinberg Equilibrium (see Section 9.1.2) for a diploid population we did not need any assumption about the fitness, because we were not considering fitness at that time.

Define the following variables:

N_t: The total number of individuals in generation t.

p_t: The relative frequency of allele A in generation t.

$q_t = 1 - p_t$: The relative frequency of allele a in generation t.

f_{AA}: The expected value of the number of gametes an adult individual with genotype AA contributes to the next generation.

f_{aa}: The expected value of the number of gametes an adult individual with genotype aa contributes to the next generation.

f_{Aa}: The expected value of the number of gametes an adult individual with genotype Aa contributes to the next generation.

v_{AA}: The probability of an individual with genotype AA surviving to adulthood.

v_{aa}: The probability of an individual with genotype aa surviving to adulthood.

v_{Aa}: The probability of an individual with genotype Aa surviving to adulthood.

Suppose that we assume a near-infinite population and start with a population in Hardy-Weinberg Equilibrium. Then, owing to Equalities 9.1 and 9.2, the relative frequency of genotype AA in the population at time t is equal to p_t^2. Similarly, the relative frequency of genotype aa is q_t^2, and the relative frequency of genotype Aa is $2p_tq_t$. Therefore, the expected value of the number of gametes contributed by individuals with genotype AA to generation $t + 1$ is

$$f_{AA}v_{AA}N_tp_t^2.$$

Similarly, the expected value of the number of gametes contributed by individuals with genotype aa to generation $t + 1$ is

$$f_{aa}v_{aa}N_tq_t^2,$$

and the expected value of the number of gametes contributed by individuals with genotype Aa to generation $t + 1$ is

$$f_{Aa}v_{Aa}N_t\left(2p_tq_t\right).$$

We then have that the expected value of the number of occurrences of each allele in generation $t + 1$ is

$$
\begin{aligned}
E_{t+1}(A) &= f_{AA}v_{AA}N_tp_t^2 + .5f_{Aa}v_{Aa}N_t\left(2p_tq_t\right) \\
&= f_{AA}v_{AA}N_tp_t^2 + f_{Aa}v_{Aa}N_tp_tq_t
\end{aligned}
$$

$$
\begin{aligned}
E_{t+1}(a) &= f_{aa}v_{aa}N_tq_t^2 + .5f_{Aa}v_{Aa}N_t\left(2p_tq_t\right) \\
&= f_{aa}v_{aa}N_tq_t^2 + f_{Aa}v_{Aa}N_tp_tq_t.
\end{aligned}
$$

Define the **absolute fitness** of the genotypes as follows:

$$
\begin{aligned}
W_{AA} &= .5 \times f_{AA} \times v_{AA} \\
W_{aa} &= .5 \times f_{aa} \times v_{aa} \\
W_{Aa} &= .5 \times f_{Aa} \times v_{Aa}.
\end{aligned}
$$

The absolute fitness is the expected value of the number of offspring for an individual with the given genotype. Note that we have the factor .5 because each gamete produces .5 individuals. The **relative fitness** is defined as follows (assuming our reference genotype is aa):

$$w_{aa} = \frac{W_{aa}}{W_{aa}} = 1$$

$$w_{AA} = \frac{W_{AA}}{W_{aa}}$$

$$w_{Aa} = \frac{W_{Aa}}{W_{aa}}.$$

We then have that the expected value of the number of occurrences of each allele in generation $t + 1$ is

$$E_{t+1}(A) = 2W_{AA}N_t p_t^2 + 2W_{Aa}N_t p_t q_t \qquad (9.5)$$

$$E_{t+1}(a) = 2W_{aa}N_t q_t^2 + 2W_{Aa}N_t p_t q_t.$$

Again assuming that the expected values are about equal to the actual numbers, the relative frequency of allele A in generation $t + 1$ is

$$
\begin{aligned}
p_{t+1} &= \frac{2W_{AA}N_t p_t^2 + 2W_{Aa}N_t p_t q_t}{2W_{AA}N_t p_t^2 + 2W_{Aa}N_t p_t q_t + 2W_{aa}N_t q_t^2 + 2W_{Aa}N_t p_t q_t} \\[2mm]
&= \frac{W_{AA}p_t^2 + W_{Aa}p_t q_t}{W_{AA}p_t^2 + 2W_{Aa}p_t q_t + W_{aa}q_t^2} \\[2mm]
&= \frac{w_{AA}p_t^2 + w_{Aa}p_t q_t}{w_{AA}p_t^2 + 2w_{Aa}p_t q_t + w_{aa}q_t^2}. \qquad (9.6)
\end{aligned}
$$

Similarly,

$$q_{t+1} = \frac{q_t^2 + w_{Aa}p_t q_t}{w_{AA}p_t^2 + 2w_{Aa}p_t q_t + w_{aa}q_t^2},$$

and therefore

$$\frac{p_{t+1}}{q_{t+1}} = \frac{w_{AA}p_t^2 + w_{Aa}p_t q_t}{w_{aa}q_t^2 + w_{Aa}p_t q_t}. \qquad (9.7)$$

Next we discuss several possible relationships among the fitnesses and investigate the consequences of each relationship.

Codominance

In general in genetics, *codominance* refers to the situation in which both alleles of a gene are expressed in the heterozygote, with neither one being dominant or recessive to the other. However, when we are discussing the fitness of the genotype, codominance ordinarily means something more specific. That is, by **codominance** we mean one of the homozygotes (the AA or aa genotype) is strongest, the other one is weakest, and the heterozygote (Aa) has a fitness level between the two. We call the allele in the stronger homozygote the **more fit allele** and the allele in the weaker homozygote the **less fit allele**. Realize, however, that it is not the allele that is fit, but rather the organism.

Next we discuss multiplicative and additive codominance.

Multiplicative Fitness   Codominance with multiplicative fitness has the relationships in the following table:

Genotype	AA	Aa	aa
Fitness	$w_{AA} = (1+s)^2$	$w_{Aa} = 1+s$	$w_{aa} = 1$

Owing to Equality 9.7, we then have

$$\frac{p_{t+1}}{q_{t+1}} = \frac{w_{AA}p_t^2 + w_{Aa}p_tq_t}{w_{aa}q_t^2 + w_{Aa}p_tq_t}$$

$$= \frac{(1+s)^2 p_t^2 + (1+s)\,p_tq_t}{q_t^2 + (1+s)p_tq_t}$$

$$= \frac{(1+s)p_t[(1+s)p_t + q_t]}{q_t(q_t + p_t + sp_t)}$$

$$= \frac{(1+s)p_t(1+sp_t)}{q_t(1+sp_t)}$$

$$= (1+s)\frac{p_t}{q_t}.$$

This recurrence is the same as Recurrence 9.3. So, the result will be the same as the one presented for a haploid population in Section 9.2.1. That is, the relative frequency of the more fit allele will fairly quickly approach 1 as the number of generations approaches infinity (depending on the value of s).

Additive Fitness   Codominance with additive fitness has the relationships in the following table:

Genotype	AA	Aa	aa
Fitness	$w_{AA} = 1+2s$	$w_{Aa} = 1+s$	$w_{aa} = 1$

Owing to Equality 9.7, we then have

$$\frac{p_{t+1}}{q_{t+1}} = \frac{w_{AA}p_t^2 + w_{Aa}p_tq_t}{w_{aa}q_t^2 + w_{Aa}p_tq_t}$$

$$= \frac{(1+2s)p_t^2 + (1+s)p_tq_t}{q_t^2 + (1+s)p_tq_t}$$

$$= \left(\frac{1+s+sp_t}{1+sp_t}\right)\frac{p_t}{q_t}.$$

Unlike the multiplicative case, this recurrence does not have a closed-form solution. However, if $s > 1$, then

$$\frac{1 + s + sp_t}{1 + sp_t} > 1,$$

which means the sequence p_t/q_t is increasing, and therefore the sequence p_t is increasing. We therefore have that

$$\begin{aligned}
\frac{p_{t+1}}{q_{t+1}} &= \frac{1 + s + sp_t}{1 + sp_t} \frac{p_t}{q_t} \\
&> \frac{1 + s + sp_0}{1 + s} \frac{p_t}{q_t}.
\end{aligned}$$

This implies that

$$\frac{p_t}{q_t} > \left(\frac{1 + s + sp_0}{1 + s}\right)^t \frac{p_0}{q_0},$$

which means

$$\lim_{t \to \infty} \frac{p_t}{q_t} = \infty.$$

So, we again conclude that the relative frequency of the more fit allele will approach 1 as the number of generations approaches infinity.

Next we investigate the convergence speed for small and large values of s.

1. Suppose s is small. Then when p_t is small,

$$\frac{1 + s + sp_t}{1 + sp_t} \approx \frac{1 + s + 0}{1 + 0} = 1 + s,$$

and when p_t is large (close to 1),

$$\frac{1 + s + sp_t}{1 + sp_t} \approx \frac{1 + s + s}{1 + s} = \frac{1 + 2s}{1 + s} \approx \frac{1 + 2s + s^2}{1 + s} = 1 + s.$$

So, when s is small, the convergence is about the same as that for the multiplicative case.

2. Suppose s is large (close to 1). Then when p_t is small,

$$\frac{1 + s + sp_t}{1 + sp_t} \approx \frac{1 + s + 0}{1 + 0} = 1 + s.$$

However, once p_t becomes large (close to 1)

$$\frac{1 + s + sp_t}{1 + sp_t} \approx \frac{1 + s + s}{1 + s} = \frac{1 + 2s}{1 + s}$$

Now

$$\lim_{s \to 1} \frac{1 + 2s}{1 + s} = \frac{3}{2} = 1 + \frac{1}{2}.$$

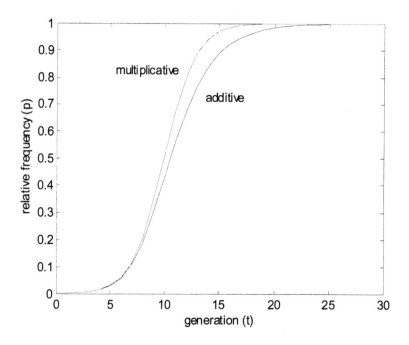

Figure 9.2: The relative frequency of allele A as a function of the generation when $p_0 = .001$ and $s = 1$ in the case of multiplicative and additive codominance.

So, when s is large and p_t is large, the multiplier in our recurrence is about equal to $1 + s/2$. That is,

$$\frac{p_{t+1}}{q_{t+1}} \approx (1 + s/2)\frac{p_t}{q_t}.$$

This means that, once p_t becomes large, convergence slows down relative to the multiplicative case.

Example 9.3 *Figure 9.2 shows p_t (the relative frequency of allele A) as a function of t (the generation) when $p_0 = .001$ and $s = 1$ in the case of multiplicative and additive codominance.*

Dominance

In **dominance**, one of the alleles is dominant as far as determining the trait (in this case, the fitness) of the heterozygote. Therefore, the heterozygote has the same fitness as the homozygote with the dominant allele.

More Fit Allele Is Dominant If the more fit allele is dominant, we have the relationships in the following table:

Genotype	AA	Aa	aa
Fitness	$w_{AA} = 1 + s$	$w_{Aa} = 1 + s$	$w_{aa} = 1$

Owing to Equality 9.7, we then have

$$\frac{p_{t+1}}{q_{t+1}} = \frac{w_{AA}p_t^2 + w_{Aa}p_t q_t}{w_{aa}q_t^2 + w_{Aa}p_t q_t}$$
$$= \frac{(1+s)p_t^2 + (1+s)p_t q_t}{q_t^2 + (1+s)p_t q_t}$$
$$= \frac{1+s}{1+sp_t}\frac{p_t}{q_t}.$$

It is not difficult to show that

$$\lim_{t \to \infty} \frac{p_t}{q_t} = \infty,$$

which means that the relative frequency of the more fit allele will approach 1 as the number of generations approaches infinity.

When p_t is small, the convergence is about the same as that of the codominant multiplicative fitness. So, a new, stronger allele will initially quickly overtake the wild-type allele. However, once p_t becomes close to 1,

$$\frac{1+s}{1+sp_t} \approx 1,$$

which means convergence will slow to a crawl. This implies that it is difficult to completely get rid of a deleterious recessive allele.

Example 9.4 *Figure 9.3 shows p_t (the relative frequency of allele A) as a function of t (the generation) when $p_0 = .001$ and $s = .01$ in the case of dominance of the more fit allele (A). Note that the relative frequency of A exceeds .9 in the 2000th generation, but in the 10,000th generation the relative frequency of A is still noticeably smaller than 1.*

Example 9.5 *Examples in humans of dominance of the more fit allele include the diseases Tay-Sachs syndrome and cystic fibrosis. The allele for each of these diseases is recessive, and the homozygote, which has two alleles for the disease, is by far less fit. However, this recessive allele is able to "hide" in the heterozygote, which is just as fit as the dominant homozygote, for a long time.*

Less Fit Allele Is Dominant If the less fit allele is dominant, we have the relationships in the following table:

Genotype	AA	Aa	aa
Fitness	$w_{AA} = 1 + s$	$w_{Aa} = 1$	$w_{aa} = 1$

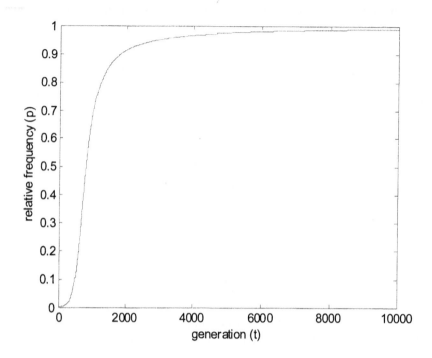

Figure 9.3: The relative frequency of allele A as a function of the generation when $p_0 = .001$ and $s = .01$ in the case of dominance of the more fit allele (A).

Owing to Equality 9.7, we then have

$$
\begin{aligned}
\frac{p_{t+1}}{q_{t+1}} &= \frac{w_{AA}p_t^2 + w_{Aa}p_tq_t}{w_{aa}q_t^2 + w_{Aa}p_tq_t} \\
&= \frac{(1+s)p_t^2 + p_tq_t}{q_t^2 + p_tq_t} \\
&= (1+sp_t)\frac{p_t}{q_t}.
\end{aligned}
$$

When p_t is small, $1 + sp_t \approx 1$, whereas when p_t is large, $1 + sp_t \approx 1 + s$. This means a less fit, dominant wild-type allele will initially persist for a long time before it is taken over.

Example 9.6 *Figure 9.4 shows p_t (the relative frequency of allele A) as a function of t (the generation) when $p_0 = .001$ and $s = .01$ in the case of dominance of the less fit allele (a). Note that the relative frequency of A increases very slowly for a long time and then shoots close to 1 near the 100,000th generation.*

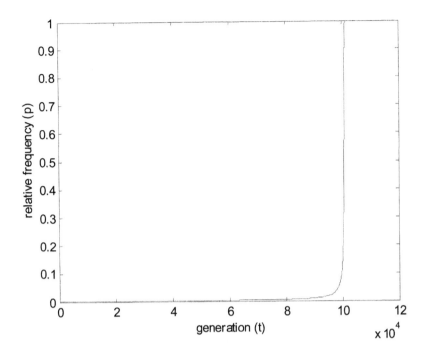

Figure 9.4: The relative frequency of allele A as a function of the generation when $p_0 = .001$ and $s = .01$ in the case of dominance of the less fit allele (a).

Overdominance

In **overdominance**, the heterozygote is more fit than either homozygote. So, overdominance is actually a special case of codominance, when the term codominance is used in its more general sense in genetics. This relationship is described by the following table, where $0 < s \leq 1$ and $0 < t \leq 1$:

Genotype	AA	Aa	aa
Fitness	$w_{AA} = 1 - s$	$w_{Aa} = 1$	$w_{aa} = 1 - r$

Owing to Equality 9.7, we then have

$$\frac{p_{t+1}}{q_{t+1}} = \frac{w_{AA}p_t^2 + w_{Aa}p_tq_t}{w_{aa}q_t^2 + w_{Aa}p_tq_t}$$

$$= \frac{(1-s)p_t^2 + p_tq_t}{(1-r)q_t^2 + p_tq_t}$$

$$= \frac{1 - sp_t}{1 - rq_t}\frac{p_t}{q_t}.$$

It is difficult to investigate this recurrence directly because we have two constants s and t. However, it seems that in this case we could reach an equilibrium in which both alleles coexist. Let's investigate this possibility. Using Equality 9.6, it is left as an exercise to show that in general

$$\Delta p = p_{t+1} - p_t$$

$$= \frac{p_t q_t \left[(w_{Aa} - w_{aa}) q_t + (w_{AA} - w_{Aa}) p_t \right]}{w_{AA} p_t^2 + 2 w_{Aa} p_t q_t + w_{aa} q_t^2}.$$

In the case of overdominance, we obtain that

$$\Delta p = \frac{p_t q_t \left[(w_{Aa} - w_{aa}) q_t + (w_{AA} - w_{Aa}) p_t \right]}{w_{AA} p_t^2 + 2 w_{Aa} p_t q_t + w_{aa} q_t^2}$$

$$= \frac{p_t q_t (r q_t - s p_t)}{(1 - s) p_t^2 + 2 p_t q_t + (1 - r) q_t^2}$$

$$= \frac{p_t q_t [r - (s + r) p_t]}{1 - s p_t^2 - r q_t^2}. \tag{9.8}$$

To approach an equilibrium, we would have to have $\Delta p \to 0$. Owing to the previous equality, $\Delta p \to 0$ if we have one of the following:

1. $p_t \to 0$

2. $q_t \to 0$

3. $r - (s + r) p_t \to 0$, which means $p_t \to r/(s + r)$.

In the first two cases, we are approaching the elimination of one of the alleles; in the third case, we are approaching coexistence.

Note that the numerator of Equality 9.8 has the same sign as

$$r - (s + r) p_t.$$

If p is close to 0, $r - (s + r) p_t$ is close to r, which is positive, which means $\Delta p > 0$ and p_t will not approach 0. If p is close to 1, $r - (s + r) p_t$ is close to $-s$, which is negative, which means $\Delta p < 0$ and p_t will not approach 1. So, unless we start with $p_0 = 0$ or $p_0 = 1$, p_t will not approach 0 or 1.

Now suppose $p_t < r/(s + r)$. Then

$$r - (s + r) p_t > r - (s + r) \left(\frac{r}{s + r} \right) = r - r = 0.$$

So, $\Delta p > 0$, which means p_t will be approaching $r/(s + r)$.

Next suppose $p_t > r/(s + r)$. Then

$$r - (s + r) p_t < r - (s + r) \left(\frac{r}{s + r} \right) = r - r = 0.$$

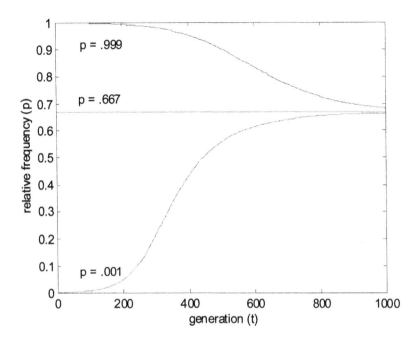

Figure 9.5: The relative frequency of allele A as a function of the generation when $s = .01$ and $r = .02$ in the case of overdominance. The initial relative frequency of A is p.

So, $\Delta p < 0$, which means p_t will be approaching $r/(s + r)$.

The possibility still exists that p_t will jump back and forth over $r/(s + r)$. It is left as an exercise to show that this does not happen. So, the equilibrium is in fact stable. Intuitively, since the heterozygote is strongest, both alleles should perpetuate in the population so as to maintain the strongest genotype.

Example 9.7 *Figure 9.5 shows p_t (the relative frequency of allele A) as a function of t (the generation) when $s = .01$ and $r = .02$ in the case of overdominance. We have that*

$$p_t \to \frac{r}{s + r} = \frac{.02}{.01 + 02} = \frac{2}{3}.$$

Regardless of the initial distribution of alleles, eventually about $\frac{2}{3}$ of the alleles will be type A.

Example 9.8 *Sickle-cell anemia is an example of overdominance in humans. The homozygote, which has two alleles for the disease, has a severe illness that often results in death. The heterozygote is essentially asymptomatic but has* a superior resistance to malaria, which is an important fitness advantage in malarial regions. In parts of Africa, the Middle East, and India, where malaria is endemic, there is a relatively high relative frequency of the sickle-cell allele.

Underdominance

In **underdominance**, the heterozygote is less fit than either homozygote. So, underdominance is actually a special case of codominance, when the term codominance is used in its more general sense in genetics. This relationship is described by the same table used for overdominance, but with $s < 0$ and $t < 0$. Recall that the table is as follows:

Genotype	AA	Aa	aa
Fitness	$w_{AA} = 1 - s$	$w_{Aa} = 1$	$w_{aa} = 1 - r$

We then obtain the same expression for Δp as the one for overdominance:

$$\Delta p = \frac{p_t q_t [r - (s + r)p_t]}{1 - sp_t^2 - rq_t^2},$$

and therefore the same equilibrium points:

1. $p_t \to 0$

2. $q_t \to 0$

3. $r - (s + r)p_t \to 0$, which means $p_t \to r/(s + r)$.

If p_t is close to 0, $r - (s + r)p_t$ is close to r, which is now negative. So, $\Delta p < 0$ and p_t will approach 0. Similarly, if p is close to 1, $r - (s+r)p_t$ is close to $-s$, which is now positive. So, $\Delta p > 0$ and p_t will approach 1.

Suppose $p_t < r/(s+r)$. Then since $s + r < 0$

$$r - (s + r)p_t < r - (s + r)\left(\frac{r}{s + r}\right) = r - r = 0.$$

So, $\Delta p < 0$, which means p_t will be moving away from $r/(s + r)$.

If $p_t > r/(s+r)$, then since $s + r < 0$

$$r - (s + r)p_t > r - (s + r)\left(\frac{r}{s + r}\right) = r - r = 0.$$

So, $\Delta p > 0$, which means p_t will be moving away from $r/(s + r)$.

We conclude that $r/(s + r)$ is an unstable equilibrium. That is, if $p_t = r/(s + r)$, the slightest move in either direction will make it move away from that point. If the nudge is to the left, p_t will approach 0; if it is to the right, p_t will approach 1. Intuitively, the heterozygote should be eliminated from the population since it is weakest, and this can only happen if one of the two alleles is eliminated.

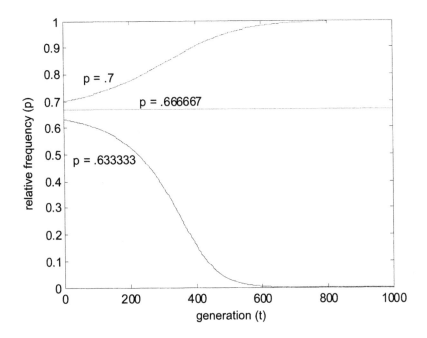

Figure 9.6: The relative frequency of allele A as a function of the generation when $s = -.01$ and $r = -.02$ in the case of underdominance. The initial relative frequency of A is p.

Example 9.9 *Figure 9.6 shows p_t (the relative frequency of allele A) as a function of t (the generation) when $s = -.01$ and $r = -.02$ in the case of underdominance. We have that*

$$p_t = \frac{-.02}{-.01 - .02} = \frac{2}{3}$$

is an unstable equilibrium. Note that the equilibrium is at a point where there is more of the less fit allele, namely A. Initially, this might seem unintuitive. However, if we go just beneath the equilibrium point of $\frac{2}{3}$, p_t will be driven to 0. So, there is a bigger range of relative frequencies, namely the interval $\left(0, \frac{2}{3}\right)$, for which A will be eliminated.

Fitness as a Product of Fitnesses

We assumed that the fitness of an organism is determined by one gene (locus). This certainly is not the case. However, if we assume that the fitness of the organism is determined by the product of the fitness contribution of each gene, we obtain the same results for individual genes as those obtained earlier. We illustrate this idea using just two genes.

Suppose we have one gene with alleles A and a and another gene with alleles B and b. Assume that the absolute fitness of an individual is the product of absolute fitness values associated with each of the two alleles. For example, if the individual's genotype is $AaBB$, then the absolute fitness of the individual is

$$W_{AaBB} = kW_{Aa}W_{BB},$$

where k is a constant. If we further assume that the alleles from the two genes mix independently in the population, and we let p_t, q_t, s_t, and r_t be, respectively, the relative frequencies of alleles A, a, B, and b in generation t, then Equality 9.5 is replaced by the following equality:

$$
\begin{aligned}
E_{t+1}(A) &= kW_{AA}W_{BB}N_t p_t^2 s_t^2 + kW_{AA}W_{Bb}N_t p_t^2 (2s_t r_t) + \\
&\quad kW_{AA}W_{bb}N_t p_t^2 r_t^2 + .5kW_{Aa}W_{BB}N_t(2p_t q_t)s_t^2 + \\
&\quad .5kW_{Aa}W_{Bb}N_t(2p_t q_t)(2s_t r_t) + .5kW_{Aa}W_{bb}N_t(2p_t q_t)r_t^2 \\
\\
&= kW_{AA}W_{BB}N_t p_t^2 s_t^2 + 2kW_{AA}W_{Bb}N_t p_t^2 2s_t r_t + \\
&\quad kW_{AA}W_{bb}N_t p_t^2 r_t^2 + kW_{Aa}W_{BB}N_t p_t q_t s_t^2 + \\
&\quad 2kW_{Aa}W_{Bb}N_t p_t q_t s_t r_t + kW_{Aa}W_{bb}N_t p_t q_t r_t^2 \\
\\
&= k(W_{BB}s_t^2 + 2W_{Bb}s_t r_t + W_{bb}r_t^2)\left(W_{AA}N_t p_t^2 + W_{Aa}N_t p_t q_t\right).
\end{aligned}
$$

Similarly,

$$E_{t+1}(a) = k(W_{BB}s_t^2 + 2W_{Bb}s_t r_t + W_{bb}r_t^2)\left(W_{Aa}N_t p_t^2 + W_{Aa}N_t p_t q_t\right).$$

So, when we compute p_{t+1} as done in Equality 9.6, the expression

$$k\left(W_{BB}s_t^2 + 2W_{Bb}s_t r_t + W_{bb}r_t^2\right)$$

will cancel out, and all our results will be the same.

How realistic is this assumption that we can multiply fitnesses? Joe Felsenstein [private correspondence] notes the following:

> It is not realistic to assume that the fitness of a diploid is a product of quantities, each determined by one locus. On the other hand most pairs of loci don't interact, and one can approximate by using a model in which we take the products. This model is not realistic, but it is useful.

Let's investigate further what the assumption of noninteraction entails. Recall that we represented this assumption mathematically as follows:

$$W_{AaBB} = kW_{Aa}W_{BB}. \tag{9.9}$$

Next we determine the value of the constant k. If we give the absolute fitness the usual interpretation that it is the expected value of the number of offspring for an individual with the given genotype, then Equality 9.9 states that

$$E(n|AaBB) = kE(n|Aa)E(n|BB),$$

where n is the number of offspring an individual has and E denotes expected value. We wrote this equality for particular allele values. The general form of it is as follows:

$$E(n|XYZW) = kE(n|XY)E(n|ZW), \qquad (9.10)$$

where X and Y are random variables whose possible values are A and a, and Z and W are random variables whose possible values are B and b. Now

$$E(n|XY) = \sum_{ZW} E(n|XYZW)P(ZW|XY)$$

$$= \sum_{ZW} E(n|XYZW)P(ZW),$$

where the second equality is owing to the fact that we assumed that the genes mix independently. Owing to Equality 9.10, we then have

$$E(n|XY) = \sum_{ZW} E(n|XYZW)P(ZW)$$

$$= k \sum_{ZW} E(n|XY)E(n|ZW)P(ZW)$$

$$= kE(n|XY)E(n),$$

which means

$$k = \frac{1}{E(n)}.$$

So, we are assuming that

$$W_{XYZW} = E(n|XYZW) = \frac{E(n|XY)E(n|ZW)}{E(n)}.$$

We illustrated the situation with only two loci. If there were k loci, it is not hard to show that denominator would be $(E(n))^{k-1}$.

Example 9.10 *Suppose*

$$W_{Aa} = E(n|Aa) = 4 \qquad W_{BB} = E(n|BB) = 3 \qquad E(n) = 2.$$

Then

$$W_{AaBB} = E(n|Aa, BB) = \frac{E(n|Aa)E(n|BB)}{E(n)} = \frac{4 \times 3}{2} = 6.$$

It might seem odd that the combination of Aa and BB is more fit than either type individually. Note that both genotypes Aa and BB are more fit than average. Note further that $E(n|Aa)$ is an expected value based on lack of knowledge of the second genotype. When we further condition on the fact that the second genotype is more fit than average, the fitness should go up.

9.3   Genetic Drift

In Section 9.1 we investigated how populations might change if there were no mutations and we had an essentially infinite population. We assumed that, owing to the law of large numbers, the actual number of individuals of a given type in each generation is equal to the expected value of that number. In Section 9.1.1 we found that this implies that, in the case of a haploid asexual population, the relative frequency of each strain will not change from generation to generation (which is no surprise). In Section 9.1.2 we found that this implies that, in the case of a diploid population, once we pass the first generation, the relative frequencies of the alleles will remain the same, which is called a Hardy-Weinberg Equilibrium. In Section 9.2, we assumed that mutations can occur, resulting in new alleles that affect the fitness of the individuals with the new alleles. We again assumed an essentially infinite population so that the actual numbers in each generation are the expected values of those numbers. We found that this assumption implies that a less fit allele will methodically be eliminated from the population, except in the case of overdominance. In this latter case, the relative frequencies of the alleles will methodically approach values determined by their relative fitnesses.

Both of these models are deterministic because we assumed that the actual numbers are equal to their expected values. In actuality, populations are finite, which means that we cannot make this assumption. For example, suppose there are N haploid individuals in each generation and the generations do not overlap. Suppose further that the relative frequency of strain A in generation t is p and the relative frequency of strain a in generation t is $1-p$. The expected value of the number of strain A individuals in generation $t+1$ is then Np. However, since the population is finite, by chance we could have more or less than this number. In the worst-case scenario for strain A, by chance none of the strain A individuals will have surviving offspring, leaving us with no strain A individuals in generation $t+1$, whereas in the best-case scenario, none of the strain a individuals will have surviving offspring, leaving us with all strain A individuals in generation $t+1$. The situation is similar in a diploid population. A given heterozygote might by chance not contribute an equal number of A alleles and a alleles to the gamete pool. Furthermore, a given individual might by chance not have any offspring. So, there is a small chance that all the alleles in the next generation may be of type A or of type a. Next we investigate **genetic drift**, which is the evolutionary consequence of random changes in strain (or allele) relative frequencies. We make the simplifying assumptions in the **Wright-Fisher Model**. For a diploid population, these assumptions are as follows:

1. There are N individuals in each generation and the generations do not overlap.

2. There are two alleles, A and a. This means that there are a total of $2N$ alleles in each generation.

3. Each individual in generation $t+1$ has each of its alleles generated as

follows: First, an individual is chosen at random from generation t; then an allele is chosen at random from that individual.

These assumptions imply the following probabilistic statement:

> The probability of each allele in generation $t+1$ being type A is p, where p is the relative frequency of allele A in generation t. The event that one allele in generation $t+1$ is type A is independent of the event that another allele in generation $t+1$ is type A.

The following example illustrates that this statement follows from the Wright-Fisher Model.

Example 9.11 *Suppose there are 100 individuals in generation t, 30 of these individuals are type AA, 60 are type Aa, and 10 are type aa. Suppose further for each individual in generation $t+1$, each of its alleles is obtained by randomly choosing an individual in generation t and then randomly choosing an allele from that individual. Then the probability of each of its alleles being type A is given by*

$$\begin{aligned}
P(allele = A) &= P(A|AA)P(AA) + P(A|Aa)P(Aa) + P(A|aa)P(aa) \\
&= 1 \times .3 + .5 \times .6 + 0 \times .1 = .6.
\end{aligned}$$

Furthermore, the relative frequency of allele A in the population is given by

$$p = \frac{2 \times 30 + 60}{2 \times 100} = .6.$$

So,

$$P(allele = A) = p,$$

as the preceding statement claims.

Notice that the Wright-Fisher Model is analogous to the following coin-tossing experiment:

1. In trial 0 we toss a coin n times, and the coin has probability p_0 of landing heads on each toss.

2. In trial $t+1$ we toss a coin n times, and the coin has probability k_t/n of landing heads on each toss, where k_t is the number of heads in trial t.

In the Wright-Fisher Model, $n = 2N$ and p_0 is the relative frequency of allele A in the initial generation. We first obtain some results for the coin-tossing experiment, and then we apply the results to the Wright-Fisher Model.

9.3.1 A Coin-Tossing Experiment

We obtain results for the coin-tossing experiment described at the end of the last subsection. The results we obtain pertain to any experiment that has two possible outcomes and for which separate trials of the experiment are probabilistically independent. We use coin-tossing terminology for the sake of concreteness. Suppose we are about to toss a possibly biased coin. Let the probability of heads be p and the probability of tails be $q = 1 - p$. That is,

$$P(heads) = p.$$

If we perform n independent trials of the experiment (i.e., toss the coin n times), the sequence of trials is called a **Bernoulli process**. If we let K be a random variable whose value is the number of heads that occur in the n trials, then the probability distribution of K is the **binomial distribution**, which is as follows:

$$P(K = k) = \binom{n}{k} p^k q^{n-k} \tag{9.11}$$

for $k = 0, 1, \ldots, n$, where

$$\binom{n}{k} = \frac{n!}{k!(n-k)!}.$$

The preceding result is obtained in most elementary probability texts, as is the proof of the following lemma.

Lemma 9.1 *For the binomial distribution, the expected value of K is given by*

$$E(K) = np,$$

and the variance of K is given by

$$Var(K) = np(1 - p).$$

Suppose now that we toss a coin n times and then choose a new coin according to the number of heads that occurred in the n tosses. Specifically, if k heads occurred in the n tosses, we choose a new coin such that

$$P(heads) = \frac{k}{n}.$$

We then toss the new coin n times and choose a third coin according to the number of heads that occurred in these n tosses. If we repeat this process indefinitely, we produce a Markov chain, as discussed in Section 6.4. The transition matrix for this chain is as follows:

$$\mathbf{P} = \begin{pmatrix} p_{00}=1 & p_{01}=0 & p_{02}=0 & p_{03}=0 & \cdots & p_{0,n}=0 \\ p_{10} & p_{11} & p_{12} & p_{13} & \cdots & p_{1n} \\ p_{20} & p_{21} & p_{22} & p_{23} & \cdots & p_{2n} \\ \vdots & \vdots & \vdots & \vdots & \ddots & \vdots \\ p_{n-1,0} & p_{n-1,1} & p_{n-1,2} & p_{n-1,3} & \cdots & p_{n-1,n} \\ p_{n0}=0 & p_{n1}=0 & p_{n2}=0 & p_{n3}=0 & \cdots & p_{nn}=1 \end{pmatrix} \tag{9.12}$$

In this matrix, p_{ij} is the probability of obtaining j heads in the next set of n tosses if i heads were obtained in the previous set of n tosses. Owing to Equality 9.11, the value of p_{ij} is given by

$$p_{ij} = \binom{n}{j} p^j q^{n-j},$$

where

$$
\begin{aligned}
p &= \frac{i}{n} \\
q &= 1 - p.
\end{aligned}
$$

Note from the first row that

$$p_{00} = 1.$$

This is because when zero heads were obtained in the previous set of tosses, we will toss a coin that is certain to land tails in the next set of tosses. Similarly, $p_{nn} = 1$.

Recall that the stationary distribution for a Markov chain is the limiting distribution of the outcomes. In the current example, intuitively it is the probability distribution of the number of heads in a set of tosses that is essentially an infinite amount of time into the future, based on the information we have when we start the experiment. To find the stationary distribution for this chain, we need to solve

$$\mathbf{r}^T = \mathbf{r}^T \mathbf{P}$$

along with

$$\sum_j r_j = 1.$$

This system does not have a unique solution. Rather, the solution is

$$
\begin{aligned}
r_0 &= 1 - x \\
r_1 &= r_2 = \cdots = r_{n-1} = 0 \\
r_n &= x.
\end{aligned}
$$

The following example illustrates this idea.

Example 9.12 *Solve*

$$
\begin{pmatrix} r_0 & r_1 & r_2 & r_3 \end{pmatrix} = \begin{pmatrix} r_0 & r_1 & r_2 & r_3 \end{pmatrix}
\begin{pmatrix}
1 & 0 & 0 & 0 \\
.296 & .444 & .222 & .037 \\
.037 & .222 & .444 & .296 \\
0 & 0 & 0 & 1
\end{pmatrix}
$$

along with

$$\sum_{j=0}^{3} r_j = 1.$$

The solution is

$$
\begin{aligned}
r_0 &= 1 - x \\
r_1 &= r_2 = 0 \\
r_3 &= x.
\end{aligned}
$$

The reason there is not a unique solution is that the chain is not reducible. That is, every state is not reachable from every other state. Once we have a set of outcomes with zero heads, we will forever toss a coin weighted to always land tails. Similarly, once we have a set of outcomes with n heads, we will forever toss a coin weighted to always land heads. These states are called **absorbing states**. The stationary distribution in this case depends on the initial probabilities. Next we show how to determine this distribution from these initial probabilities. However, first note that for the stationary distribution

$$
\begin{aligned}
P(K = 0) &= 1 - x \\
P(K = n) &= x.
\end{aligned}
\tag{9.13}
$$

Recall that K is a random variable whose value is the number of heads. This means that eventually we will toss a coin that always lands heads or always lands tails.

Theorem 9.1 *Suppose we toss a coin n times and then choose a new coin according to the number of heads that occurred in the n tosses. We then repeat this process indefinitely producing the Markov chain whose transition matrix is in Equality 9.12. Let*

$$
p_0 = P(heads)
$$

for the first set of tosses, and K_t be a random variable representing the number of occurrences of heads in the set of tosses occurring at time t. Then for all t

$$
E(K_t) = np_0.
\tag{9.14}
$$

Proof. *The proof is by induction. Owing to Lemma 9.1, we have that*

$$
E(K_1) = np_0.
$$

By way of induction, assume that

$$
E(K_t) = np_0.
$$

Then

$$
\begin{aligned}
E(K_{t+1}) &= \sum_{k=0}^{n} E(n_{t+1} | K_t = k) P(K_t = k) \\
&= \sum_{k=0}^{n} n \left(\frac{k}{n} \right) P(K_t = k) \\
&= \sum_{k=0}^{n} k P(K_t = k) \\
&= E(K_t) = np_0.
\end{aligned}
$$

To obtain the second equality above, first we note that when k heads occur at time t, we toss a coin with $P(heads) = k/n$ at time $t+1$; then we apply Lemma 9.1. The last equality is due to the induction hypothesis. ∎

Now owing to Equality 9.13, for the stationary distribution

$$E(K) = 0(1 - x) + nx = nx.$$

This equality along with Equality 9.14 implies that

$$nx = np_0,$$

which means

$$x = p_0.$$

We see that the probability of eventually always throwing heads is equal to the probability of throwing heads on each individual toss in the first set of tosses. Furthermore, for the stationary distribution

$$E(K^2) = 0^2(1 - p_0) + n^2 p_0 = n^2 p_0.$$

So,

$$\begin{aligned} Var(K) &= E(K^2) - (E(K))^2 \\ &= n^2 p_0 - (np_0)^2 \\ &= n^2 p_0(1 - p_0). \end{aligned}$$

Owing to Lemma 9.1, for the first set of tosses

$$Var(K_1) = np_0(1 - p_0).$$

Example 9.13 *Suppose $n = 1000$ and*

$$p_0 = P(heads) = .3$$

for our first set of tosses. Then for every set of tosses

$$E(K_t) = np_0 = 1000 \times .3 = 300.$$

For the first set of tosses

$$\begin{aligned} Var(K_1) &= np_0(1 - p_0) \\ &= 1000 \times .3(1 - .3) = 210. \end{aligned}$$

This means that the standard deviation σ_1 for the first set of tosses is given by

$$\sigma_1 = \sqrt{210} = 14.491.$$

For the stationary distribution (the limit as t goes to infinity)

$$\begin{aligned} Var(K) &= n^2 p_0(1 - p_0) \\ &= 1000^2 \times .3(1 - .3) = 210{,}000. \end{aligned}$$

This means that the standard deviation σ for the stationary distribution is given by

$$\sigma = \sqrt{210{,}000} = 458.257.$$

Note that for the first set of tosses we are pretty confident that we will throw the expected value of the number of heads (namely 300). However, when t is very large, we have no confidence at all that we will throw this number. In fact, in the limit we know we will not. That is, in the limit we will either throw all heads or all tails.

Felsenstein [2007] shows that the variance in the number of heads at time t is given by

$$Var(K_t) = n^2 p_0 (1 - p_0) \left[1 - \left(1 - \frac{1}{n} \right)^t \right]. \tag{9.15}$$

It is not hard to see that this means that, although as we proceed in time the expected value of the number of heads remains fixed at np_0, we become increasingly less confident that we will actually throw that number of heads. Note that when n is large, the time t must be large for the variance to become large relative to the expected value.

Example 9.14 *Suppose $n = 1000$ and*

$$p_0 = P(heads) = .3$$

for our first set of tosses. Then

$$
\begin{aligned}
Var(K_{10}) &= n^2 p_0 (1 - p_0) \left[1 - \left(1 - \frac{1}{n} \right)^{10} \right] \\
&= 1000^2 \times .3(1 - .3) \left[1 - \left(1 - \frac{1}{1000} \right)^{10} \right] \\
&= 2090.58,
\end{aligned}
$$

which means

$$\sigma_{10} = \sqrt{2090.58} = 45.72,$$

and therefore

$$\frac{\sigma_{10}}{E(K_{10})} = \frac{45.72}{300} = .1524.$$

However, if $n = 1{,}000{,}000$,

$$
\begin{aligned}
Var(K_{10}) &= n^2 p_0 (1 - p_0) \left[1 - \left(1 - \frac{1}{n} \right)^{10} \right] \\
&= 1{,}000{,}000^2 \times .3(1 - .3) \left[1 - \left(1 - \frac{1}{1{,}000{,}000} \right)^{10} \right] \\
&= 209{,}990.55,
\end{aligned}
$$

which means

$$\sigma_{10} = \sqrt{209{,}990.55} = 458.25,$$

and therefore

$$\frac{\sigma_{10}}{E(K_{10})} = \frac{458.25}{300{,}000} = .00153.$$

Finally, note that these expected values and variances are all relative to our information at time 0. That is, they are relative to our knowledge that we are starting with a coin whose $P(heads) = p_0$.

9.3.2 Application to the Wright-Fisher Model

The coin-tossing experiment just discussed describes the Wright-Fisher Model as follows:

1. $2N$, the number of alleles in each generation, is analogous to n, the number of coin tosses in each set of tosses.

2. $k_0/2N$, where k_0 is the number of alleles of type A in the initial generation, is analogous to p_0, the probability of heads in the first set of tosses.

So, at time zero, the relative frequency of type A alleles is $k_0/2N$, and of course we are certain that there are k_0 type A alleles in the population. For each future generation, the expected value of the number of type A alleles remains fixed at $(k_0/2N)\,2N = k_0$, but we become increasingly less confident that this is the actual number. In the limit, we become certain that all alleles are either type A or type a. The probability of the former event is $k_0/2N$, whereas the probability of the latter event is $1 - k_0/2N$. The larger the value of N (the population size), the longer it will take for the variance to become large (relative to the mean), and therefore the longer it will take for one of the alleles to be eliminated from the population.

A good way to view these results is in terms of possible worlds. Suppose at a given point in time, in a given world, k_0 of the $2N$ alleles are type A. Suppose next that $1{,}000{,}000$ different worlds evolve from this world according to the Wright-Fisher Model. Then eventually, in about $k_0/2N$ of these worlds, allele A will be the only allele left in the population, whereas in the remaining worlds allele a will be the only allele left in the population.

Gene substitution is defined as the complete replacement of a predominant (wild-type) allele by a mutant allele. The **fixation probability** of a mutant allele is the probability that it will replace a wild-type allele. We see that if the initial number of occurrences of mutant A is k_0, its fixation probability is as follows:

$$P(\text{fixation of } A) = \frac{k_0}{2N}. \tag{9.16}$$

Example 9.15 *Suppose that the population size $N = 1{,}000{,}000$, and there are initially 800,000 occurrences of allele A in the population. Then we have that*

$$P(\text{fixation of } A) = \frac{k_0}{2N} = \frac{800{,}000}{2 \times 1{,}000{,}000} = .4.$$

Ordinarily, a mutant allele A arises as a single copy in a population. So, ordinarily $k_0 = 1$, and

$$P(\text{fixation of } A) = \frac{1}{2N}.$$

Therefore, in a small population a mutant allele has a good chance of replacing a wild type purely by chance, whereas in a large population the chance is small.

Note that the results obtained in this section are in direct opposition to a Hardy-Weinberg Equilibrium (Section 9.1.2). When we replace the assumption that the actual number of occurrences of an allele is exactly equal to its expected value by the more realistic assumption that it has a probability distribution, instead of the alleles remaining indefinitely in equilibrium we find that one of the alleles is eventually eliminated from the population. This result is obtained without any assumption concerning fitness or natural selection. A mutant allele can replace a wild-type allele simply by chance.

9.3.3   Difficulties with the Wright-Fisher Model

There is a difficulty with the Wright-Fisher Model when applied to populations such as humans. Recall that for an individual in generation $t + 1$ we assumed that each of its alleles is generated as follows: First, an individual is chosen at random from generation t; then an allele is chosen at random from that individual. However, this assumption entails the following two assumptions, which are violated in human populations:

1. The population is polygamous. That is, to generate a given individual in generation $t + 1$, we randomly select two individuals from generation t.

2. The population is hermaphroditic. A hermaphrodite is an individual that has two sets of sex organs and can therefore breed with itself. We assume not only polygamy but that the randomly chosen second parent for an individual can be the same as the first parent.

Felsenstein [2007] shows that our results are unchanged if we replace polygamy by monogamy, and that they are significantly altered in nonhermaphroditic populations only when the number of females is significantly different from the number of males. In this latter case, he obtains a formula for an effective population size that would be used to compute the variance using Equality 9.15. That formula is as follows:

$$N_e \approx \frac{4N_f N_M}{N_f + N_m},$$

where N_f is the number of females, and N_m is the number of males, and N_e is the effective population size.

Example 9.16 *Suppose that*

$$
\begin{aligned}
N &= 1{,}000{,}000 \\
N_f &= 200{,}000 \\
N_m &= 800{,}000.
\end{aligned}
$$

Then

$$N_e \approx \frac{4N_f N_M}{N_f N_m} + \frac{1}{2} = \frac{4 \times 200{,}000 \times 800{,}000}{200{,}000 + 800{,}000} = 640{,}000.$$

Note that the effective population is smaller than the actual population, which means that the variance computed using Equality 9.15 will be smaller relative to the mean, and one allele will be eliminated more quickly from the population.

9.4 Natural Selection and Genetic Drift

The natural selection model we presented is a deterministic one. It entailed that an allele with a fitness advantage will definitely replace a less fit allele, except in the case of overdominance of the less fit allele. In this latter case, we will reach an equilibrium in which both alleles coexist. Our genetic drift model entails that one allele can replace another allele simply by chance. Next we discuss results obtained when we assume that the two forces are acting simultaneously. We don't develop the formula for the fixation probability. You are referred to [Felsenstein, 2007] for that development. Rather, we just give the result for the case in which we have codominance and additive fitness. Recall that in this case we have the following relative fitnesses:

Genotype	AA	Aa	aa
Fitness	$w_{AA} = 1 + 2s$	$w_{Aa} = 1 + s$	$w_{aa} = 1$

In this case,

$$P(\text{fixation of } A) = \frac{1 - e^{-4N_e sp}}{1 - e^{-4N_e s}}, \tag{9.17}$$

where p is the initial relative frequency of allele A, and N_e is the effective population size. Now when $s = 0$ both the numerator and denominator of Equality 9.17 are 0. So, to investigate this formula for $s = 0$ (the case of no selection), we take the limit as follows:

$$P(\text{fixation of } A) = \lim_{s \to 0} \frac{1 - e^{-4N_e sp}}{1 - e^{-4N_e s}} = \lim_{s \to 0} \frac{4N_e pe^{-4N_e sp}}{4N_e e^{-4N_e s}} = \frac{4N_e p}{4N_e} = p. \tag{9.18}$$

The first equality above is due to L'Hôpital's Rule. Since $p = k_0/2N$, we see that Equality 9.17 reduces to Equality 9.16 when there is no selection advantage (as it should).

Ordinarily, a mutant allele arises as a single copy in a population so that $p = 1/2N$. In this case if we assume $N = N_e$, Equality 9.17 becomes

$$P(\text{fixation of } A) = \frac{1 - e^{-4Ns(1/2N)}}{1 - e^{-4Ns}} = \frac{1 - e^{-2s}}{1 - e^{-4Ns}}.$$

If s is small, and we take the power series expansion of the exponential function, we then have

$$P(\text{fixation of } A) = \frac{1 - e^{-2s}}{1 = e^{-4Ns}}$$

$$= \frac{1 - [1 + (-2s) + (-2s)^2/2 + \cdots]}{1 - e^{-4Ns}}$$

$$\approx \frac{2s}{1 - e^{-4Ns}}. \tag{9.19}$$

Example 9.17 *Suppose $N = 100$ and a single copy of mutant A arises in the population. Then we have the following probabilities of fixation:*

1. *Neutral mutation: Using Equality 9.18,*

$$P(\text{fixation of } A) = \frac{k_0}{2N} = \frac{1}{2N} = \frac{1}{2 \times 100} = .005.$$

2. *$s = .01$ (advantageous mutation): Using Equality 9.19,*

$$P(\text{fixation of } A) \approx \frac{2s}{1 - e^{-4Ns}} = \frac{2 \times .01}{1 - e^{-4 \times 100 \times .01}} = .0204.$$

3. *$s = -.01$ (disadvantageous mutation): Using Equality 9.19,*

$$P(\text{fixation of } A) \approx \frac{2s}{1 - e^{-4Ns}} = \frac{2 \times (-.01)}{1 - e^{-4 \times 100 \times (-.01)}} = .00037.$$

We see that when N is small, even a disadvantageous mutation has a reasonable chance of becoming fixed in the population.

Example 9.18 *Suppose $N = 1,000,000$ and a single copy of mutant A arises in the population. Then we have the following probabilities of fixation:*

1. *Neutral mutation: Using Equality 9.18,*

$$P(\text{fixation of } A) = \frac{1}{2N} = \frac{1}{2 \times 1,000,000} = .0000005.$$

2. *$s = .01$ (advantageous mutation): Using Equality 9.19,*

$$P(\text{fixation of } A) \approx \frac{2s}{1 - e^{-4Ns}} = \frac{2 \times .01}{1 - e^{-4 \times 1,000,000 \times .01}} = .02.$$

3. *$s = -.01$ (disadvantageous mutation): Using Equality 9.19,*

$$P(\text{fixation of } A) \approx \frac{2s}{1 - e^{-4Ns}} = \frac{2 \times (-.01)}{1 - e^{-4 \times 1,000,000 \times (-.01)}} = 3.3 \times 10^{-17374}.$$

We see that when N is large, the disadvantageous mutation has essentially no chance of becoming fixed in the population, whereas the advantageous mutation has about the same chance it has when N is small.

Note that when s is small and positive,

$$\lim_{N\to\infty} P(\text{fixation of } A) \approx \lim_{N\to\infty} \frac{2s}{1 - e^{-4Ns}} = 2s. \qquad (9.20)$$

So, as the previous two examples indicated, an advantageous mutation has about the same probability of fixation regardless of the population size (once the population size exceeds being very small). If $s = .01$, this probability is about .02.

There is some disagreement in the scientific community as to whether natural selection or genetic drift is more responsible for evolutionary change. Li [1997] discusses this matter and provides ample references.

9.5 Rate of Substitution

We define the **rate of substitution** for a given gene as the expected value of the number of mutant alleles of the gene reaching fixation per generation. First, consider neutral mutations. Let R_{neu} be the rate of substitution of neutral mutations for a given gene, and let μ_{neu} be the probability of one allele of the gene undergoing a neutral mutation in a given individual. Then in a population of size N, the expected value of the number of neutral mutant alleles of the gene in each generation is equal to

$$2N\mu_{neu}.$$

Since the probability of each of these mutant alleles becoming fixed is $1/2N$, we have that

$$R_{neu} = 2N\mu_{neu} \times \frac{1}{2N} = \mu_{neu}.$$

Now let R_{adv} be the rate of substitution of advantageous mutations for a given gene, and let μ_{adv} be the probability of one allele of the gene undergoing an advantageous mutation in a given individual. If we make the assumptions leading to Equality 9.20, then the probability of each of the mutant alleles becoming fixed is $2s$, which means

$$R_{adv} = 2N\mu_{asv} \times 2s = 4N\mu_{adv}s.$$

Example 9.19 *Suppose in a population of size $N = 1,000,000,000$,*

$$\mu_{neu} = \mu_{adv} = 5 \times 10^{-9},$$

and $s = .01$. Then

$$R_{neu} = \mu_{neu} = 5 \times 10^{-9}$$

$$R_{adv} = 4N\mu_{adv}s = 4 \times 1,000,000,000 \times 5 \times 10^{-9} \times .01 = .2.$$

So, in this large population, neutral mutations of a given gene will rarely become fixed, but an advantageous mutation will become fixed on the average once in every five generations.

EXERCISES

Section 9.1

Exercise 9.1 Suppose that we have a diploid population and the following values of the variables in Section 9.1.2 for generation $t = 1$:

Genotype	AA	Aa	aa
Relative frequency	$p_{AA} = .5$	$p_{Aa} = .3$	$p_{aa} = .2$

Compute $P(genotype = AA)$, $P(genotype = aa)$, and $P(genotype = Aa)$ for individuals in generation $t = 2$ and for individuals in generation $t = 3$.

Section 9.2

Exercise 9.2 Suppose that we have a haploid population with the values in the table in Example 9.1. Suppose further that $p_0 = .1$.

1. Compute p_1.

2. Compute p_{100}.

Exercise 9.3 Suppose that we have a diploid population, codominance, and multiplicative fitness. Suppose further that $s = .02$ and $p_0 = .01$.

1. Compute p_1.

2. Compute p_{100}.

Exercise 9.4 Name a disease that is an example of dominance. Why is it difficult for a population to rid itself of this disease?

Exercise 9.5 Suppose that we have a diploid population, overdominance, and $s = .05$ and $r = .02$ in the following table:

Genotype	AA	Aa	aa
Fitness	$w_{AA} = 1 - s$	$w_{Aa} = 1$	$w_{aa} = 1 - r$

What is the approximate value of $p_{10,000}$?

Exercise 9.6 Name a disease that is an example of overdominance.

Exercise 9.7 Suppose that we have the variables in Example 9.10 and the following values of these variables:

$$W_{Aa} = E(n|Aa) = 2 \qquad W_{BB} = E(n|BB) = 3 \qquad E(n) = 4.$$

Compute

$$W_{AaBB} = E(n|Aa, BB).$$

Is the combination of Aa and BB more fit ore less fit than either type individually?

Section 9.3

Exercise 9.8 Suppose that we toss a coin 1000 times and then choose a new coin according to the number of heads that occurred in the 1000 tosses. We then repeat this process indefinitely. Let

$$p_0 = P(heads) = .4$$

for the first set of tosses and K_t be a random variable representing the number of occurrences of heads in the set of tosses occurring at time t.

1. Compute $E(K_t)$ for all t.

2. Compute $Var(K_1)$.

3. Compute $Var(K_{20})$.

4. Compute $Var(K)$ for the stationary distribution.

Exercise 9.9 Suppose that a population has size $N = 1000$, and allele A arises as a single mutation in one individual. Assuming the Wright-Fisher Model, what is the probability of fixation of A?

Section 9.4

Exercise 9.10 Assume that natural selection and genetic drift are both occurring. Further assume that codominance and additive fitness are described by the following table:

Genotype	AA	Aa	aa
Fitness	$w_{AA} = 1 + 2s$	$w_{Aa} = 1 + s$	$w_{aa} = 1$

Given that the size of the population is $N = 10{,}000$, compute P(fixation of A) in each of the following situations:

1. Neutral mutation

2. $s = .02$ (advantageous mutation)

3. $s = -.02$ (disadvantageous mutation)

Section 9.5

Exercise 9.11 Assume that codominance and additive fitness. Suppose that in a population of size $N = 1,000,000$,

$$\mu_{neu} = \mu_{adv} = 4 \times 10^{-6},$$

and $s = .02$, where μ_{neu} and μ_{adv} are, respectively, the probabilities of one allele of a gene undergoing a neutral mutation and an advantageous mutation in a given individual. Compute the substitution rates R_{neu} and R_{adv}.

Chapter 10

Molecular Evolutionary Genetics

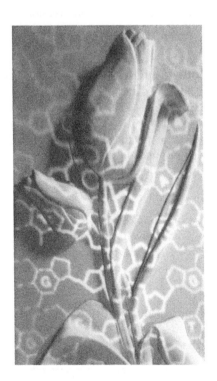

In the previous chapter, we took the gene (locus) as our basic unit and investigated ways in which allele composition can change over time when alleles undergo mutations. An allele that becomes fixed in a population is composed of a sequence that is different from the one it replaces. That is, the substitution of a new allele for an old allele consists of the substitution of a new nucleotide sequence for an old nucleotide sequence. In this chapter, we investigate how nucleotide sequences change over time. We take the nucleotide as our basic unit and develop models for nucleotide substitutions over time. First, in Section 10.1 we present some of these models of nucleotide substitution. Next Section 10.2 shows how, based on these models, we can measure the evolutionary distance between two species by comparing homologous DNA sequences from individuals who are members of each of the species. To compare homologous sequences, we must first align them. Finally, Section 10.3 presents a method for aligning homologous sequences.

10.1   Models of Nucleotide Substitution

We start with the simplest model of nucleotide substitution, namely the Jukes-Cantor Model.

10.1.1   The One-Parameter Jukes-Cantor Model

Here we develop both discrete and continuous versions of the model.

The Discrete Model

The **Jukes-Cantor Model** assumes that each base has the same probability of being substituted by each of the other bases. For example, an A is equally likely to be substituted by a G, C, or T. In the discrete version of the model, we define the **substitution rate** μ, where $\mu \leq 1/3$, to be the probability that, in one unit of time, a nucleotide at a specific site is substituted by a distinct different nucleotide throughout the population (at least in enough of the population that it shows up in our sampled sequence). The probability that the site still has the same base after one unit of time is clearly $1 - 3\mu$.

Example 10.1 *Suppose our unit of time is one year and $\mu = 5 \times 10^{-9}$. Then if a given site has base A at a given point in time, the probability that it has base G one year later is 5×10^{-9}, the probability that it has base C is 5×10^{-9}, the probability that it has base T is 5×10^{-9}, and the probability that it still has base A is $1 - 3\left(5 \times 10^{-9}\right)$.*

A discrete model of nucleotide substitution establishes a Markov chain. In the case of the Jukes-Cantor Model, the values in the transition matrix $\mathbf{P}$ for the chain are given by the following table:

	A	G	C	T
A	$1 - 3\mu$	μ	μ	μ
G	μ	$1 - 3\mu$	μ	μ
C	μ	μ	$1 - 3\mu$	μ
T	μ	μ	μ	$1 - 3\mu$

For example,

$$p_{CG} = \mu,$$

which is the entry in the third row and second column, is the probability of a C being substituted by G.

Based on the value of μ, we can compute the probability that t time units into the future the base at a given site will be the same as the original base. The following theorem obtains this probability.

Theorem 10.1 *Let $P(y|x,t)$ be the probability that a site has base y at time t given that it has base x at time 0. Then, assuming the discrete Jukes-Cantor Model,*

$$P(x|x,t) = \frac{1}{4} + \frac{3}{4}(1 - 4\mu)^t$$

and for $y \neq x$,

$$P(y|x,t) = \frac{1}{4} - \frac{1}{4}(1 - 4\mu)^t.$$

Proof. *If we let $p(t) \equiv P(x|x,t)$, we then have that*

$$
\begin{aligned}
p(t+1) &= \sum_y P(site = x \text{ at time } t+1 | site = y \text{ at time } t) P(site = y \text{ at time } t) \\
&= \mu\frac{1 - p(t)}{3} + \mu\frac{1 - p(t)}{3} + \mu\frac{1 - p(t)}{3} + (1 - 3u)p(t) \\
&= p(t)(1 - 4\mu) + \mu.
\end{aligned}
\tag{10.1}
$$

If we let $c = 1 - 4\mu$, we then have that

$$p(t+1) = cp(t) + \mu.$$

We solve this recurrence as follows:

$$
\begin{aligned}
p(0) &= 1 \\
p(1) &= c(1) + \mu \\
&= c + \mu \\
p(2) &= c(c + \mu) + \mu \\
&= c^2 + (c + 1)\mu \\
&\cdots \\
p(t) &= c^t + (c^{t-1} + c^{t-2} + \cdots + 1)\mu.
\end{aligned}
$$

So,

$$
\begin{aligned}
P(x|x,t) &= p(t) \\
&= c^t + \frac{c^t - 1}{c - 1}\mu \\
&= (1 - 4\mu)^t + \frac{(1 - 4\mu)^t - 1}{1 - 4\mu - 1}\mu \\
&= (1 - 4\mu)^t + \frac{(1 - 4\mu)^t}{-4} + \frac{1}{4} \\
&= \frac{1}{4} + \frac{3}{4}(1 - 4\mu)^t.
\end{aligned}
$$

This establishes the formula for $P(x|x,t)$. The formula for $P(y|x,t)$ is obtained by noting that $P(y|x,t) = (1 - P(x|x,t))/3$. ∎

The Continuous Model

Since μ is a probability, it might seem that we should call it the *substitution probability* rather than the *substitution rate*. However, in the continuous version of the model, we define 4μ as a rate parameter in a Poisson process. Before giving this definition, we review Poisson processes.

A Poisson process models the number of "arrivals" of some entity in intervals of time. In a **Poisson process** with rate parameter r, if $N(\triangle t)^1$ is the number of arrivals in time interval $\triangle t$, then $N(\triangle t)$ has the Poisson distribution with parameter $r\triangle t$. That is,

$$P(N(\triangle t) = k) = \frac{e^{-r\triangle t}(r\triangle t)^k}{k!}. \tag{10.2}$$

Note that the expected value of $N(\triangle t)$ is given by

$$E(N(\triangle t)) = r\triangle t.$$

So, r is the expected value of the number of arrivals in one time unit.

We have the following lemma.

Lemma 10.1 *In a Poisson process with rate parameter r, the probability of one arrival in an infinitesimally small time interval dt is equal to $r \cdot dt$.*

Proof. *Owing to Equality 10.2,*

$$
\begin{aligned}
P(N(dt) = 0) &= \frac{e^{-r\cdot dt}(r \cdot dt)^0}{0!} \\
&= e^{-r\cdot dt} \\
&= 1 - r \cdot dt + \frac{(-r \cdot dt)^2}{2!} + \cdots \\
&\approx 1 - r \cdot dt
\end{aligned}
$$

$$
\begin{aligned}
P(N(dt) = 1) &= \frac{e^{-r\cdot dt}(r \cdot dt)^1}{1!} \\
&= r \cdot dt\, e^{-r\cdot dt} \\
&= r \cdot dt \left[1 - (r \cdot dt) + \frac{(-r \cdot dt)^2}{2!} + \cdots \right] \\
&\approx r \cdot dt.
\end{aligned}
$$

This completes the proof. ∎

---

[1] The notation $N(\triangle t)$ means that N depends on $\triangle t$; it does not denote multiplication.

The result in the previous lemma is actually an alternative definition of a Poisson process. That is, if we assume the result, we can prove we must have a Poisson process.

We now apply the Poisson process model to nucleotide substitution. In the Jukes-Cantor Model, we define μ to be the expected value of the number of substitutions of one nucleotide by a distinct different nucleotide in one time unit. We assume that $r = 4\mu$ is the rate parameter in a Poisson process. We have the factor 4 instead of the factor 3 in the expression for r because a Poisson process models the number of "arrivals" of some entity. In the case of a nucleotide substitution model, the arriving entity is an A, G, C, or T. If an A is currently present at the site and an A "arrives," the site is simply unchanged.

We now have the following theorem.

Theorem 10.2 *Let $P(y|x,t)$ be the probability that a site has base y at time t, given that it has base x at time 0. Then, assuming the continuous Jukes-Cantor Model,*

$$P(x|x,t) = \frac{1}{4} + \frac{3}{4}e^{-4\mu t} \tag{10.3}$$

and for $y \neq x$,

$$P(y|x,t) = \frac{1}{4} - \frac{1}{4}e^{-4\mu t}. \tag{10.4}$$

Proof. *Since we are assuming that $r = 4\mu$ is the rate parameter in a Poisson process, owing to Lemma 10.1, $r \cdot dt$ is equal to the probability of an "arrival" in an infinitesimally small time period dt. Therefore, the probability of a specific substitution (e.g., an A by a G) is $\mu \cdot dt$.*

So, if we let $p(t) = P(x|x,t)$, we have that

$$
\begin{aligned}
p(t + dt) &= p(t)(1 - 3\mu \cdot dt) + \frac{1 - p(t)}{3}\mu \cdot dt + \frac{1 - p(t)}{3}\mu \cdot dt + \frac{1 - p(t)}{3}\mu \cdot dt \\
&= p(t) - 4p(t)\mu \cdot dt + \mu \cdot dt,
\end{aligned}
$$

which means that

$$\frac{p(t + dt) - p(t)}{dt} = -4\mu p(t) + \mu.$$

We therefore have that

$$\frac{dp(t)}{dt} = -4\mu p(t) + \mu. \tag{10.5}$$

It is left as an exercise to show that the solution to this differential equation when $p(0) = 1$ is

$$p(t) = \frac{1}{4} + \frac{3}{4}e^{-4\mu t},$$

which proves the theorem. ∎

Let's investigate what happens when the substitution rate (probability) in the discrete model is about equal to the substitution rate in the continuous model. Owing to Theorems 10.1 and 10.2, this would be the case if and only if

$$(1 - 4\mu)^t \approx e^{-4\mu t},$$

which holds if and only if

$$1 - 4\mu \approx e^{-4\mu}.$$

Now if μ is small,

$$\begin{aligned} e^{-4\mu} &= 1 + (-4\mu) + \frac{(-4\mu)^2}{2} + \cdots \\ &\approx 1 - 4\mu. \end{aligned}$$

So, as long as μ is small, the substitution rates in the two models are about the same.

Henceforth, we will consider μ a rate and use Equalities 10.3 and 10.4. Once we assume μ is a rate, we can arbitrarily change our time scale and adjust μ accordingly. We will see in Section 10.2.2 that it is sometimes convenient to do this. For example, owing to Equality 10.4, we have for $y \neq x$ that

$$P(y|x.t) = \frac{1}{4} - \frac{1}{4}e^{-4\mu t}.$$

Suppose we change our time scale, using an arbitrary positive constant k, by setting

$$t = kt'.$$

We then have that

$$P(y|x,t') = \frac{1}{4} - \frac{1}{4}e^{-4\mu k t'}.$$

So, we can set

$$\mu' = k\mu$$

and use the substitution rate μ' when using time scale t'.

Example 10.2 *Suppose the unit in time scale t is one year and the rate in time scale t is μ. Suppose further that*

$$t = 10t'.$$

Then the unit in time scale t' will be 10 years and the rate μ' in time scale t' will be 10μ. Note that $3\mu'$ is the expected value of the number of substitutions in 10 years, whereas 3μ is the expected value of the the the number of substitutions in one year.

When we allow μ' to have an arbitrary value, we can no longer call it a probability (indeed, it may exceed 1), and the probability that one base will change to another in one time unit will no longer necessarily be μ'.

Example 10.3 *Suppose the unit in time scale t is one year and $\mu = .00002$ for time scale t. Then $3 \times \mu = .00006$ is the expected value of the number of substitutions in one year. Suppose further that the unit in time scale t' is 100,000 years. We have that*

$$\mu' = 100{,}000\mu = 2,$$

and $3\mu' = 6$ *is the expected value of the number of substitutions in 100,000 years.*

If we use time scale t to compute the probability of one base changing to another base in one year, we have

$$
\begin{aligned}
P(y|x, t = 1) &= \frac{1}{4} - \frac{1}{4}e^{-4\mu \times 1} \\
&= \frac{1}{4} - \frac{1}{4}e^{-4(.00002)1} \\
&= .019999,
\end{aligned}
$$

which is about equal to μ. The reason is that μ is small.

If we use time scale t' to compute the probability of one base changing to another in 100,000 years, we have

$$
\begin{aligned}
P(y|x, t' = 1) &= \frac{1}{4} - \frac{1}{4}e^{-4\mu' \times 1} \\
&= \frac{1}{4} - \frac{1}{4}e^{-4 \times 2 \times 1} \\
&= .24992,
\end{aligned}
\tag{10.6}
$$

which is not about equal to μ'.

Note that .24992 is the substitution rate (probability) that we would use for time scale t' in our discrete model (see Theorem 10.1). That is,

$$
\frac{1}{4} - \frac{1}{4}(1 - 4 \times .24992)^1 = .24992.
$$

However, .24992 is not the rate of substitution.

10.1.2 Equilibrium Frequencies

A model of nucleotide substitution determines an irreducible ergodic Markov chain (see Section 6.4), which means that the chain will have a stationary distribution. Biologists call this stationary distribution the **equilibrium frequencies**. These relative frequencies are the probabilities of a site containing each of the bases at a distant point in the future. Next we compute the equilibrium frequencies for the Jukes-Cantor Model.

Recall that to find the stationary distribution, we need to solve (we now denote the stationary distribution by π because that is the notation used by biologists in this context):

$$
\boldsymbol{\pi}^T = \boldsymbol{\pi}^T \mathbf{P}
$$

along with

$$
\sum_j \pi_j = 1.
$$

Therefore, to find the stationary distribution for the Jukes-Cantor Model, we need to solve the following equation:

$$(\pi_A \quad \pi_G \quad \pi_C \quad \pi_T)$$

$$= (\pi_A \quad \pi_G \quad \pi_C \quad \pi_T) \begin{pmatrix} 1-3\mu & \mu & \mu & \mu \\ \mu & 1-3\mu & \mu & \mu \\ \mu & \mu & 1-3\mu & \mu \\ \mu & \mu & \mu & 1-3\mu \end{pmatrix}$$

along with

$$\sum_x \pi_x = 1.$$

The solution is

$$(\pi_A \quad \pi_G \quad \pi_C \quad \pi_T) = (1/4 \quad 1/4 \quad 1/4 \quad 1/4).$$

This result is not surprising, since the model is symmetrical in terms of the four bases.

10.1.3 Other Models

The Jukes-Cantor Model assumes that a given base is equally likely to be substituted by any other base. It seems that this is not the case. Rather transitions are more likely than transversions. To model this situation, Kimura [1980] proposed **Kimura's two-parameter model**. The substitution rates for this model are as follows:

	A	G	C	T
A	$1 - \alpha - 2\beta$	α	β	β
G	α	$1 - \alpha - 2\beta$	β	β
C	β	β	$1 - \alpha - 2\beta$	α
T	β	β	α	$1 - \alpha - 2\beta$

Since transitions are more likely than transversions, $\alpha > \beta$.

Similar to the way we obtained our continuous results for the Jukes-Cantor Model (see Equalities 10.3 and 10.4), Kimura [1980] derives the following results for the continuous version of Kimura's two-parameter model. If we let $P(y|x,t)$ be the probability that a site has base y at time t, given that it has base x at time 0, then

$$P(x|x,t) = \frac{1}{4} + \frac{1}{4}e^{-4\beta t} + \frac{1}{2}e^{-2(\alpha+\beta)t},$$

for $y \neq x$, when x to y is a transition,

$$P(y|x,t) = \frac{1}{4} + \frac{1}{4}e^{-4\beta t} - \frac{1}{2}e^{-2(\alpha+\beta)t}, \tag{10.7}$$

and for $y \neq x$, when x to y is a transversion,

$$P(y|x,t) = \frac{1}{4} - \frac{1}{4}e^{-4\beta t}. \tag{10.8}$$

Next we determine the equilibrium frequencies for Kimura's two-parameter model. We need to solve

$$\begin{pmatrix} \pi_A & \pi_G & \pi_C & \pi_T \end{pmatrix} =$$

$$\begin{pmatrix} \pi_A & \pi_G & \pi_C & \pi_T \end{pmatrix} \begin{pmatrix} 1-\alpha-2\beta & \alpha & \beta & \beta \\ \alpha & 1-\alpha-2\beta & \beta & \beta \\ \beta & \beta & 1-\alpha-2\beta & \alpha \\ \beta & \beta & \alpha & 1-\alpha-2\beta \end{pmatrix}$$

along with

$$\sum_x \pi_x = 1.$$

The solution is

$$\begin{pmatrix} \pi_A & \pi_G & \pi_C & \pi_T \end{pmatrix} = \begin{pmatrix} 1/4 & 1/4 & 1/4 & 1/4 \end{pmatrix}.$$

Again, this result is not surprising, owing to the symmetry in the model.

There are more general models. For example, **Blaisdell's four-parameter model** [1985] has the following substitution rates:

	A	G	C	T
A	$1-\alpha-2\gamma$	α	γ	γ
G	β	$1-\beta-2\gamma$	γ	γ
C	δ	δ	$1-\beta-2\delta$	β
T	δ	δ	α	$1-\alpha-2\delta$

The equilibrium frequencies for this model depend on the values of the parameters.

Example 10.4 *Suppose $\alpha = .3, \beta = .4, \gamma = .1$, and $\delta = .2$ in Blaisdell's four-parameter model. It is left as an exercise to show that the equilibrium frequencies are*

$$\begin{pmatrix} \pi_A & \pi_G & \pi_C & \pi_T \end{pmatrix} = \begin{pmatrix} .182 & .370 & .296 & .152 \end{pmatrix}.$$

The most general model would have 12 parameters. That is, there would be a unique free parameter for every entry in the transition matrix other than the entries in the diagonal.

10.1.4 Time Reversibility

A model of nucleotide substitution is called **time-reversible** if for all x and y and all t

$$P(x)P(y|x,t) = P(y)P(x|y,t),$$

where $P(x) = \pi_x$ is the equilibrium frequency and $P(y|x,t)$ is the probability of changing from base x to base y in time t. **Time reversibility** means that the probability of starting with base x and ending with base y in time t is the same as the probability of starting with base y and ending with base x in time t.

It is not hard to see that if the equilibrium frequencies are $\pi_A = \pi_G = \pi_C = \pi_T = \frac{1}{4}$, and if, in the transition matrix for the model, we have for all x and y that

$$p_{xy} = p_{yx},$$

the model is time-reversible. So, both the Jukes-Cantor Model and Kimura's two-parameter model are time reversible. Blaisdell's four-parameter model is not time-reversible.

It is possible to show that a model is time-reversible if and only if its substitution rates have the following form (see [Lanave et al., 1984]):

	A	G	C	T
A	$1 - \pi_G\alpha$ $-\pi_C\beta - \pi_T\gamma$	$\pi_G\alpha$	$\pi_C\beta$	$\pi_T\gamma$
G	$\pi_A\alpha$	$1 - \pi_A\alpha$ $-\pi_C\delta - \pi_T\epsilon$	$\pi_C\delta$	$\pi_T\epsilon$
C	$\pi_A\beta$	$\pi_G\delta$	$1 - \pi_A\beta$ $-\pi_G\delta - \pi_T\eta$	$\pi_T\eta$
T	$\pi_A\gamma$	$\pi_G\epsilon$	$\pi_C\eta$	$1 - \pi_A\gamma$ $-\pi_G\epsilon - \pi_C\eta$

where

$$\pi_A + \pi_G + \pi_C + \pi_T = 1.$$

Furthermore, it can be shown that π_A, π_G, π_C, and π_T are the equilibrium frequencies for the model.

10.2   Evolutionary Distance

Consider the same section of DNA in every individual in a particular population (species). In each generation, each site in the sequence has a chance of undergoing a mutation in each gamete that produces an individual for the next generation. One result is substitution at a given site of one base by another base in the entire population (or much of it). Another result would be that a speciation event eventually occurs, which means that the members of the species separate into two separate species. (See the introduction to Chapter 11 for more on this.) In this case, the substitutions that occurred in one of the species can be quite different from those that occurred in the other species. This means that the sequences in individuals, taken from each of the two species, may be quite different. We say that the sequences have **diverged**. This idea is illustrated in Figure 10.1. The corresponding sequences from the two species are called **homologous sequences**. In endeavors such as phylogenetic tree inference, we are interested in comparing homologous sequences from different species and

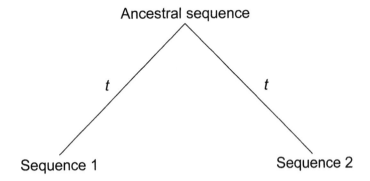

Figure 10.1: Divergence of two sequences from a common ancestral seqeuence t time units ago.

estimating their distance in an evolutionary sense. Specifically, we define the **genetic distance** D between two homologous sequences as the average of the total number of substitutions in both of them since they diverged.

Example 10.5 *Suppose two homologous sequences have length 100, that sequence 1 has undergone 80 substitutions since they diverged, and that sequence 2 has undergone 60 substitutions since they diverged. Then*

$$D = \frac{80 + 60}{100} = 1.4.$$

We could never actually know how many substitutions each of the sequences has undergone. Next we discuss estimating D from the two sequences.

10.2.1 Assuming Uniform Rates

In noncoding regions of DNA (introns), it is reasonable to assume that all sites in the sequence follow the same substitution model and evolve at the same rate (that is, the values of the parameters in the model are the same). We make this assumption in what follows. First, we discuss the Jukes-Cantor Model.

Jukes-Cantor Model

We start with the following theorem.

Theorem 10.3 *If we assume the Jukes-Cantor Model and μ is the substitution rate for a given site, then the probability $p(t)$ that two homologous sequences are different at that site at time t is given by*

$$p(t) = \frac{3}{4} \left(1 - e^{-8\mu t} \right). \tag{10.9}$$

Proof. *If the ancestor sequence has base x at a given site, then the two descendent sequences would be the same at that site if the base did not change in both of them or if both bases changed to the same base. For example, if the ancestor sequence has an A at a site, the two descendent sequences will have the same base at the site if they both still have an A at the site, if they both have the site change to G, if they both have the site change to C, or if they both have the site change to T. So, the probability r(t) that the two sequences are the same at a given site at time t is given by*

$$r(t) = \sum_y [P(y|x,t)]^2 .$$

From Equalities 10.3 and 10.4, we have that

$$
\begin{aligned}
[P(x|x,t)]^2 &= \left(\frac{1}{4} + \frac{3}{4}e^{-4\mu t}\right)\left(\frac{1}{4} + \frac{3}{4}e^{-4\mu t}\right) \\
&= \frac{1}{16} + \frac{3}{8}e^{-4\mu t} + \frac{9}{16}e^{-8\mu t}
\end{aligned}
$$

and for $x \neq y$

$$
\begin{aligned}
[P(y|x,t)]^2 &= \left(\frac{1}{4} - \frac{1}{4}e^{-4\mu t}\right)\left(\frac{1}{4} - \frac{1}{4}e^{-4\mu t}\right) \\
&= \frac{1}{16} - \frac{1}{8}e^{-4\mu t} + \frac{1}{16}e^{-8\mu t}.
\end{aligned}
$$

Therefore,

$$
\begin{aligned}
r(t) &= \sum_y [P(y|x,t)]^2 \\
&= \frac{1}{16} + \frac{3}{8}e^{-4\mu t} + \frac{9}{16}e^{-8\mu t} + 3\left(\frac{1}{16} - \frac{1}{8}e^{-4\mu t} + \frac{1}{16}e^{-8\mu t}\right) \\
&= \frac{1}{4} + \frac{3}{4}e^{-8\mu t}.
\end{aligned}
$$

So,

$$
\begin{aligned}
p(t) &= 1 - r(t) \\
&= 1 - \left(\frac{1}{4} + \frac{3}{4}e^{-8\mu t}\right) \\
&= \frac{3}{4}\left(1 - e^{-8\mu t}\right),
\end{aligned}
$$

which completes the proof. ∎

Theorem 10.4 *If we assume the Jukes-Cantor Model and that all sites in two homologous sequences have the same substitution rate, then the expected value of the genetic distance D at time t is given by*

$$E_t(D) = -\frac{3}{4}\ln\left(1 - \frac{4}{3}p(t)\right), \qquad (10.10)$$

where $p(t)$ is the probability that the sequences are different at a given site at time t.

Proof. *Recall that the substitution rate μ is the expected value of the number of substitutions of one nucleotide by a distinct different nucleotide in one unit of time. So, the expected value of the number of substitutions in t units of time at each site in each sequence is $3\mu t$, which means that if the sequences have length n,*

$$E_t(D) = 2\frac{3\mu t \times n}{n} = 6\mu t. \tag{10.11}$$

Solving Equality 10.9 for μ, we obtain

$$\mu = \frac{-\ln\left(1 - \frac{4}{3}p(t)\right)}{8t}.$$

Combining the last two equalities yields

$$\begin{aligned}
E_t(D) &= 6\mu t \\
&= 6\frac{-\ln\left(1 - \frac{4}{3}p(t)\right)}{8t}t \\
&= -\frac{3}{4}\ln\left(1 - \frac{4}{3}p(t)\right).
\end{aligned}$$

This completes the proof. ∎

Theorem 10.5 *Suppose we have two homologous sequences of length n that are different at m sites. Assume that the probability p that they are different at a site is the same for all sites, and that the event that they are different at one site is independent of the event that they are different at another site. Then the maximum likelihood estimate of p is given by*

$$\hat{p} = \frac{m}{n}.$$

Proof. *Let DATA be our data concerning the n sites. For a given value of p, the likelihood $L(p)$ is given by*

$$L(p) = P(\text{DATA}|p) = p^m(1 - p)^{n-m}.$$

Owing to Example 3.14, we therefore have that

$$\hat{p} = \frac{m}{n}.$$

This completes the proof. ∎

We use the result in the previous theorems to estimate the evolutionary distance D as follows: First, note that we do not know t. However, regardless of the value of t, we can obtain the maximum likelihood estimate of $p(t)$ using Theorem 10.5. If we use this estimate in Equality 10.10, we obtain an estimate of the $E_t(D)$. We can then use this estimate of the expected value of D as an estimate of its actual value.

Example 10.6 *Suppose two homologous sequences have length 100 and they differ on 70 sites. Then our estimate of $p(t)$ is*

$$\hat{p} = \frac{70}{100} = .7,$$

and our estimate of D is

$$
\begin{aligned}
\hat{D} &= -\frac{3}{4}\ln\left(1 - \frac{4}{3}\hat{p}\right) \\
&= -\frac{3}{4}\ln\left(1 - \frac{4}{3}(.7)\right) \\
&= 2.031.
\end{aligned}
$$

Kimura's Two-Parameter Model

Next we obtain the expected value of the genetic distance in the case of Kimura's two-parameter model.

Theorem 10.6 *If we assume Kimura's two-parameter model, and α and β are the substitution rates for a given site, then the probability $p(t)$ that two homologous sequences are different at that site by a transition at time t is given by*

$$p(t) = \frac{1}{4}\left(1 + e^{-8\beta t}\right) - \frac{1}{2}e^{-4(\alpha+\beta)t}, \tag{10.12}$$

and the probability $q(t)$ that two homologous sequences are different at that site by a transversion at time t is given by

$$q(t) = \frac{1}{2}\left(1 - e^{-8\beta t}\right). \tag{10.13}$$

Proof. *The proof is left as an exercise.* ■

Theorem 10.7 *If we assume Kimura's two-parameter model and that all sites in two homologous sequences have the same substitution rates, then the expected value of the genetic distance D at time t is given by*

$$E_t(D) = -\frac{1}{2}\ln(1 - 2p(t) - q(t)) - \frac{1}{4}\ln(1 - 2q(t)).$$

Proof. *Recall that α is the probability that the nucleotide at a specific site undergoes a transition in one unit of time, and β is the probability that it undergoes each of the possible transversions. So, the expected value of the total number of substitutions at each site in each sequence is $(\alpha + 2\beta)t$, which means, if the sequences have length n, that*

$$E_t(D) = 2\frac{(\alpha + 2\beta)t \times n}{n} = (2\alpha + 4\beta)t. \tag{10.14}$$

It is left as an exercise to show that if we solve Equalities 10.12 and 10.13 for α and β, we obtain that

$$\alpha = -\frac{1}{4t} \ln(1 - 2p(t) - q(t)) + \frac{1}{8t} \ln(1 - 2q(t))$$

$$\beta = -\frac{1}{8t} \ln(1 - 2q(t)).$$

Combining the last three equalities yields

$$
\begin{aligned}
E_t(D) &= (2\alpha + 4\beta)t \\
&= -\frac{1}{2} \ln(1 - 2p(t) - q(t)) - \frac{1}{4} \ln(1 - 2q(t)).
\end{aligned}
$$

This completes the proof. ∎

Similar to what we did in the case of the Jukes-Cantor Model, we can use Theorem 10.7 to estimate the value of D in the case of Kimura's two-parameter model.

Example 10.7 *Suppose two homologous sequences have length 100 and they differ by a transition at 22 sites and by a transversion at 48 sites. Then our estimates of $p(t)$ and $q(t)$ are*

$$\hat{p} = \frac{22}{100} = .22$$

$$\hat{q} = \frac{48}{100} = .48,$$

and our estimate of D is

$$
\begin{aligned}
\hat{D} &= -\frac{1}{2} \ln(1 - 2\hat{p} - \hat{q}) - \frac{1}{4} \ln(1 - 2\hat{q}) \\
&= -\frac{1}{2} \ln(1 - 2 \times .22 - .48) - \frac{1}{4} \ln(1 - 2 \times .48) \\
&= 2.0676.
\end{aligned}
$$

Example 10.8 *When we compare homologous sequences from two individuals in two different species, we first align the sequences because one or both of the sequences may have undergone insertions and/or deletions since they diverged. Sequence alignment is discussed in Section 10.3. An alignment of the first introns of the human and owl monkey insulin genes results in a 196-nucleotide sequence in which 163 of the sites do not have a gap in either sequence. See [Li, 1997] for the alignment. There are 14 sites that differ by a transition and 4 sites that differ by a transversion. Therefore, according to the Jukes-Cantor Model, our estimate of p is*

$$\hat{p} = \frac{18}{163} = .11043,$$

and our estimate of D is

$$
\begin{aligned}
\hat{D} &= -\frac{3}{4}\ln\left(1-\frac{4}{3}\hat{p}\right) \\
&= -\frac{3}{4}\ln\left(1-\frac{4}{3}(.11043)\right) \\
&= .11846.
\end{aligned}
$$

According to Kimura's two-parameter model, our estimates of $p(t)$ and $q(t)$ are

$$
\hat{p} = \frac{14}{163} = .08589
$$

$$
\hat{q} = \frac{4}{163} = .02454,
$$

and our estimate of D is

$$
\begin{aligned}
\hat{D} &= -\frac{1}{2}\ln(1-2\hat{p}-\hat{q}) - \frac{1}{4}\ln(1-2\hat{q}) \\
&= -\frac{1}{2}\ln(1-2\times.08589-.02454) - \frac{1}{4}\ln(1-2\times.02454) \\
&= .12186.
\end{aligned}
$$

Other Models

Explicit analytical estimates of the evolutionary distance have been obtained for several other models. See [Zharkikh, 1994] for a discussion of these models.

10.2.2   Assuming Nonuniform Rates

In coding regions of DNA (exons), it is not reasonable to assume that all sites in the sequence evolve at the same rate. Furthermore, Yang [1996] notes that rates vary considerably among amino acid sites. Next we discuss modeling nonuniform rates. One technique researchers have used to model this situation is to place probability distributions on the substitution rates. For example, if we assume the Jukes-Cantor Model, for each site i we place a probability distribution $\rho(\mu)$ on the substitution rate μ at that site, and then, relative to that distribution, we compute the expected value $\bar{p}(t)$ of the probability $p(t)$ that two homologous sequences are different at a given site at time t. Rather than assuming μ is the same for all sites, we assume $\rho(\mu)$ is the same for all sites.

The gamma distribution is often used to quantify our uncertainty concerning rates. Furthermore, it is often tractable to compute genetic distances when this distribution is used. However, Felsenstein [2004] cautions that "There is nothing about the gamma distribution that makes it more biologically realistic than any other distribution, such as lognormal. It is used because of its mathematical tractability." In any case, biologists (see e.g., [Yang, 1994]) have used the gamma distribution for $\rho(\mu)$, and we discuss the use of this distribution.

Before proceeding, let us review the gamma distribution. Recall that the **gamma density function** with parameters a and b, where $a > 0$ and $b > 0$, is

$$\rho(x) = \frac{b^{\alpha}}{\Gamma(a)} x^{a-1} e^{-bx} \qquad x > 0$$

and is denoted GammaDen$(x; a, b)$. If the random variable X has the gamma density function, then

$$E(X) = \frac{a}{b} \qquad \text{and} \qquad V(X) = \frac{a}{b^2}. \qquad (10.15)$$

Jukes-Cantor Model

Next we obtain the expected value of the genetic distance when we use the Jukes-Cantor Model and we place a gamma distribution on μ. We have the following theorem.

Theorem 10.8 *Assume the Jukes-Cantor Model and that all sites in two homologous sequences have the following gamma density function of μ:*

$$\rho(\mu) = \frac{b^a}{\Gamma(a)} \mu^{a-1} e^{-b\mu} \qquad \mu > 0.$$

Then the expected value of the genetic distance D at time t is given by

$$E_t(D) = \frac{3a}{4} \left[\left(1 - \frac{4}{3} \bar{p}(t) \right)^{-1/a} - 1 \right], \qquad (10.16)$$

where $\bar{p}(t)$ is the expected value of the probability that two sites are different at time t.[2]

Proof. *Owing to Theorem 10.3, if we assume the Jukes-Cantor Model and μ is the substitution rate for a given site, then the probability $p(t, \mu)$ that two homologous sequences are different at time t at the site is given by*

$$p(t, \mu) = \frac{3}{4} \left(1 - e^{-8\mu t} \right),$$

where now we have explicitly shown the dependence of the probability on μ.

---

[2] It might seem odd that we denote the expected value of a probability rather than just showing a probability since, in the Bayesian framework, the expected value of a probability is our belief (probability). However, we are considering $p(t)$ a relative frequency, and $\bar{p}(t)$ is the expected value of $p(t)$ relative to our belief, which is represented by a gamma distribution.

Therefore, the expected value of p(t) is given by

$$
\begin{aligned}
\bar{p}(t) &= \int_0^\infty p(t,\mu)\rho(\mu)d\mu \\
&= \int_0^\infty \frac{3}{4}\left(1 - e^{-8\mu t}\right)\frac{b^a}{\Gamma(a)}\mu^{a-1}e^{-b\mu}d\mu \\
&= \frac{3}{4}\int_0^\infty \frac{b^a}{\Gamma(a)}\mu^{a-1}e^{-b\mu}d\mu - \frac{3}{4}\int_0^\infty e^{-8\mu t}\frac{b^a}{\Gamma(a)}\mu^{a-1}e^{-b\mu}d\mu \\
&= \frac{3}{4} - \frac{3}{4}\int_0^\infty e^{-8\mu t}\frac{b^a}{\Gamma(a)}\mu^{a-1}e^{-b\mu}d\mu \\
&= \frac{3}{4} - \frac{3}{4}\int_0^\infty \frac{b^a}{\Gamma(a)}\mu^{a-1}e^{(-b+8t)\mu}d\mu \\
&= \frac{3}{4} - \frac{3}{4}\left(\frac{b}{b+8t}\right)^a \int_0^\infty \frac{(b+8t)^a}{\Gamma(a)}\mu^{a-1}e^{-(b+8t)\mu}d\mu \\
&= \frac{3}{4} - \frac{3}{4}\left(\frac{b}{b+8t}\right)^a.
\end{aligned}
\tag{10.17}
$$

The fourth and last equalities are owing to the fact that we are integrating over a gamma density function, which means the integral is 1. Now since $\bar\mu = a/b$,

$$ b = \frac{a}{\bar\mu}. $$

Substituting this expression for b in Equality 10.17 yields

$$ \bar{p}(t) = \frac{3}{4} - \frac{3}{4}\left(\frac{a}{a+8\bar\mu t}\right)^a. $$

It is left as an exercise to solve this equation for $\bar\mu t$ to obtain that

$$ 6\bar\mu t = \frac{3a}{4}\left[\left(1 - \frac{4}{3}\bar{p}(t)\right)^{-1/a} - 1\right]. $$

The proof now follows from Equality 10.11. ∎

We estimate the genetic distance as follows: We first estimate $\bar{p}(t)$ by dividing the number of sites at which the sequences differ by the total number of sites. We then estimate $E_t(D)$ using Equality 10.16 and use this estimate as our estimate of D.

It is left as an exercise to show that

$$ \lim_{a\to\infty} \frac{3a}{4}\left[\left(1 - \frac{4}{3}\bar{p}(t)\right)^{-1/a} - 1\right] = -\frac{3}{4}\ln\left(1 - \frac{4}{3}\bar{p}(t)\right), $$

which is the same formula obtained in Theorem 10.4 when we considered the substitution rates to be the same at all sites. This result is not surprising since for $\bar\mu$ to be finite, we would need to have b approach infinity as a approaches infinity, which means that the variance, which is a/b^2, is approaching zero.

The question remains as to how to choose a. Jin and Nei [1990] performed experiments testing various values of a and using Kimura's two-parameter model. In these experiments, they obtained the best results with $a = 1$, but they found that the results were fairly robust as to the choice of a.

Example 10.9 *Suppose two homologous sequences have length 100 and they differ on 70 sites. Suppose further that we assume the Jukes-Cantor Model and a gamma distribution with $a = 1$ of the substitution rate μ. Then our estimate of $\bar{p}(t)$ is*

$$\hat{p} = \frac{60}{100} = .7,$$

and our estimate of D is

$$
\begin{aligned}
\hat{D} &= \frac{3a}{4}\left[\left(1 - \frac{4}{3}\hat{p}\right)^{-1/a} - 1\right] \\
&= \frac{3 \times 1}{4}\left[\left(1 - \frac{4}{3}(.7)\right)^{-1/1} - 1\right] \\
&= 10.5.
\end{aligned}
$$

Notice that the result is quite different from that obtained in Example 10.6.

Kimura's Two-Parameter Model

Next we show the expected value of the genetic distance when we use Kimura's two-parameter model and we assume nonuniform rates. The variation in α and β at different sites is primarily caused by the variation in mutation rate, which means α and β are not independent. The model commonly used assumes that

$$R = \frac{\alpha}{2\beta},$$

which is the ratio of the expected number of transitions to the expected number of transversions, has the same value for every site. We have the following theorem, which originally appeared in [Jin and Nei, 1990].

Theorem 10.9 *Assume Kimura's two-parameter model and that all sites in two homologous sequences have the following gamma density functions of α:*

$$\rho(\alpha) = \frac{b^a}{\Gamma(a)}\alpha^{a-1}e^{-b\alpha} \qquad \alpha > 0.$$

Assume further that $\alpha = 2R\beta$ for a constant R. Then the expected value of the genetic distance D at time t is given by

$$E_t(D) = \frac{a}{2}\left[(1 - 2\bar{p}(t) - \bar{q}(t))^{-1/a} - 1\right] + \frac{a}{4}\left[(1 - 2\bar{q}(t))^{-1/a} - 1\right],$$

where $\bar{p}(t)$ is the expected value of the probability that two sites differ by a transition at time t, and $\bar{q}(t)$ is the expected value of the probability that two sites differ by a transversion at time t.

Proof. *Owing to Theorem 10.6, the probability $p(t, \alpha, \beta)$ that two homologous sequences are different at a given site by a transition at time t is given by*

$$p(t, \alpha, \beta) = \frac{1}{4}\left(1 + e^{-8\beta t}\right) - \frac{1}{2}e^{-4(\alpha+\beta)t},$$

and the probability $q(t, \alpha, \beta)$ that two homologous sequences are different at that site by a transversion at time t is given by

$$q(t, \alpha, \beta) = \frac{1}{2}\left(1 - e^{-8\beta t}\right).$$

Using the technique used in Theorem 10.8, you should show the following as an exercise:

$$
\begin{aligned}
\bar{p}(t) &= \int_0^\infty \frac{1}{4}\left(1 + e^{-8\beta t}\right)\rho(\alpha)d\alpha - \int_0^\infty \frac{1}{2}e^{-4(\alpha+\beta)t}\rho(\alpha)d(\alpha) \\
&= \frac{1}{4} + \frac{1}{4}\left(\frac{a}{a + 8\bar{\beta}t}\right)^a - \frac{1}{2}\left(\frac{a}{a + 4\left(\bar{\alpha}+\bar{\beta}\right)t}\right)^a
\end{aligned}
\tag{10.18}
$$

$$
\begin{aligned}
\bar{q}(t) &= \frac{1}{2}\left(1 - e^{-8\beta t}\right)\rho(\beta)d\beta \\
&= \frac{1}{2} - \frac{1}{2}\left(\frac{a}{a + 8\bar{\beta}t}\right)^a.
\end{aligned}
\tag{10.19}
$$

Using Equalities 10.18 and 10.19, you should also show that

$$2(\bar{\alpha}+\bar{\beta})t = \frac{a}{2}\left[(1 - 2\bar{p}(t) - \bar{q}(t))^{-1/a} - 1\right]$$

$$2\bar{\beta}t = \frac{a}{4}\left[(1 - 2\bar{q}(t))^{-1/a} - 1\right].$$

Owing to Equality 10.14

$$E_t(D) = (2\bar{\alpha} + 4\bar{\beta})t.$$

Therefore,

$$
\begin{aligned}
E_t(D) &= 2(\bar{\alpha}+\bar{\beta})t + 2\bar{\beta}t \\
&= \frac{a}{2}\left[(1 - 2\bar{p}(t) - \bar{q}(t))^{-1/a} - 1\right] + \frac{a}{4}\left[(1 - 2\bar{q}(t))^{-1/a} - 1\right].
\end{aligned}
$$

This completes the proof. ■

It is left as an exercise to show that

$$\lim_{a\to\infty} \frac{a}{2}\left[(1 - 2\bar{p}(t) - \bar{q}(t))^{-1/a} - 1\right] + \frac{a}{4}\left[(1 - 2\bar{q}(t))^{-1/a} - 1\right]$$

$$= -\frac{1}{2}\ln(1 - 2\bar{p}(t) - \bar{q}(t)) - \frac{1}{4}\ln(1 - 2\bar{q}(t)),$$

which is the same formula obtained in Theorem 10.7 when we considered the substitution rates to be the same at all sites.

Similar to what we did in the case of the Jukes-Cantor Model, we can use Theorem 10.9 to estimate the value of D in the case of Kimura's two-parameter model.

Example 10.10 *Suppose two homologous sequences have length 100 and they differ by a transition at 22 sites and by a transversion at 48 sites. Suppose further we assume Kimura's two-parameter model, and we set $a = 1$ in our gamma distribution. Then our estimates of $\bar{p}(t)$ and $\bar{q}(t)$ are*

$$\hat{p} = \frac{22}{100} = .22$$

$$\hat{q} = \frac{48}{100} = .48,$$

and our estimate of D is

$$
\begin{aligned}
E_t(D) &= \frac{a}{2}\left[(1 - 2\bar{p}(t) - \bar{q}(t))^{-1/a} - 1\right] + \frac{a}{4}\left[(1 - 2\bar{q}(t))^{-1/a} - 1\right] \\
&= \frac{1}{2}\left[(1 - 2(.22) - .48)^{-1/1}\right] - 1 + \frac{1}{4}\left[(1 - 2(.48))^{-1/1} - 1\right] \\
&= 11.75.
\end{aligned}
$$

Notice that the result is close to that obtained in Example 10.9 but quite different from that obtained in Example 10.7.

10.3   Sequence Alignment

Recall from Section 10.2 that in endeavors such as phylogenetic tree inference, we are interested in comparing homologous sequences from individuals in different species and estimating their distance in an evolutionary sense. Recall further from Example 10.8 that when we compare homologous sequences from two different individuals, we first align the sequences because one or both of the sequences may have undergone insertions and/or deletions since they diverged.

Example 10.11 *Suppose we have the following homologous sequences:*

$$A \quad A \quad C \quad A \quad G \quad T \quad T \quad A \quad C \quad C$$

$$T \quad A \quad A \quad G \quad G \quad T \quad C \quad A \; .$$

We could align them in many ways. The following shows two possible alignments:

$$
\begin{array}{cccccccccc}
- & A & A & C & A & G & T & T & A & C & C \\
T & A & A & - & G & G & T & - & - & C & A
\end{array}
$$

$$
\begin{array}{ccccccccc}
A & A & C & A & G & T & T & A & C & C \\
T & A & - & A & G & G & T & - & C & A
\end{array} .
$$

When we include a dash $(-)$ in an alignment, this is called inserting a gap. *It indicates that either the sequence with the gap has undergone a deletion or the other sequence has undergone an insertion.*

Which alignment in the previous example is better? Both have five matching base pairs. The top alignment has two mismatched base pairs, but at the expense of inserting four gaps. On the other hand, the bottom alignment has three mismatched base pairs but at the expense of inserting only two gaps. In general, it is not possible to say which alignment is better without first specifying the penalty for a mismatch and the penalty for a gap. For example, suppose we say a gap has a penalty of 1 and a mismatch has a penalty of 3. We call the sum of all the penalties in an alignment the **cost** of the alignment. Given these penalty assignments, the top alignment in Example 10.11 has a cost of 10, whereas the bottom one has a cost of 11. So, the top one is better. On the other hand, if we say a gap has a penalty of 2 and a mismatch has a penalty of 1, it is not hard to see that the bottom alignment has a smaller cost and is therefore better.

Once we do specify the penalties for gaps and mismatches, it is possible to determine the optimal alignment. However, doing this by checking all possible alignments is an intractable task. Sometimes we can efficiently solve an optimization problem, without checking all candidate solutions, using a technique called **dynamic programming**. Next we develop a dynamic programming algorithm for the sequence alignment problem. An understanding of recursion and dynamic programming, which can be obtained from a standard algorithms textbook such as [Neapolitan and Naimipour, 2004], would help you follow this development.

For the sake of concreteness, in our discussion we assume the following:

- The penalty for a mismatch is 1.

- The penalty for a gap is 2.

First, we represent the two sequences in arrays as follows:

A	A	C	A	G	T	T	A	C	C
$x[0]$	$x[1]$	$x[2]$	$x[3]$	$x[4]$	$x[5]$	$x[6]$	$x[7]$	$x[8]$	$x[9]$

T	A	A	G	G	T	C	A
$y[0]$	$y[1]$	$y[2]$	$y[3]$	$y[4]$	$y[5]$	$y[6]$	$y[7]$

Let $opt(i, j)$ be the cost of the optimal alignment of the subsequences $x[i..9]$ and $y[j..7]$. Then $opt(0, 0)$ is the cost of the optimal alignment of $x[0..9]$ and $y[0..7]$, which is the alignment we want to perform. This optimal alignment must start with one of the following.

1. $x[0]$ is aligned with $y[0]$. If $x[0] = y[0]$ there is no penalty at the first alignment site, whereas if $x[0] \neq y[0]$ there is a penalty of 1.

2. $x[0]$ is aligned with a gap and there is a penalty of 2 at the first alignment site.

3. $y[0]$ is aligned with a gap and there is a penalty of 2 at the first alignment site.

Suppose the optimal alignment A_{opt} of $x[0..9]$ and $y[0..7]$ has $x[0]$ aligned with $y[0]$. Then this alignment contains within it an alignment B of $x[1..9]$ and $y[1..7]$. Suppose this is not the optimal alignment of these two subsequences. Then there is some other alignment C of them that has a smaller cost. However, then alignment C, along with aligning $x[0]$ with $y[0]$, will yield an alignment of $x[0..9]$ and $y[0..7]$ that has a smaller cost than A_{opt}. So, alignment B must be the optimal alignment of $x[1..9]$ and $y[1..7]$. Similarly, if the optimal alignment of $x[0..9]$ and $y[0..7]$ has $x[0]$ aligned with a gap, then that alignment contains within it the optimal alignment of $x[1..9]$ and $y[0..7]$, and if the optimal alignment of $x[0..9]$ and $y[0..7]$ has $y[0]$ aligned with a gap, then that alignment contains within it the optimal alignment of $x[0..9]$ and $y[1..7]$.

Example 10.12 *Suppose the following is an optimal alignment of $x[0..9]$ and $y[0..7]$:*

$x[0]$	$x[1]$	$x[2]$	$x[3]$	$x[4]$	$x[5]$	$x[6]$	$x[7]$	$x[8]$	$x[9]$
A	A	C	A	G	T	T	A	C	C

T	A	–	A	G	G	T	–	C	A
$y[0]$	$y[1]$		$y[2]$	$y[3]$	$y[4]$	$y[5]$		$y[6]$	$y[7]$

Then the following must be an optimal alignment of $x[1..9]$ and $y[1..7]$:

$x[1]$	$x[2]$	$x[3]$	$x[4]$	$x[5]$	$x[6]$	$x[7]$	$x[8]$	$x[9]$
A	C	A	G	T	T	A	C	C

A	–	A	G	G	T	–	C	A
$y[1]$		$y[2]$	$y[3]$	$y[4]$	$y[5]$		$y[6]$	$y[7]$

If we let *penalty* $= 0$ if $x[0] = y[0]$ and 1 otherwise, we have established the following recursive property:

$$opt(0,0) = \min(opt(1,1) + penalty, opt(1,0) + 2, opt(0,1) + 2).$$

Although we illustrated this recursive property starting at the 0th position in both sequences, it clearly holds if we start at arbitrary positions. So, in general,

$$opt(i,j) = \min(opt(i+1,j+1) + penalty, opt(i+1,j) + 2, opt(i,j+1) + 2).$$

To complete the development of a recursive algorithm, we need terminal conditions. Now if we have passed the end of sequence **x** ($i = m$), and we are the jth position in sequence **y**, where $j < n$, then we must insert $n - j$ gaps. So, one terminal condition is

$$opt(m,j) = 2(n - j).$$

Similarly, if we have passed the end of sequence **y** ($j = n$), and we are the ith position in sequence **x**, where $i < m$, then we must insert $m - i$ gaps. So, another terminal condition is

$$opt(i,n) = 2(m - i).$$

We now have the following algorithm:

```
void opt (int i, j)
{
  if i == m
    opt = 2(n − j);
  else if j == n
    opt = 2(m − i);
  else {
    if x[i] == x[j]
       penalty = 0;
    else
       penalty = 1;
    opt = min(opt(i + 1, j + 1) + penalty, opt(i + 1, j) + 2, opt(i, j + 1) + 2);
  }
}
```

The top-level call to the algorithm is

$$optimal\_cost = opt(0, 0);$$

Note that this algorithm only gives the cost of an optimal alignment; it does not produce one. It could be modified to also produce one. However, we will not pursue that here because the algorithm is very inefficient and is therefore not the one we would actually use. It is left as an exercise to show that it has exponential time complexity.

The problem is that many subinstances are evaluated more than once. For example, to evaluate $opt(0, 0)$ at the top level, we need to evaluate $opt(1, 1)$, $opt(1, 0)$, and $opt(0, 1)$. To evaluate $opt(1, 0)$ in the first recursive call, we need to evaluate $opt(1, 2)$, $opt(2, 0)$, and $opt(1, 1)$. The two evaluations of $opt(1, 1)$ will unnecessarily be done independently. Often when a recursive algorithm for a problem has exponential time complexity, we can solve the problem efficiently using a dynamic programming algorithm that exploits the same recursive property as the recursive algorithm. The reason that dynamic programming yields an efficient algorithm is that the solution is constructed from the bottom up instead of from the top down. Next we illustrate how to do this for the current problem instance.

To solve the problem using dynamic programming, we create an $m + 1$ by $n + 1$ table, as shown in Figure 10.2. Note that we include one extra character in each sequence which is a gap. The purpose of this is to give our upward iteration scheme a starting point. We want to compute and store $opt(i, j)$ in the ijth slot of this table. Recall that we have the following:

$$opt(i, j) = \min(opt(i+1, j+1) + penalty, opt(i+1, j) + 2, opt(i, j+1) + 2) \quad (10.20)$$

$$opt(10, j) = 2(8 − j) \quad (10.21)$$

$$opt(i, 8) = 2(10 − i). \quad (10.22)$$

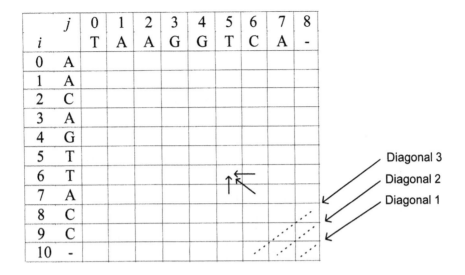

	j	0	1	2	3	4	5	6	7	8
i		T	A	A	G	G	T	C	A	-
0	A									
1	A									
2	C									
3	A									
4	G									
5	T									
6	T									
7	A									
8	C									
9	C									
10	-									

Figure 10.2: This table is used to find the optimal alignment when we use dynamic programming.

Note that we have put the current instance's values of m and n in the formulas. If we are in the bottom row of the table, we use Equality 10.21 to compute $opt(i, j)$; if we are in the right-most column, we use Equality 10.22; otherwise, we use Equality 10.20. Note in Equality 10.20 that each array item's value can be computed from the value of the array item to the right of it, the value of the array item underneath it, and the value of the array item underneath it and to the right. For example, we illustrate in Figure 10.2 that $opt(6, 5)$ is computed from $opt(7, 6)$, $opt(7, 5)$, and $opt(6, 6)$.

Therefore, we can compute all values in the table in Figure 10.2 by first computing all values on Diagonal 1, then computing all values on Diagonal 2, then computing all values on Diagonal 3, and so on. We illustrate the computations for the first three diagonals:

- Diagonal 1:
$$opt(10, 8) = 2(10 - 10) = 0.$$

- Diagonal 2:
$$opt(9, 8) = 2(10 - 9) = 2$$
$$opt(10, 7) = 2(8 - 7) = 2.$$

- Diagonal 3:
$$opt(8, 8) = 2(10 - 8) = 4$$

$opt(9, 7)$

$$= \min(opt(9 + 1, 7 + 1) + penalty, opt(9 + 1, 7) + 2, opt(9, 7 + 1) + 2)$$
$$= \min(0 + 1, 3 + 2, 2 + 2) = 1$$

j	0	1	2	3	4	5	6	7	8
i	T	A	A	G	G	T	C	A	-
0 A	**7**	8	10	12	13	15	16	18	20
1 A	6	**6**	8	10	11	13	14	16	18
2 C	6	5	**6**	8	9	11	12	14	16
3 A	7	5	**4**	6	7	9	11	12	14
4 G	9	7	5	**4**	5	7	9	10	12
5 T	8	8	6	4	**4**	5	7	8	10
6 T	9	8	7	5	3	**3**	5	6	8
7 A	11	9	7	6	4	2	**3**	4	6
8 C	13	11	9	7	5	3	**1**	3	4
9 C	14	12	10	8	6	4	2	**1**	2
10 -	16	14	12	10	8	6	4	2	**0**

Figure 10.3: The completed table used to find the optimal alignment.

$$opt(10,6) = 2(8-6) = 4.$$

Figure 10.3 shows the table after all the values are computed. The value of the optimal alignment is $opt(0,0)$, which is 7.

Next we show how the optimal alignment can be retrieved from the completed table. First we must obtain the path that led to $opt(0,0)$. We do this by starting in the upper-left corner of the table and retracing our steps. We look at the three table items that could have led to $opt(0,0)$ and choose the one that gives the correct value. We then repeat this procedure with the chosen item. Ties are broken arbitrarily. We do this until we arrive in the lower-right corner. The path obtained is highlighted in Figure 10.3. We show how the first few values in the path were obtained. First denote the array slot that occupies the ith row and the jth column by $[i,j]$. Then proceed as follows:

1. Choose array slot $[0,0]$.

2. Find second array item in path.

   (a) Check array slot $[0,1]$: Since from this slot we move to the left to arrive at array slot $[0,0]$, a gap is inserted, which means 2 is added to the cost. We have that

   $$opt(0,1) + 2 = 8 + 2 = 10 \neq 7.$$

   (b) Check array slot $[1,0]$: Since from this slot we move up to arrive at array slot $[0,0]$, a gap is inserted, which means 2 is added to the cost.

We have that

$$opt(1,0) + 2 = 6 + 2 = 8 \neq 7.$$

(c) Check array slot $[1,1]$: Since from this slot we move diagonally to arrive at array slot $[0,0]$, the value of *penalty* is added to the cost. Since $x[0] = A$ and $y[0] = T$, *penalty* $= 1$. We have that

$$opt(1,1) + 1 = 6 + 1 = 7.$$

So, the second array slot in the path is $[1,1]$.

Alternatively, we could create the path while storing the entries in the table. That is, each time we store an array element we create a pointer back to the array element that determined its value.

Once we have the path, we retrieve the alignment as follows (note that the sequences are generated in reverse order).

1. Starting in the bottom-right corner of the table, we follow the highlighted path.

2. Every time we make a diagonal move to arrive at array slot $[i,j]$, we place the character in the ith row into the **x** sequence and we place the character in the jth column into the **y** sequence.

3. Every time we make a move directly up to arrive at array slot $[i,j]$, we place the character in the ith row into the **x** sequence, and we place a gap into the **y** sequence.

4. Every time we make a move directly to the left to arrive at array slot $[i,j]$, we place the character in the jth row into the **y** sequence, and we place a gap into the **x** sequence.

If you follow this procedure for the table in Figure 10.3, you will obtain the following optimal alignment:

$$\begin{array}{ccccccccc} A & A & C & A & G & T & T & A & C & C \\ T & A & - & A & G & G & T & - & C & A \end{array}.$$

Note that if we assigned different penalties, we might obtain a different optimal alignment. Li [1997] discusses the assignment of penalties.

It is left as an exercise to write the dynamic programming algorithm for the sequence alignment problem. This algorithm for sequence alignment, which is developed in detail in [Waterman, 1984], is one of the most widely used sequence alignment methods. It is used in sophisticated sequence alignment systems such as BLAST [Bedell et al., 2003] and DASH [Gardner-Stephen and Knowles, 2004].

EXERCISES

Section 10.1

Exercise 10.1 Suppose the substitution rate in the continuous version of the Jukes-Cantor Model is given by

$$\mu = .0000007.$$

1. Compute the probability that the nucleotide at a site will not change in 1 year, 100 years, 10,000 years, and 100,000 years.

2. Compute the probability that the nucleotide at a site will change to a specific different nucleotide in 1 year, 100 years, 10,000 years, and 100,000 years.

Exercise 10.2 Suppose the parameters in the continuous version of Kimura's model are given by

$$\alpha = .000006$$

$$\beta = .0000007$$

1. Compute the probability that the nucleotide at a site will not change in 1 year, 100 years, 10,000 years, and 100,000 years.

2. Compute the probability that the nucleotide at a site will change by a transition in 1 year, 100 years, 10,000 years, and 100,000 years.

3. Compute the probability that the nucleotide at a site will change by a transversion in 1 year, 100 years, 10,000 years, and 100,000 years.

Exercise 10.3 In Example 10.4 it was left as an exercise to show that the equilibrium frequencies are

$$\left(\begin{array}{cccc} \pi_A & \pi_G & \pi_C & \pi_T \end{array} \right) = \left(\begin{array}{cccc} .182 & .370 & .296 & .152 \end{array} \right).$$

Show this.

Section 10.2

Exercise 10.4 Suppose two homologous sequences have length 500, sequence 1 has undergone 50 substitutions since they diverged, and sequence 2 has undergone 80 substitutions since they diverged. Compute their evolutionary distance D.

Exercise 10.5 Suppose two homologous sequences have length 500 and they differ on 150 sites. Assuming the Jukes-Cantor Model, estimate their evolutionary distance D.

Exercise 10.6 Prove Theorem 10.6.

Exercise 10.7 Suppose two homologous sequences have length 500, they differ by a transition on 90 sites, and they differ by a transversion on 60 sites. Assuming Kimura's two-parameter model, estimate their evolutionary distance D.

Exercise 10.8 At the end of Theorem 10.8, it was left as an exercise to show that

$$6\bar{\mu}t = \frac{3a}{4}\left[\left(1 - \frac{4}{3}\bar{p}(t)\right)^{-1/a} - 1\right].$$

Show this.

Exercise 10.9 Following Theorem 10.8, it was left as an exercise to show that

$$\lim_{a\to\infty}\frac{3a}{4}\frac{\left(1 - \left(1 - \frac{4}{3}\bar{p}(t)\right)^{1/a}\right)}{\left(1 - \frac{4}{3}\bar{p}(t)\right)^{1/a}} = -\frac{3}{4}\ln\left(1 - \frac{4}{3}\bar{p}(t)\right).$$

Show this. Hint: Use the fact that

$$\ln x = \lim_{k\to 0}\frac{x^k - 1}{k}.$$

Exercise 10.10 Following Theorem 10.9, it was left as an exercise to show that

$$\lim_{a\to\infty}\frac{a}{2}\left[(1 - 2\bar{p}(t) - \bar{q}(t))^{-1/a} - 1\right] + \frac{a}{4}\left[(1 - 2\bar{q}(t))^{-1/a} - 1\right]$$

$$= -\frac{1}{2}\ln(1 - 2\bar{p}(t) - \bar{q}(t)) - \frac{1}{4}\ln(1 - 2\bar{q}(t)).$$

Show this.

Exercise 10.11 Suppose two homologous sequences have length 500, and they differ on 150 sites. Assuming the Jukes-Cantor Model, estimate their evolutionary distance D for the following values of a in our gamma distribution of the substitution rate:

1. $a = 1$

2. $a = 2$

3. $a = 10$

Compare these results to those obtained in Exercise 10.5.

Exercise 10.12 Suppose two homologous sequences have length 500, they differ by a transition on 90 sites, and they differ by a transversion on 60 sites. Assuming Kimura's two-parameter model and that there is a constant R such that $\alpha = 2R\beta$, estimate their evolutionary distance D for the following values of a in our gamma distribution of the parameter α:

1. $a = 1$

2. $a = 2$

3. $a = 10$

Compare these results to those obtained in Exercise 10.7.

Section 10.3

Exercise 10.13 Analyze the time complexity of Algorithm *opt*, which appears in Section 10.3.

Exercise 10.14 Write the dynamic programming algorithm for the sequence alignment problem.

Exercise 10.15 Assuming a penalty of 1 for a mismatch and a penalty of 2 for gap, use the dynamic programming algorithm for the sequence alignment problem to find an optimal alignment of the following sequences:

$$C \quad C \quad G \quad G \quad G \quad T \quad T \quad A \quad C \quad C \quad A$$

$$G \quad G \quad A \quad G \quad T \quad T \quad C \quad A \quad .$$

Exercise 10.16 The sequence alignment package BLAST (Basic Local Alignment Search Tool) and its successor PSI-BLAST are available at

http://blast.ncbi.nlm.nih.gov/Blast.cgi.

Furthermore, a tutorial on using these packages can be viewed at

www.geospiza.com/outreach/BLAST/.

1. Study the tutorial.

2. When you use BLAST, you can enter a nucleotide sequence. Then BLAST will search a database of sequences looking for close matches to the one you entered. Alternatively, you can enter an accession number that corresponds to a sequence in the database. For example, the accession number corresponding to the PAX-6 gene in humans is U63833. The PAX-6 gene is a master regulatory gene controlling a great deal of eye development in many species, including the human, the house mouse, the octopus, and the fly *Drosophila*. Mutations of this gene in humans cause the condition aniridia, a defect in which the iris of the eye is absent or deformed. A mutation of this gene in *Drosophila* can cause the fly to develop without eyes.

   Enter accession number U63833 and investigate the sequences that are found to be close matches. Is the house mouse or *Drosophila* closer to humans based on this analysis?

3. You can also align protein sequences using BLAST. The accession code for the protein sequence corresponding to the PAX-6 gene in humans is P26367. Enter this number and again investigate the sequences that are found to be close matches. Compare the results to those obtained in Part (2) above when nucleotide sequences were compared.

Chapter 11

Molecular Phylogenetics

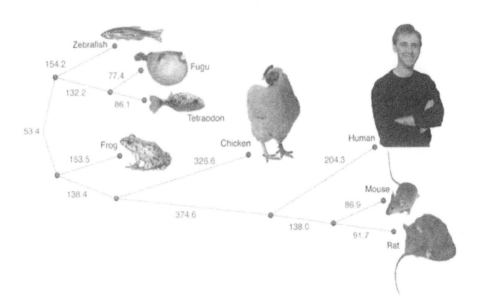

The theory of evolution entails that changes in traits in populations occur over time through natural selection, genetic drift, or both. At some point we say that a **speciation event** occurs, which means that a new species arises in a population. For the most part, this happens at some fuzzy point in time when the set of members of the species breaks into two subsets such that the members of one subset cannot mate with the members of the other. We try to reconstruct the history of these events with a **phylogenetic tree**. Each internal node in the tree is called a **taxonomic unit**. External nodes (leaves) represent **extant** taxonomic units. That is, they represent species that exist today. They are called **operational taxonomic units (OTUs)**. Internal nodes are called **hypothetical taxonomic units (HTUs)** because they cannot be directly observed. The weights of the edges in a phylogenetic tree in some sense are a measure of the amount of difference between the two nodes connected by the edge. The figure at the beginning of this chapter shows a phylogenetic tree.

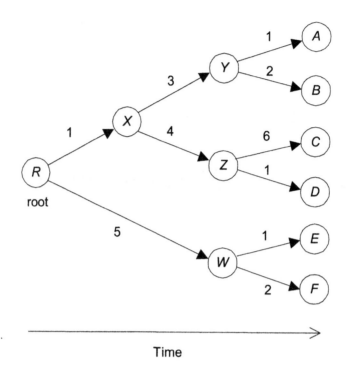

Figure 11.1: A rooted phylogenetic tree.

The field of **molecular phylogenetics** uses molecular structure to construct phylogenetic trees. For example, we could compare homologous DNA sequences from individuals in different species, or we could compare protein sequences. Various methods have been developed to perform these tasks. We concentrate on the method that represents a phylogenetic tree as a Bayesian network and learns the most probable tree using a Bayesian network learning algorithm. This method was actually developed independently by researchers in biology before Bayesian networks had been fully developed and were commonly known. These researchers therefore did not refer to the trees as Bayesian networks; they called the method the **maximum likelihood method**. Before discussing this method in detail, we show some simpler methods to illustrate molecular phylogenetic inference. We make no effort to discuss all methods that have been developed for learning phylogenetic trees. See [Felsenstein, 2004] for a comprehensive coverage of the field.

11.1 Phylogenetic Trees

We start by distinguishing rooted and unrooted phylogenetic trees.

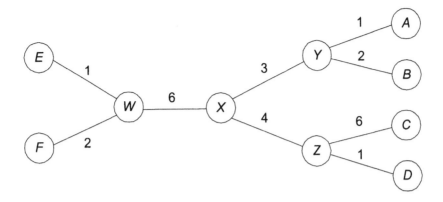

Figure 11.2: An unrooted phylogenetic tree.

11.1.1 Rooted Trees

A **rooted phylogenetic tree** contains a unique node called the **root** that has two or more children, and each of those children has two more children, and so on. If the tree is binary (i.e., each node has exactly two children), it is called a **bifurcating** tree. Otherwise, it is called a **multifurcating** tree. We will only consider bifurcating trees. Rooted trees entail an arrow of time, and show the evolutionary paths culminating in the OTUs. The root occurred longest ago and is the common ancestor of all the OTUs in the tree. The parent of each pair of nodes in the tree is the most recent common ancestor of the two nodes. It occurs at the point in time when the two species diverged (the parent species set divided into two subsets). The weights of the edges in the tree are a measure of the amount of molecular change since the species diverged, but ordinarily we do not draw the edges proportional to their weights. Rather, we line up all the leaves.

Figure 11.1 shows a rooted phylogenetic tree. In the tree in Figure 11.1, the OTUs A and B have the common ancestor Y; C and D have the common ancestor Z; Y and Z have the common ancestor X; E and F have the common ancestor W; and finally, X and W have the common ancestor R. Sometime authors assume that the HTUs must be extinct species. That is, they assume that the time since the speciation event has occurred is sufficiently long that neither of the OTUs could mate with the parent HTU in a rooted tree. However, there is no reason that this must be the case, and there is no way of actually knowing whether it is the case. For example, species A and Y in Figure 11.1 could be the same species.

11.1.2 Unrooted Trees

An **unrooted phylogenetic tree** contains no root and no direction. It only specifies the relationships among the OTUs and does not show the evolutionary

298   CHAPTER 11. MOLECULAR PHYLOGENETICS

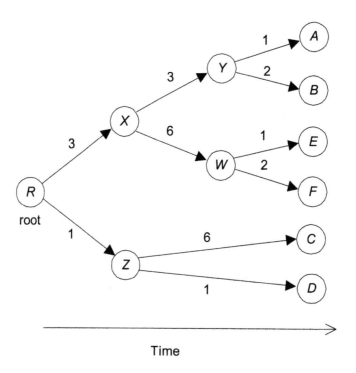

Figure 11.3: Another rooted phylogenetic tree.

paths. Figure 11.2 shows the tree from Figure 11.1 as an unrooted tree. Notice that the distance between W and X is 6, which is the sum of their distances to the root in Figure 11.1. To every unrooted tree there correspond many rooted trees. That is, we can place the root on the edge between any two nodes and then assign distances to the two new edges, which connect the nodes and the root, that sum to the original distance between the two nodes. Figure 11.1 shows the rooted tree resulting from the unrooted tree in Figure 11.2 when we choose the edge between X and W; Figure 11.3 shows the resultant rooted tree when we choose the edge between X and Z. Notice that the evolutionary path can depend on which edge we choose. For example, in the tree in Figure 11.1, A has a more recent common ancestor with C than it does with E, whereas in the tree in Figure 11.3 the reverse is true.

Most methods for learning phylogenetic trees learn unrooted trees. To root the tree, we ordinarily need an outgroup. When comparing several OTUs, an **outgroup** is an OTU for which there is evidence (e.g., paleontological information) that the other OTUs are more closely related to each other than each of them is to the outgroup. An outgroup is the closest thing to a current-day ancestor of all the other OTUs, and indeed it could be their ancestor. As an example, if we have reason to believe that A, B, D, E, and F are all more

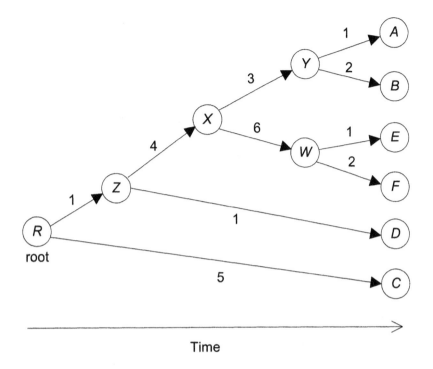

Figure 11.4: A third rooted phylogenetic tree.

closely related to each other than each of them is to C, then C is an outgroup relative to these six OTUs. In this case, we would place the root as shown in Figure 11.4.

11.2   Distance Matrix Learning Methods

As we mentioned, in molecular phylogenetics we could compare homologous DNA sequences from individuals in different species, or we could compare protein sequences. We will illustrate learning methods using DNA sequences. We assume that the sequences have already been aligned as discussed in Section 10.3. In distance matrix methods, the evolutionary distance D (see Section 10.2) is computed from the aligned sequences for every pair of OTUs, and then the algorithm tries in some way to make two OTUs, which have a relatively small distance between them, close together in the phylogenetic tree. When we create a tree using a distance matrix learning method, the weights on the edges represent evolutionary distances, and we call them **distances**.

11.2.1   UPGMA

The simplest distance matrix method is the **unweighted pair-group method with arithmetic mean (UPGMA)**. The method proceeds as follows: We first determine the two OTUs A and B with minimum distance between them and make them siblings in the tree, breaking ties arbitrarily. Next we set the distance on the edge connecting each of them to their common ancestor equal to

$$\frac{D_{AB}}{2},$$

where D_{AB} is the distance between A and B. Then we treat A and B as a unit Y, and for each remaining OTU X we set

$$D_{YX} = \frac{D_{AX} + D_{BX}}{2}.$$

Replacing A and B by Y in the table, we then find the two table elements that have minimum distance between them and group them in the tree. Again, the distance between each of them and their common ancestor is set equal to half the distance between them. We then group these two elements as a unit and repeat this procedure. We do this until there are no ungrouped elements. In general, when computing the distance between two elements Y and Z, each of which represents a subset of OTUs, we set

$$D_{YZ} = \frac{\sum\limits_{i \in Y, j \in Z} D_{ij}}{n_Y \times n_Z},$$

where n_Y and n_Z are the number of OTUs in Y and Z, respectively. The next example illustrates this method.

Example 11.1 *Suppose we have four OTUs A, B, C, and D and the distances in the following table:*

OTU	A	B	C
B	8		
C	7	9	
D	12	14	11

Since $D_{AC} = 7$ is the minimum distance, we group A and C and set the distance from each of them to their common ancestor equal to

$$\frac{D_{AC}}{2} = \frac{7}{2} = 3.5.$$

See Figure 11.5 to view the assignments of this value in the tree.

   Next we compute

$$D_{\{A,C\}B} = \frac{D_{AB} + D_{CB}}{2} = \frac{8 + 9}{2} = 8.5$$

$$D_{\{A,C\}D} = \frac{D_{AD} + D_{CD}}{2} = \frac{12 + 11}{2} = 11.5.$$

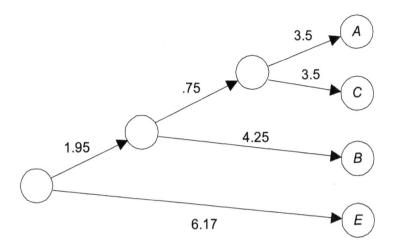

Figure 11.5: The tree constructed in Example 11.1.

We then have the following table:

OTU	$\{A,C\}$	B
B	8.5	
D	11.5	14

Since $D_{\{A,C\}B} = 8.5$ is the minimum distance, we group $\{A,C\}$ and B and set the distance from each of them to their common ancestor equal to

$$\frac{D_{\{A,C\}B}}{2} = \frac{8.5}{2} = 4.25.$$

Again, see Figure 11.5.
  Next we compute

$$D_{\{A,C,B\}D} = \frac{D_{AD} + D_{CD} + D_{BD}}{3} = \frac{12 + 11 + 14}{3} = 12.333.$$

Since $\{A,C,B\}$ and D are the only remaining elements, we group them and set the distance from each of them to their common ancestor equal to

$$\frac{D_{\{A,C,B\}D}}{2} = \frac{12.333}{2} = 6.17.$$

See Figure 11.5 for the final tree. Note that this method produces a rooted tree.

The UPGMA method assumes a molecular clock. The **molecular clock hypothesis** is that the rate of evolutionary change is approximately constant over time in all lineages (edges). If this hypothesis were correct, then the total distance of each OTU to the root would be the same. Inherent in the UPGMA

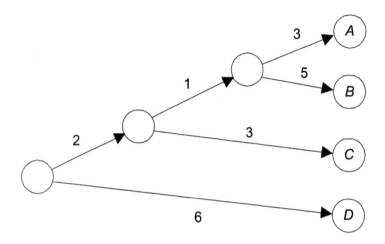

Figure 11.6: If the distances are those shown in the table in Example 11.1 and this is the true tree, UPMGA will learn the wrong tree.

method is the assumption that these distances are the same, and it therefore produces a rooted tree in which this is the case. Notice that the distance of each OTU to the root is 6.17 for the tree in Figure 11.5.

The molecular clock hypothesis is controversial. See [Li, 1997] for a discussion. The problem with UPGMA is that it can give seriously inaccurate results if the distances do not reflect a molecular clock. For example, suppose the true tree is the one in Figure 11.6. Suppose further that the distances between the OTUs are those shown in the table in Example 11.1. Notice that the molecular clock hypothesis does not hold for the tree in Figure 11.6. For example, the distance from B to the root is 8, whereas the distance from C to the root is 5. In this case, the UPGMA method will incorrectly learn the tree in Figure 11.5. The **transformed distance method** (see [Li, 1997]), which is a variation of the UPGMA method, can often find the correct tree when there is no molecular clock. This method, however, requires that there be a known outgroup among the OTUs.

11.2.2   Neighbors-Relation Method

A phylogenetic tree, whose weights represent distances, is called **additive** if the distance between any two OTUs is equal to the sum of the distances on the edges that connect them. If the distances between OTUs are those shown in the table in Example 11.1, the tree in Figure 11.5 is not additive. For example, $D_{AB} = 8$, but the sum of the distances on the edges connecting A and B is equal to $3.5 + .75 + 4.25 = 8.5$. However, the tree in Figure 11.6 is additive. Rather than assuming a molecular clock, the **neighbors-relation method** assumes that the tree is additive, or at least is close to being additive. If this

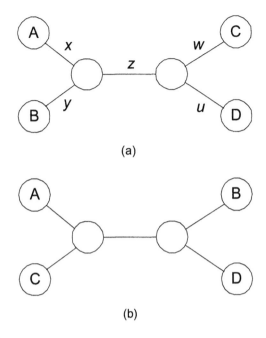

Figure 11.7: The unrooted tree that pairs A with B is in (a); the one that pairs A with C is in (b).

is so, the method can recover the correct unrooted tree, when there are four OTUs A, B, C, and D, as follows: The unrooted tree must pair A with B, C, or D. So, there are three possible unrooted trees. The one that pairs A with B and the one that pairs A with C are shown in Figure 11.7. If the tree in Figure 11.7 (a) were additive, we would have

$$D_{AC} + D_{BD} = D_{AD} + D_{BC} = x + z + w + y + z + u = D_{AB} + D_{CD} + 2z.$$

Therefore,

$$D_{AB} + D_{CD} < D_{AC} + D_{BD} \tag{11.1}$$

$$D_{AB} + D_{CD} < D_{AD} + D_{BC}. \tag{11.2}$$

The requirement of satisfying Equalities 11.1 and 11.2 is called the **four-point condition** [Buneman, 1971]. If they do not hold, the tree could not be additive. Therefore, we choose a tree for which the condition holds.

Example 11.2 *Suppose we have four OTUs A, B, C, and D and the distances in Example 11.1. That is, we have the following distances:*

OTU	A	B	C
B	8		
C	7	9	
D	12	14	11

Then

$$D_{AB} + D_{CD} = 8 + 11 = 19 < 21 = 7 + 14 = D_{AC} + D_{BD}$$

$$D_{AB} + D_{CD} = 8 + 11 = 19 < 21 = 12 + 9 = D_{AD} + D_{BC}.$$

This means that the tree in Figure 11.7 (a) is additive. It is left as an exercise to show that the conditions do not both hold for the other two trees. So, we choose the tree in Figure 11.7 (a). Note that this unrooted tree correctly represents the rooted tree in Figure 11.6.

Equalities 11.1 and 11.2 are necessary but not sufficient conditions for additivity. There are five variables (x, y, z, w, and u) in the tree in Figure 11.7 (a), and these five variables need to satisfy six linear constraints (one for each distance). So, we have six equations in five unknowns. In the case of the previous example, we have

$$
\begin{aligned}
x + z + w &= 7 \\
x + z + u &= 12 \\
y + z + w &= 9 \\
y + z + u &= 14 \\
x + y &= 8 \\
w + u &= 11.
\end{aligned}
$$

The solution to this system of equations is $x = 3, y = 5, z = 1, w = 3, u = 8$. However, in general there is no solution. For example, if we change D_{AD} to 20, Equalities 11.1 and 11.2 will still hold for the tree in Figure 11.7 (a), but there will be no solution to the system of equations, which means additivity cannot hold.

Sattath and Tversky [1977] developed a method for extending the neighbors-relation method to handle more than four OTUs.

11.3   Maximum Likelihood Method

We presented a couple of simple methods for learning phylogenetic trees to introduce you to the problem. Our real purpose is to present the probabilistic method in detail. As mentioned at the beginning of this chapter, this method represents a phylogenetic tree as a Bayesian network, even though the researchers who developed the method did not call it that. However, as we present it, we will use Bayesian network terminology. First we show the Bayesian network representation of a phylogenetic tree.

11.3.1   A Phylogenetic Tree as a Bayesian Network

Given a set of related OTUs, we assume that some rooted phylogenetic tree resulted in the OTUs. For the sake of concreteness, let's say it is the tree in Figure 11.8. We further assume that a model for nucleotide substitution

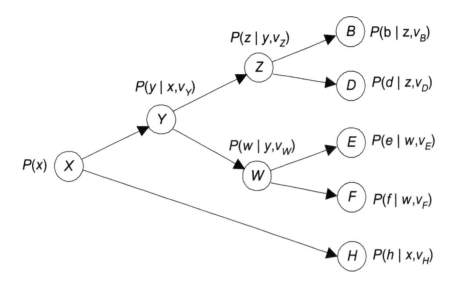

Figure 11.8: A Bayesian network representation of a phylogenetic tree.

(as discussed in Section 10.1) describes the probabilistic relationship between each nucleotide site for the species at a given node in the tree and the same nucleotide site for the species at the node's parent. For example, if we assume the continuous version of the Jukes-Cantor Model, we assume that for each site in the sequences being compared, if the species at the parent node has base i at the site, and the edge from the parent to the child represents time equal to t, then the probability that the species at the child has each of the bases at the site is given by

$$P(x|x,t) = \frac{1}{4} + \frac{3}{4}e^{-4\mu t}$$

and for $x \neq y$

$$P(y|x,t) = \frac{1}{4} - \frac{1}{4}e^{-4\mu t},$$

where μ is the mutation rate when the parent species evolves to the child species.

In what follows, we will assume the Jukes-Cantor Model, but any other model could be used. The probability distribution at node Y, for example, depends both on the time t_Y since Y diverged from its parent and the mutation rate μ. Rather than represent both these parameters in the tree, we set

$$v_Y = \mu t_Y.$$

The parameter v_Y is the expected value of the number of substitutions along the edge (branch) and is called the **branch length**. We denote the probability distribution at Y as

$$P(y|x, v_Y).$$

We then have the Bayesian network representation of a phylogenetic tree shown in Figure 11.8. Each node in the tree not only represents a species, but is a random variable whose values are the possible bases the species could have at a given site. That is, its four possible values are A, G, C, and T.

Example 11.3 *Suppose $v_Y = .3$ in the tree in Figure 11.8. We have that the conditional probability distribution of Y given $X = A$ is as follows:*

$$P(Y = A|X = A, v_Y) = \frac{1}{4} + \frac{3}{4}e^{-4(.3)} = .4759$$

$$P(Y = G|X = A, v_Y) = \frac{1}{4} - \frac{1}{4}e^{-4(.3)} = .1747$$

$$P(Y = C|X = A, v_Y) = \frac{1}{4} - \frac{1}{4}e^{-4(.3)} = .1747$$

$$P(Y = T|X = A, v_Y) = \frac{1}{4} - \frac{1}{4}e^{-4(.3)} = .1747.$$

The prior probabilities are usually set equal to the equilibrium frequencies in the model that is used. So, for the Jukes-Cantor Model we set

$$\begin{aligned}
P(X = A) &= \pi_A = .25 \\
P(X = G) &= \pi_G = .25 \\
P(X = C) &= \pi_C = .25 \\
P(X = T) &= \pi_T = .25.
\end{aligned}$$

Our assumption is that the true tree existed in nature and gave rise to the observed bases at the leaves for each site in the sequence. We can think in terms of possible worlds. In any given world, for each site a value x of X is generated according to probability distribution $P(X)$. Then a value of y of Y is generated according to probability distribution $P(Y|X = x)$, and so on until the leaves are reached. So, in any particular world (e.g., our world) each random variable in the tree has a precise value.

Our observed data consist of the bases in homologous sequences from all the OTUs in the tree. Our task is to do model selection (i.e. learn the tree) based on these data. To accomplish this task we need to score candidate trees. Next we discuss how to do that.

11.3.2   Scoring a Tree Assuming Uniform Rates

Here we discuss scoring a tree assuming that all sites evolve at the same rate.

The Scoring Criterion

One data item $data_i$ consists of the bases at the ith site for every OTU. Given that each of our sequences has m sites, our set of data then is

$$\mathsf{D} = \{data_1, data_2, \ldots, data_m\}.$$

Denote an assignment of values to the parameters in the tree by P. For example, for the tree in Figure 11.8, P denotes a set of values of the following set of parameters:

$$\{v_Y, v_Z, v_W, v_B, v_D, v_E, v_F, v_H\}.$$

Let a tree $\mathbb{T}$ be given. We will score the tree using the Bayesian information criterion (BIC) score. Recall that this score is given by

$$BIC\,(\mathbb{T}:\mathsf{D}) = \ln\left(P(\mathsf{D}|\hat{\mathsf{P}},\mathbb{T})\right) - \frac{d}{2}\ln m, \qquad (11.3)$$

where $\hat{\mathsf{P}}$ is the maximum likelihood estimate of the parameters, and d is the size of the model.

All our candidate phylogenetic trees are full binary trees that have the same number leaves. A binary tree is called **full** if every nonleaf has two children. We have the following theorem concerning such trees.

Theorem 11.1 *If two full binary trees have the same number of leaves, they have the same number of edges.*

Proof. *The proof is left as an exercise.* ∎

Owing to the previous theorem, all our candidate trees have the same size, and so we can ignore the second term in Equality 11.3. Therefore, our score will simply be

$$P(\mathsf{D}|\hat{\mathsf{P}},\mathbb{T}).$$

We assume that each site evolved independently. So, given a tree $\mathbb{T}$ and values of the parameters, the data items are independent. We therefore finally have that our score is

$$P(\mathsf{D}|\hat{\mathsf{P}},\mathbb{T}) = \prod_{i=1}^{m} P(data_i|\hat{\mathsf{P}},\mathbb{T}). \qquad (11.4)$$

The Inference Algorithm

We can compute the probability of each data item $data_i$ for a given tree and set of values of its parameters using a Bayesian network inference algorithm. If we knew the maximum likelihood value $\hat{\mathsf{P}}$, to score the tree we would simply do this computation once with the parameters having the values in $\hat{\mathsf{P}}$. However, we do not know it. So, we must do a heuristic search over values of P, each time computing the probability of each $data_i$ given the values in P. Before showing the heuristic search algorithm, we present the Bayesian network inference algorithm we use.

We illustrate the algorithm using the tree in Figure 11.8 as an example. For that tree, one data item $data_i$ is a set of values b_i, d_i, e_i, f_i, and h_i of the random variables B, D, E, F, and H. So, we need to compute

$$P(data_i|\mathsf{P},\mathbb{T}) = P(b_i, d_i, e_i, f_i, h_i|\mathsf{P},\mathbb{T}).$$

Since the evidence is all at the leaves, the computation is quite easy using an efficient factorization and eliminating variables. For example, for the tree in Figure 11.8 (for the sake of simplicity, we do not show the conditioning on P and $\mathbb{T}$),

$$P(b_i, d_i, e_i, f_i, h_i)$$

$$= \sum_{x,y,z,w} [P(b_i|z)P(d_i|z)P(e_i|w)P(f_i|w)P(h_i|x)P(z|y)P(w|y)P(y|x)P(x)]$$

$$= \sum_x P(x)P(h_i|x)$$

$$\left(\sum_y P(y|x) \left(\sum_w P(w|y)P(e_i|w)P(f_i|w) \right) \left(\sum_z P(z|y)P(b_i|z)P(d_i|z) \right) \right).$$

In general, we sum over all parents of leaves first, then all their parents, and so on until we reach the root, which is summed over last. The algorithm is most easily expressed in terms of recursion as follows: For the root X, the **likelihood** of the tree originating at X is (note that all our likelihoods depend on $\mathbb{T}$, P, and the data D; for simplicity we do not show any of these dependencies):

$$P(data_i|\mathsf{P}, \mathbb{T}) = L_i = \sum_x P(x)L_i^X(x).$$

If Y is the parent of Z and W, we recursively define the **conditional likelihood** of the subtree originating at Y given $Y = y$ as follows

$$L_i^Y(y) = \left(\sum_w P(w|y)L_i^Y(w) \right) \left(\sum_z P(z|y)L_i^Z(z) \right).$$

The terminal condition is when we reach a leaf. For a leaf B in the tree with value b_i, we define

$$L_i^B(b_i) = 1$$
$$L_i^B(b) = 0 \quad \text{for } b \neq b_i.$$

The algorithm proceeds in a dynamic programming fashion, computing the likelihoods of leaves first and then proceeding up the tree until the root is reached. The following example illustrates the algorithm.

Example 11.4 *Suppose we have the tree $\mathbb{T}$ in Figure 11.8 with a set of values* P *of the parameters, and*

$$data_i = \{b_i, d_i, e_i, f_i, h_i\}$$

contains the values of the random variables B, D, E, F, and H at the ith site.
    For each value z of node Z we compute

$$L_i^Z(z) = \left(\sum_b P(b|z)L_i^B(b) \right) \left(\sum_d P(d|z)L_i^D(d) \right)$$

$$= P(b_i|z)P(d_i|z).$$

For the sake of illustration, we show the computation of one of the conditional probabilities in the last expression. For $z = b_i$,

$$P(b_i|z) = \frac{1}{4} + \frac{3}{4}e^{-4v_B},$$

and for $z \neq b_i$,

$$P(b_i|z) = \frac{1}{4} - \frac{1}{4}e^{-4v_B}.$$

For each value w of node W we compute

$$
\begin{aligned}
L_i^W(w) &= \left(\sum_e P(e|w)L_i^E(e)\right)\left(\sum_f P(f|w)L_i^F(f)\right) \\
&= P(e_i|w)P(f_i|w).
\end{aligned}
$$

Once we compute these values for all values z of Z and w of W, for each value y of Y we compute

$$L_i^Y(y) = \left(\sum_w P(w|y)L_i^W(w)\right)\left(\sum_z P(z|y)L_i^Z(z)\right).$$

Then for each value x of X we compute

$$
\begin{aligned}
L_i^X(x) &= \left(\sum_y P(y|x)L_i^Y(y)\right)\left(\sum_h P(h|x)L_i^H(h)\right) \\
&= \left(\sum_y P(y|x)L_i^Y(y)\right)P(h_i|x).
\end{aligned}
$$

Finally,

$$P(data_i|\mathsf{P}, \mathbb{T}) = L_i = \sum_x P(x)L_i^X(x),$$

and

$$P(\mathsf{D}|\mathsf{P}, \mathbb{T}) = \prod_{i=1}^m P(data_i|\mathsf{P}, \mathbb{T}).$$

Before showing our heuristic search algorithm for the maximum likelihood values of P, we show that we are actually scoring unrooted trees.

Scoring Unrooted Trees

We discussed a method for scoring rooted trees. However, the method scores unrooted trees also. For example, consider the rooted tree in Figure 11.9 (b). It can be obtained from the unrooted tree in Figure 11.9 (a) by placing the root on the edge between nodes Y and Z. The unrooted tree in Figure 11.9 (a) represents a family of rooted trees, each one obtained by placing the root anywhere on one of the six edges in the unrooted tree. We will show that,

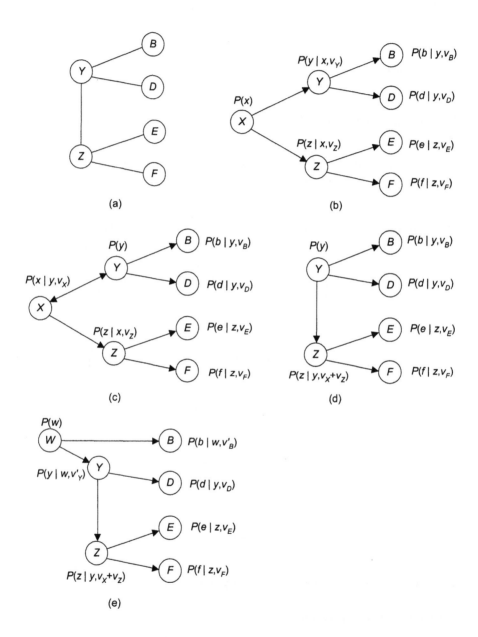

Figure 11.9: By following the sequence of trees in (b), (c), (d), and (e), we can show that the trees in (b) and (e) have the same score.

as long as our model of nucleotide substitution is time-reversible and we use the equilibrium frequencies as the prior probabilities, each of these rooted trees has the same score, and therefore we can apply the score to the unrooted tree. Although we illustrate the matter using the trees in Figure 11.9, the argument clearly applies to every family of rooted trees described by an unrooted tree.

First we need to review some preliminary concepts. A **probabilistic model** for a set V of random variables is a set of joint probability distributions of the variables. Ordinarily, a probabilistic model is specified using a parameter set and combinatoric rules for determining the joint probability distribution from the parameter set. Each member of the model is then obtained by assigning values to the members of the parameter set and applying the rules. A rooted tree, along with a model for nucleotide substitution (e.g. the Jukes-Cantor Model) that specifies the conditional probability distribution at each node, is a probabilistic model, called a **rooted tree model**. For example, the tree in Figure 11.9 (b), along with the Jukes-Cantor Model, is a rooted tree model. Each assignment of values to the parameters v_B, v_D, v_E, v_F, v_Y, and v_Z yields a probability distribution in the model, and these are the only probability distributions in the model.

To show that all the rooted tree models described by the unrooted tree in Figure 11.9 (a) have the same score, we need to show that, for every pair of such models $\mathbb{T}_1$ and $\mathbb{T}_2$, for each probability distribution included in $\mathbb{T}_1$ (that is, an assignment P_1 of values to the parameters in the tree), there is a probability distribution included in $\mathbb{T}_2$ (that is, an assignment P_2 of values to the parameters in the tree) such that for all values b, d, e, and f of the random variables B, D, E, and F,

$$P(b, d, e, f | \mathsf{P}_1, \mathbb{T}_1) = P(b, d, e, f | \mathsf{P}_2, \mathbb{T}_2). \tag{11.5}$$

To accomplish this task, we start with one distribution, which is included in the tree model in Figure 11.9 (b), and show that there is a distribution that is included in the tree model in which we put the root on the edge between nodes Y and B, such that Equality 11.5 holds. Once this is done, our task is accomplished because obviously we can then iteratively move the root to any edge.

To that end, first let

$$\mathsf{P}_1 = \{v_B, v_D, v_E, v_F, v_Y, v_Z\}.$$

We then have that

$$P(x, y, z, b, d, e, f | \mathsf{P}_1, \mathbb{T}_1)$$

$$= P(b|y, v_B)P(d|y, v_D)P(e|z, v_E)P(f|z, v_F)P(z|x, v_Z)P(y|x, v_Y)P(x).$$

Owing to the time reversibility of the nucleotide substitution model,

$$P(y|x, v_Y)P(x) = P(x|y, v_Y)P(y).$$

Therefore,

$P(x, y, z, b, d, e, f | \mathsf{P}_1, \mathbb{T}_1)$

$$= P(b|y, v_B)P(d|y, v_D)P(e|z, v_E)P(f|z, v_F)P(z|x, v_Z)P(x|y, v_Y)P(y). \quad (11.6)$$

If we set $v_X = v_Y$, this is the same joint probability distribution as the one described by the tree in Figure 11.9 (c). So, the trees in Figure 11.9 (b) and (c) describe the same set of joint probability distributions.

Continuing, since $v_X = v_Y$ we have, owing to Equality 11.6, that

$P(b, d, e, f | \mathsf{P}_1, \mathbb{T}_1)$

$$= \sum_{z,y,x} P(b|y, v_B)P(d|y, v_D)P(e|z, v_E)P(f|z, v_F)P(z|x, v_Z)P(x|y, v_X)P(y)$$

$$= \sum_{z,y} P(b|y, v_B)P(d|y, v_D)P(e|z, v_E)P(f|z, v_F)$$

$$\left(\sum_x P(z|x, v_Z)P(x|y, v_X) \right) P(y). \quad (11.7)$$

Next we create the tree $\mathbb{T}'$ in Figure 11.9 (d) with parameter set

$$\mathsf{P}' = \{v_B, v_D, v_E, v_F, v_Z'\},$$

where the parameter values are the same as those in Figure 11.9 (c) except that $v_Z' = v_X + v_Z$. It is left as an exercise to show that for the Jukes-Cantor Model,

$$P(z|y, v_X + v_Z) = \sum_x P(z|x, v_Z)P(x|y, v_X).$$

Owing to the Chapman-Kolmogorov Equation, this equality actually holds for all the models we are considering. Therefore, due to Equality 11.7, we have that

$P(b, d, e, f | \mathsf{P}_1, \mathbb{T}_1)$

$$= \sum_z \sum_y P(b|y, v_B)P(d|y, v_D)P(e|z, v_E)P(f|z, v_F)P(z|y, v_X + v_Z)P(y)$$

$$= P(b, d, e, f | \mathsf{P}', \mathbb{T}').$$

Our last step is to move the root to the edge connecting nodes Y and B as shown in Figure 11.9 (e). In the same way that we obtained parameter values that enabled us to move from the tree in Figure 11.9 (c) to the one in Figure 11.9 (d), we can obtain parameter values that enable us to move from the tree in Figure 11.9 (d) to the one in Figure 11.9 (e). We need merely choose v_Y' and v_B' such that

$$v_B = v_Y' + v_B'.$$

It is left as an exercise to show this. So, if our parameter set for the tree $\mathbb{T}_2$ in Figure 11.9 (e) is

$$\mathsf{P}_2 = \{v_B', v_D, v_E, v_F, v_Y', v_Z' = v_X + v_Z\},$$

then

$$P(b, d, e, f | \mathsf{P}_1, \mathbb{T}_1) = P(b, d, e, f | \mathsf{P}_2, \mathbb{T}_2).$$

This same argument enables us to go from the tree in Figure 11.9 (e) to the one in Figure 11.9 (b). Therefore,

$$\prod_{i=1}^{m} P(b_i, d_i, e_i, f_i | \mathsf{P}_1, \mathbb{T}_1) = \prod_{i=1}^{m} P(b_i, d_i, e_i, f_i | \mathsf{P}_2, \mathbb{T}_2),$$

where b_i, d_i, e_i, and f_i are the values of B, D, E, and F for the ith site and m is the number of sites. This means that the two trees must have the same score. Clearly, we can iteratively repeat this argument, thereby showing that all rooted trees, obtained by putting the root on one of the edges in the tree in Figure 11.9 (a), must have the same score.

A Heuristic Search Algorithm

We need the maximum likelihood value of the parameters to compute the score in Equality 11.4. There is no analytical solution to this problem. We could do a near-exhaustive[1] search over all combinations of values of the parameters in the tree, but this is computationally unfeasible. We will show a heuristic search algorithm. In this algorithm, we first assign arbitrary values to all the parameters. Then we develop an iterative method for visiting one edge at a time and trying to maximize the likelihood relative to the value of the parameter on that edge, but without changing the contribution to the likelihood of the other edges. We visit all the edges in sequence. However, each time we change the parameter at one edge, it is possible that there may be new assignments to the parameters at the other edges which would increase the likelihood more. So, we take several passes through the tree until we cannot increase the likelihood more. Even so, we are not guaranteed to find the optimal value. Rather, we could be stuck at a local optimum. However, Felsenstein [2004] notes that in practice this rarely happens.

The parameters we want to estimate are weights on the edges in an unrooted tree. This is illustrated in Figure 11.10 (a). Once we have estimates for these parameters, we can create a rooted tree, which is represented by the unrooted tree, by placing the root on one of the edges in the unrooted tree and then assigning values to the parameters in the rooted tree. Figure 11.10 (b) shows a rooted tree when we place the root on the edge between node Y and node Z.

---

[1] We say near-exhaustive because the domain of each parameter is the positive real numbers, which is an uncountable infinity. So, we would have to discretize the range when we're searching.

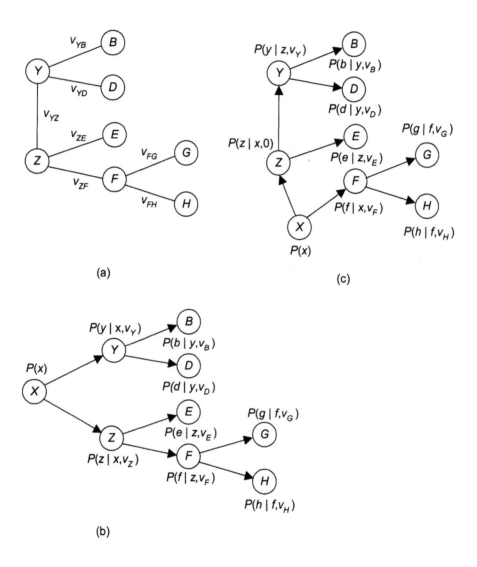

Figure 11.10: An unrooted tree is in (a). A rooted tree represented by that unrooted tree is in (b). A rooted tree represented by that unrooted tree and in which the weight on the edge from X to Z is 0 is in (c).

In Figure 11.10 we have

$$
\begin{aligned}
v_B &= v_{YB} \\
v_D &= v_{YD} \\
v_E &= v_{ZE} \\
v_F &= v_{ZF} \\
v_G &= v_{FG} \\
v_H &= v_{FH} \\
v_Y + v_Z &= v_{YZ}.
\end{aligned}
$$

To estimate maximum-likelihood values of the parameters in the tree in Figure 11.10 (a), we first assign arbitrary values to these parameters. Then we iterate over the edges, and, when we visit an edge, we try to estimate the maximum-likelihood value of the parameter on the edge. To accomplish this, for a given edge, we need to obtain a formula for the overall likelihood in terms of the edge's parameter and the likelihoods computed from the other edges. Next we illustrate how to do this using the tree in Figure 11.10 (a).

Suppose we are currently visiting the edge between nodes Z and F so that we are looking for a maximum-likelihood value of v_{ZF}. We construct the rooted tree in Figure 11.10 (c), where the weight v_Z on the edge from X to Z is 0. In this way,

$$v_F = v_{ZF}.$$

Therefore, our task becomes that of estimating the maximum-likelihood value of v_F for the tree in Figure 11.10 (c). The likelihood of $data_i$ (the data at the ith site) for that tree is as follows (note that the likelihood depends on the tree $\mathbb{T}$, the set P of values of the parameters, and the data D, but for simplicity we do not show these dependencies):

$$
\begin{aligned}
L_i &= \sum_x P(x) L_i^X(x) \\
&= \sum_x P(x) \left(\sum_z P(z|x, v_Z) L_i^Z(z) \right) \left(\sum_f P(f|x, v_F) L_i^F(f) \right).
\end{aligned}
$$

Since $v_Z = 0$, regardless of the model

$$
\begin{aligned}
P(x|x, v_Z) &= 1 \\
P(z|x, v_Z) &= 0 \qquad \text{for } z \neq x.
\end{aligned}
$$

We therefore have that

$$L_i = \sum_x P(x) L_i^Z(x) \left(\sum_f P(f|x, v_F) L_i^F(f) \right). \tag{11.8}$$

The only value in this expression that depends on v_F is $P(f|x, v_F)$. The values of $L_i^Z(x)$ can be computed from the subtree emanating from Z using the current values of the parameters on that tree; the values of $L_i^F(f)$ can be computed from the subtree emanating from F using the current values of the parameters on that tree.

The way we proceed using Equality 11.8 can depend on the nucleotide substitution model. We show how to proceed using the Jukes-Cantor Model. Recall that in this model we have

$$
\begin{aligned}
P(x|x, v_F) &= \frac{1}{4} + \frac{3}{4}e^{-4v_F} \equiv p \\
P(f|x, v_F) &= \frac{1}{4} - \frac{1}{4}e^{-4v_F} \equiv \frac{q}{3} \qquad \text{for } f \neq x.
\end{aligned}
\tag{11.9}
$$

Note that $q = 1 - p$. We can estimate a maximum-likelihood value of p and then compute the corresponding maximum-likelihood value of v_F using Equality 11.9.

To that end, we have for the ith site that

$$
\begin{aligned}
L_i &= \sum_x P(x)L_i^Z(x) \left(\sum_f P(f|x, v_F)L_i^F(f) \right) \\
&= \sum_x P(x)L_i^Z(x) \left(pL_i^F(x) + \frac{q}{3}\sum_{f \neq x} L_i^F(f) \right) \\
&= p \sum_x \left(P(x)L_i^Z(x)L_i^F(x) \right) + \frac{q}{3}\sum_x \left(P(x)L_i^Z(x)\sum_{f \neq x} L_i^F(f) \right) \\
&= a_i p + b_i q,
\end{aligned}
$$

where

$$
\begin{aligned}
a_i &= \sum_x \left(P(x)L_i^Z(x)L_i^F(x) \right) \\
b_i &= \frac{1}{3}\sum_x \left(P(x)L_i^Z(x)\sum_{f \neq x} L_i^F(f) \right).
\end{aligned}
$$

We then have

$$
L_i = a_i p + b_i q.
\tag{11.10}
$$

We want to estimate the maximum-likelihood value of the parameters for the products of the likelihoods. That is, we want to maximize

$$
\prod_i L_i = \prod_i \left(a_i p + b_i q \right).
$$

Taking logarithms, we have

$$
\ln L = \sum_i \ln \left(a_i p + b_i q \right).
$$

Equating the derivative to 0, we then have

$$\frac{d\ln L}{dp} = \sum_i \frac{b_i - a_i}{a_i p + b_i q} = \sum_i \frac{b_i - a_i}{a_i p + b_i (1 - p)} = 0.$$

This equation can be solved for p using numerical methods (e.g., Newton's method). Once we have a value of p, we solve

$$\frac{1}{4} + \frac{3}{4} e^{-4v_F} = p$$

for v_F.

As mentioned previously, we iterate this procedure over all edges. Then we repeat the iteration until we cannot improve the score.

11.3.3 Scoring a Tree Assuming Nonuniform Rates

In Section 10.2.2, we noted that in coding regions of DNA (exons), it is not reasonable to assume that all sites in the sequence evolve at the same rate. In this case, the true tree (Bayesian network), which describes the evolutionary path of the species, has the same structure for all sites, but its parameters may vary from site to site. As also noted in that section, one technique researchers have used to model this situation is to place probability distributions on the substitution rates. For example, if we assume the Jukes-Cantor Model, we place a probability distribution $\rho(\mu)$ on the parameter μ. We then have that

$$P(data_i | \mathsf{P}, \mathbb{T}) = L_i = \int_0^\infty L_{i,\mu} \rho(\mu) d\mu \tag{11.11}$$

and

$$P(\mathsf{D} | \mathsf{P}, \mathbb{T}) = L = \prod_{i=1}^m L_i,$$

where $L_{i,\mu}$ is the likelihood computed using μ as the substitution rate, L_i is the likelihood for the ith site, and there are m sites. Note that now the weight on the edge to a node Y is a time t_Y rather than $v_Y = \mu t_Y$, and P is a set of values of these times for all nodes. We approximate the integrals using numerical methods, and we try to estimate the value of P that maximizes L. The value of $L_{i,\mu}$ for a given μ can be computed as illustrated in Example 11.4.

As also discussed in Section 10.2.2, ordinarily a gamma density function is used for $\rho(\mu)$. So, that we have only one parameter in the density function, we set

$$E(\mu) = 1.$$

Owing to the equalities labeled 10.15 in Section 10.2.2, we then have that $a = b$, which means our density function is

$$\rho(\mu) = \frac{a^a}{\Gamma(a)} \mu^{a-1} e^{-a\mu} \qquad \mu > 0.$$

We still need the value of a. Gu, Fu, and Li [1995] developed methods for maximizing the likelihood relative to this parameter also.

11.3.4 Scoring a Tree Assuming Context-Dependent Rates

Nucleotide substitution can be context-dependent. That is, the substitution rate for a given site may depend on the bases at neighboring sites. The models we developed do not consider this. See [Siepel and Haussler, 2004] for a discussion of such models and the development of a model that uses an expectation-maximization (EM) algorithm.

11.3.5 Scoring a Tree Using Other Substitution Models

We illustrated scoring a tree using the Jukes-Cantor Model, which only has one parameter μ. Other models have more than one parameter, which complicates the matter. Next we illustrate how to handle such models using Kimura's two-parameter model.

Assuming Uniform Rates

Consider Kimura's two-parameter model, which has two parameters α and β, where α is the substitution rate for a transition and 2β is the substitution rate for a transversion. We could search over values of α, β, and values of t_Y for all nodes Y in the tree. However, we can reduce our effort to searching over only two parameters by proceeding as follows: First, note that the total substitution rate is

$$r = \alpha + 2\beta.$$

We choose a time scale t so that $r = 1$, which means

$$1 = \alpha + 2\beta. \tag{11.12}$$

In this way, the expected value of the number of substitutions in time t is $1 \times t = t$, which means t_Y itself is the branch length on the edge leading to node Y.

Next, we set

$$R = \frac{\alpha}{2\beta}, \tag{11.13}$$

which is the ratio of the expected number of transitions to the expected number of transversions. It is left as an exercise to simultaneously solve Equations 11.12 and 11.13 to obtain that

$$\alpha = \frac{R}{R+1}$$
$$\beta = \frac{1}{2(R+1)}.$$

Owing to Equalities 10.7 and 10.8, we then have that the probability of a specific transition is

$$P(y|x, t_Y) = \frac{1}{4} + \frac{1}{4}e^{-\frac{2}{R+1}t_Y} - \frac{1}{2}e^{-\frac{2R+1}{R+1}t_Y},$$

and the probability of a specific transversion is

$$P(y|x, t_Y) = \frac{1}{4} - \frac{1}{4}e^{-\frac{2}{R+1}t_Y}.$$

By eliminating one parameter, we need only search over values of R and values of t_Y for all nodes Y.

Assuming Nonuniform Rates

When modeling the possibility that the substitution rates at different sites need not be the same, rather than setting the overall substitution rate r to 1, we place a gamma distribution on the overall rate r and we set

$$E(r) = 1.$$

Equality 11.11 then becomes

$$P(data_i|\mathsf{P}, \mathbb{T}) = L_i = \int_0^\infty L_{i,r}\rho(r)dr,$$

where P is a set of values of t_Y for all nodes Y in the tree. Suppose that, for example, we use Kimura's two-parameter model. When computing $L_{i,r}$, we must use values of α and β, but a given value of r does not determine unique values of α and β. So, we need to search for the the maximum likelihood values of α and β satisfying $\alpha + 2\beta = r$.

11.3.6 Searching over the Space of Trees

Our goal is to find the unrooted tree(s) with the highest score. To accomplish this task, in general we would need to check every unrooted tree. However, when the number of species is not small, the number of trees is prohibitively large. First we develop a formula for the number of bifurcating, unrooted trees with n leaves, thereby showing that there are indeed too many for an exhaustive search to be tractable. Then we develop heuristic search methods that avoid searching over all the trees, but that, of course, are not guaranteed to find the optimal tree.

The Number of Unrooted Trees with n Leaves

Suppose we have n nodes, which we label $S_1, S_2, \ldots$ and S_n (for species). We want to determine the number of different unrooted bifurcating trees that have these n nodes as leaves. For a given n, all the trees will have the same structure. Two trees are deemed different if some of the nodes labeled $S_1, S_2, \ldots$ and S_n have different relative positions to each other. Figure 11.11 (a) shows the single tree when $n = 2$. If we moved S_1 to the right and moved S_2 to the left, the nodes would still have the same relative position to each other, so this would not be a new tree.

Figure 11.11 (b) shows the single tree when $n = 3$. If we interchanged the values of any of the nodes (e.g., move S_3 to the left and move S_1 to the top),

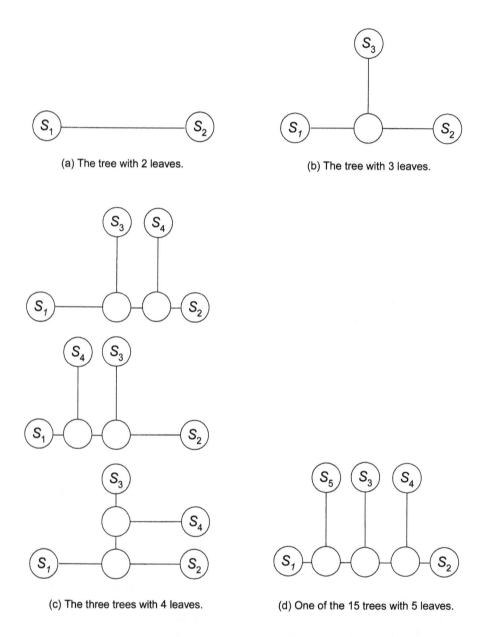

(a) The tree with 2 leaves.

(b) The tree with 3 leaves.

(c) The three trees with 4 leaves.

(d) One of the 15 trees with 5 leaves.

Figure 11.11: Unrooted bifurcating trees.

the nodes S_1, S_2, and S_3 would still have the same relative positions to each other, so this would not be a new tree. Notice that we can obtain the tree in Figure 11.11 (b) by placing a hidden node on the edge connecting S_1 and S_2 in the tree in Figure 11.11 (a), then adding an edge from this hidden node to S_3.

Figure 11.11 (c) shows the three trees when $n = 4$. Notice, for example, that the top two trees are different because in the very top tree S_4 is closest to S_2, whereas in the tree under it S_4 is closest to S_1. Notice further that each of these trees can be obtained from the tree in Figure 11.11 (b) by placing a hidden node on one of the edges in that tree and then adding an edge from this hidden node to S_4. It is not hard to see that this result is true in general. That is, each tree with the leaves $S_1, S_2, \ldots$ and S_{n+1} can be obtained from a unique tree $\mathbb{T}$ with the leaves $S_1, S_2, \ldots$ and S_n by placing a hidden node on one of the edges in $\mathbb{T}$ and then adding an edge from this hidden node to S_{n+1}.

Figure 11.11 (d) shows one of the trees with five leaves that can be obtained from the top tree in Figure 11.11 (c). Since there are three trees with four leaves and each of these trees has five edges, we conclude that there are $3 \times 5 = 15$ trees with five leaves.

The previous considerations enable us to prove a theorem concerning the number of unrooted bifurcating trees with n leaves. First we need the following lemma.

Lemma 11.1 *If a bifurcating unrooted tree has n leaves, its number of edges is equal to*

$$2n - 3.$$

Proof. *The proof is left as an exercise.* ∎

Theorem 11.2 *Suppose we have n nodes $S_1, S_2, \ldots$ and S_n, where $n \geq 3$. Then the number of different bifurcating unrooted trees that have these nodes as leaves is equal to*

$$1 \times 3 \times 5 \times 7 \times \cdots \times (2n - 5).$$

Proof. *The proof is by induction.*

Induction base: *For $n = 3$ we've already shown there is only one tree (see Figure 11.11 (b). Furthermore,*

$$1 \times (2(3) - 5) = 1.$$

Induction hypothesis: *Suppose the number of unrooted bifurcating trees that have the nodes $S_1, S_2, \ldots$ and S_n as leaves is equal to*

$$1 \times 3 \times 5 \times 7 \times \cdots \times (2n - 5).$$

Induction step: *Suppose we have the nodes $S_1, S_2, \ldots$ and S_{n+1}, and we have created all the unrooted bifurcating trees that have the first n of these nodes as leaves. As mentioned, each unrooted bifurcating tree, which has all $n + 1$ nodes as leaves, can be obtained from a unique unrooted bifurcating tree $\mathbb{T}$, which has the first n of the nodes as leaves, by placing a hidden node on a unique edge in $\mathbb{T}$*

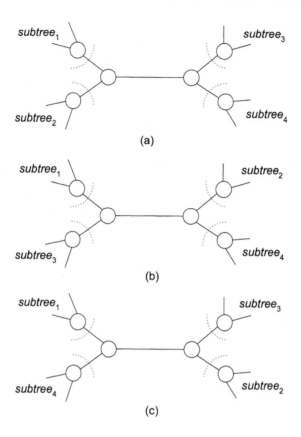

Figure 11.12: The two trees that can be obtained from the tree in (a) using a nearest-neighbor interchange appear in (b) and (c).

and creating an edge from that hidden node to S_{n+1}. Due to Lemma 11.1, there are $2n - 3$ edges in each unrooted bifurcating tree with n leaves. So, there are $2n - 3$ trees with $n + 1$ leaves that can be created from each unrooted bifurcating tree that has the first n nodes as leaves. Therefore, owing to the induction hypothesis, the number of unrooted bifurcating trees with $n + 1$ leaves is equal to

$$1 \times 3 \times 5 \times 7 \times \cdots \times (2n - 5)(2n - 3) = 1 \times 3 \times 5 \times 7 \times \cdots \times (2(n + 1) - 5).$$

This completes the proof. ∎

Example 11.5 Suppose $n = 25$. Then the number of unrooted bifurcating trees is

$$1 \times 3 \times 5 \times 7 \times \cdots \times (2(25) - 5) = 2.5374 \times 10^{28}.$$

We see that for n not small, the number of trees is prohibitively large. Next we investigate heuristic search algorithms.

Nearest-Neighbor Interchange Search

In all heuristic search strategies for finding the optimal tree, we start with either an arbitrary tree or a tree that we suspect might be the correct tree based on expert judgment. We then investigate various trees that can be obtained from the original tree, and we choose the one that most increases the score. We repeat this procedure until we cannot increase the score further.

An **exterior edge** in an unrooted tree is an edge connecting a leaf to a nonleaf; an **interior edge** in an unrooted tree is an edge connecting two nonleaves. It is left as an exercise to show that an unrooted tree with n leaves has $n - 3$ interior edges. In a **nearest-neighbor interchange search**, we look at each interior edge and investigate the trees that can be formed from the current tree by swapping two of the subtrees touching the edge. This is illustrated in Figure 11.12. Given the tree and interior edge shown in Figure 11.12 (a), we can swap *subtree₂* and *subtree₃*, resulting in the new tree in Figure 11.12 (b), or we can swap subtrees *subtree₂* and *subtree₄*, resulting in the new tree in Figure 11.12 (c). These are the only distinct new trees that can be obtained from the tree in Figure 11.12 (a) using a nearest-neighbor interchange. (The proof of this fact is analogous to the proof that there are only three different trees containing four leaves.) Since there are $n - 3$ interior edges, we investigate $2(n - 3)$ new trees in each iteration of the search.

Subtree Pruning and Regrafting Search

In a subtree pruning and regrafting search, we cut off a branch emanating from a nonleaf node, eliminate the nonleaf node resulting in a new edge connecting the two nodes to which the eliminated node was connected, insert a new nonleaf node on another edge, and attach the removed branch to that new node. This is illustrated in Figure 11.13. We first eliminate a branch from the tree in Figure 11.13 (a) resulting in the tree in Figure 11.13 (b). Note that the nonleaf node to which the branch was attached has been eliminated. We then insert a new nonleaf node on the edge touching node J, and we attach the removed branch to that node, resulting in the tree in Figure 11.13 (c). The most thorough version of the algorithm would consider every possible insertion and choose the new tree that has a maximum score. We would then repeat this procedure until we cannot improve the score. However, if we considered every possible insertion before choosing a new tree, we would have to look at $\Theta(n^2)$ number of trees in each iteration. The following theorem obtains this result.

Theorem 11.3 *The number of different trees that can be created from an unrooted bifurcating tree with n leaves, using subtree pruning and regrafting, is in $\Theta(n^2)$.*

Proof. *First, we show that the number of such trees is in $\Omega(n^2)$. To that end, consider the trees obtained when we remove an edge connected to a leaf L. If we insert that edge on the edge connected to any other leaf M (other than L's sibling if L's sibling is a leaf), L will be closer to M in the new tree than it was to M in the original tree. Since there are at least $n - 2$ leaves that are not*

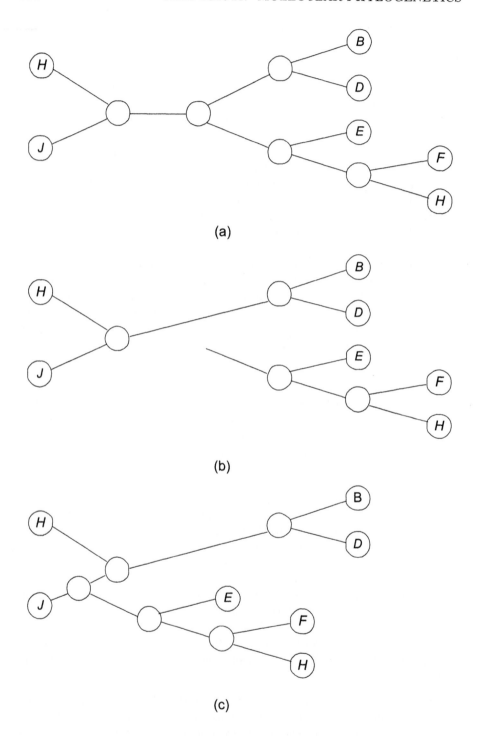

(a)

(b)

(c)

Figure 11.13: A branch is removed from the tree in (a), resulting in the separate tree and branch in (b). The branch is then reinserted, resulting in the tree in (c).

L and are not L's sibling, there are $n - 2$ new trees that can be obtained from L this way. Since there are n leaves, there are $n(n - 2)$ total trees that can be obtained this way. In some cases, inserting L on the edge connected to M will produce the same tree as the one obtained by inserting M on the edge connected to L. However, these are the only pairs that could be the same. So, there are at least $n(n - 2)/2$ different trees that can be obtained this way, which means the number of different trees is in

$$\Omega(n^2).$$

Next, we show that the number of such trees is in $O(n^2)$. Due to Lemma 11.1 there are $2n - 3$ edges, which means there are no more than $2(2n - 3)$ branches we could choose (we have the factor of 2 because when an edge does not touch a leaf, the branch can start with either node touching the edge). Furthermore, there are no more than $2n - 3$ edges at which we could insert each of them. So, the number of different trees is no more than $2(2n - 3)(2n - 3)$, which means that the number of such trees is in

$$O(n^2).$$

This completes the proof. ■

Notice that any tree obtained using a nearest-neighbor interchange can be obtained using subtree pruning and regrafting. For example, the tree in Figure 11.12 (b) can be obtained from the tree in Figure 11.12 (a) by cutting the branch containing $subtree_2$ and inserting it on the edge leading to $subtree_4$. So, if we consider all possible trees that can be obtained from the current tree before choosing a new tree, this method will definitely find a tree whose score is at least as large as the tree found using a nearest-neighbor interchange. However, this method is much more costly in that it requires $\Theta(n^2)$ time in each iteration rather than $\Theta(n)$.

If looking at every possible new tree proves to be too time-consuming, we could stop our search early. However, in this particular application only a very large amount of time should be unacceptable. Whether we consider the computation time too large depends on the nature of the application. For example, computation time must be no more than a fraction of a second in an on-board tracking system used by a pilot in a dogfight. On the other hand, computation time could be around an hour or so in an outbreak disease biosurveillance system that rechecks for a disease outbreak every hour. Finally, in our application there is no immediacy with which we need the answer. We are only trying to obtain a result for science. So, we could tolerate computation time that takes days or perhaps even weeks, but of course years would not be acceptable.

A Structural EM Algorithm

Friedman et al. [2002] developed a structural EM algorithm for learning the structure of a phylogenetic tree. In the expectation step, they use the current tree structure and edge lengths to compute expected sufficient statistics that summarize the data. In the maximization step, they search for a tree structure that maximizes the likelihood with respect to these expected sufficient statistics. They show that searching for better tree structures inside the maximization step can be done efficiently, and they note that the structural EM algorithm is "dramatically faster" than existing methods for searching maximum likelihood phylogenies.

11.3.7   Discussion

We've presented a Bayesian network approach to learning phylogenetic trees from sequence data. There are numerous other approaches. We discussed two of them, namely UPGMA and the neighbors-relation method, which are distance matrix methods. The least-squares method is another well-known distance matrix method. Maximum parsimony is a classic method that is not a distance matrix method.

The question remains as to how the performance of the Bayesian network approach compares to that of the other methods. We do not have the space to give this matter justice. However, many researchers have argued that this approach is superior in terms of both accuracy and statistical justification. You are referred to [Fukami-Kobayashi and Tateno, 1991], [Hasegawa and Fujiwara, 1993], [Kuhner and Felsenstein, 1994], [Tateno et al., 1994], and [Huelsenbeck, 1995].

The Bayesian network approach to learning phylogenetic trees has been implemented as freeware in PHYLIP Version 3.6. You can link to PHYLIP and to other packages for learning phylogenetic trees at the following site: http://evolution.genetics.washington.edu/phylip/software.html.

11.4   Distance Matrix Methods Using ML

The maximum likelihood method we presented becomes computationally intractable for large datasets because of the size of the tree search space. Recall that the number of possible trees grows worse than exponentially with the number of sequences. Currently, this method cannot handle more than a few hundred sequences. However, we are developing datasets that have thousands of homologous sequences. In contrast to the maximum likelihood method, the distance matrix methods, which we briefly introduced in Section 11.2, require computational time that is only quadratic in terms of the number of sequences. Recall that these methods require that we compute the genetic distance D for each pair of homologous sequences, which is the average of the total number of substitutions in both of them since they diverged.

Recall further that in Section 10.2 we developed analytical solutions for estimates of the genetic distance in the case of the Jukes-Cantor Model and Kimura's two-parameter model. In the case of more complex models, we cannot obtain analytical solutions. In particular, since amino acids have an alphabet size of 20 instead of size 4, a model of amino acid substitutions requires many more parameters, and an analytical solution is not possible (see, for example, [Jones et al., 1992]). In such cases we can estimate the genetic distance between two sequences using maximum likelihood techniques similar to those used in the maximum likelihood method for learning the entire tree. We discuss these techniques next.

11.4.1 Rates Not varying Among Sites

Suppose the nucleotide substitution model is time-reversible. Then the genetic distance between the two sequences can be estimated by assuming that we started with one of the sequences and ended with the other one, and all the substitutions took place on the latter one. For $1 \leq i \leq m$, let x_i be the base at site i in one of the sequences and y_i be the base at site i in the other sequence. Then for a given value D of the genetic distance, we compute the likelihood of D as follows:

$$
\begin{aligned}
L(D) &= \prod_{i=1}^{m} P(x_i, y_i | D) \\
&= \prod_{i=1}^{m} P(x_i) P(y_i | x_i, D) \\
&= \prod_{i=1}^{m} \pi_{x_i} P(y_i | x_i, D).
\end{aligned}
$$

We then search for the maximum likelihood value of D.

Example 11.6 *Assume the Jukes-Cantor Model. Suppose that two homologous sequences have length 100 and they differ on 11 sites. Then our estimate of $p(t)$, the probability that the sequences are different at a given site at time t, is*

$$
\hat{p} = \frac{11}{100} = .11,
$$

and, using the formula in Theorem 10.4, our analytical estimate of D is

$$
\begin{aligned}
\hat{D} &= -\frac{3}{4} \ln \left(1 - \frac{4}{3} \hat{p} \right) \\
&= -\frac{3}{4} \ln \left(1 - \frac{4}{3} (.11) \right) \\
&= .118\,95.
\end{aligned}
$$

Let's compute the likelihood of the analytical estimate. Let μ be the substitution rate for the two sequences. Owing to Equality 10.11,

$$
E_t(D) = 6\mu t, \tag{11.14}
$$

where t is the time since the sequences diverged. Since we are doing our computation as though we started with one sequence and ended with the other, the time in our computation is actually $2t$, and we need to estimate $2\mu t$. Owing to Equality 11.14, our estimate of $2\mu t$ for a given value of D is

$$2\mu t = \frac{D}{3}.$$

Therefore if $D = .11895$, our estimate is

$$2\mu t = \frac{D}{3} = \frac{0.118\,95}{3} = .039\,65.$$

We then have that

$$
\begin{aligned}
L(.11895) &= \prod_{i=1}^{100} \pi_{x_i} P(y_i | x_i, .11895) \\
&= \left(\frac{1}{4}\right)^{100} \left(\frac{1}{4} - \frac{1}{4}e^{-4\times 2\mu t}\right)^{11} \left(\frac{1}{4} + \frac{3}{4}e^{-4\times 2\mu t}\right)^{89} \\
&= \left(\frac{1}{4}\right)^{100} \left(\frac{1}{4} - \frac{1}{4}e^{-4\times .039\,65}\right)^{11} \left(\frac{1}{4} + \frac{3}{4}e^{-4\times .039\,65}\right)^{89} \\
&= 3.\,138\,319\,759 \times 10^{-81}.
\end{aligned}
$$

The following table shows the likelihoods for values of D around .11895:

D	$L(D)$
.11	$3.041079424 \times 10^{-81}$
.11894	$3.138319554 \times 10^{-81}$
.11895	$3.138\,319\,759 \times 10^{-81}$
.11896	$3.138319554 \times 10^{-81}$
.12	$3.137050282 \times 10^{-81}$

We see that this method yields the same estimate as the analytical solution.

Example 11.7 Suppose we use Kimura's two-parameter model. If α and β are the substitute rates for the sequences, then 2α and 2β are the rates if we are doing our computation as if we are starting with one sequence and ending with the other. Owing to Equality 10.14,

$$E_t(D) = (2\alpha + 4\beta)t.$$

Therefore, for a given value of D we estimate that

$$(2\alpha + 4\beta)t = D. \tag{11.15}$$

We have that

$$
\begin{aligned}
L(D) &= \prod_{i=1}^{m} \pi_{x_i} P(y_i | x_i, D) \\
&= \left(\frac{1}{4}\right)^{m} \left(\frac{1}{4} + \frac{1}{4}e^{-4\times 2\beta t} - \frac{1}{2}e^{-2(2\alpha t + 2\beta t)}\right)^{k} \\
&\quad \left(\frac{1}{4} - \frac{1}{4}e^{-4\times 2\beta t}\right)^{l} \left(\frac{1}{4} + \frac{1}{4}e^{-4\times 2\beta t} + \frac{1}{2}e^{-2(2\alpha t + 2\beta t)}\right)^{m-k-l},
\end{aligned}
$$

where k is the number of transitions and l is the number of transversions. To compute this likelihood, we search for the maximum likelihood values of αt and βt satisfying Equality 11.15.

As an alternative to the procedure outlined in the previous example, we could proceed as outlined in Section 11.3.5. That is, we choose a time scale t so that the total substitution rate $r = 1$, which means

$$1 = \alpha + 2\beta,$$

and we set

$$R = \frac{\alpha}{2\beta}.$$

We then need to only search over one parameter R for a given value of D. It is left as an exercise to develop the formula for this method.

We illustrated the maximum likelihood technique for estimating the genetic distance using the Jukes-Cantor Model and Kimura's two-parameter model because of the simplicity of the model. In practice we would not do this because we have analytical solutions for the estimate. We would use the technique when employing a more complex model with more parameters. In this case, we would proceed as shown in Example 11.7.

11.4.2   Rates Varying among Sites

As discussed in Section 11.3.5, if we allow the rates to vary among sites, we place a gamma distribution $\rho(r)$ on the overall rate parameter r such that $\bar{r} = 1$. The time t is the time separating the two sequences, which means it is twice the time since the species diverged. For given values of r and t, we therefore have that the estimate of the evolutionary distance is

$$D = rt.$$

We search over values of t and integrate over values of r. For a given value of t the likelihood is as follows:

$$L(t) = \prod_{i=1}^{m} \int_0^\infty \pi_{x_i} P(y_i|x_i, rt)\rho(r)dr.$$

Example 11.8 *Assume the Jukes-Cantor Model. Suppose two homologous sequences have length 100 and they differ on 11 sites. Suppose further that we have the following gamma density function of r:*

$$\rho(\mu) = \frac{a^a}{\Gamma(a)}\mu^{a-1}e^{-a\mu} \qquad \mu > 0.$$

For a site i at which the sequences are different, we have the following (the following integrals are evaluated in the same way as the one in Theorem 10.8):

$$\int_0^\infty \pi_{x_i} P(y_i|x_i, tr)\rho(r)dr = \frac{1}{4}\int_0^\infty \left(\frac{1}{4} - \frac{1}{4}e^{-4rt}\right)\frac{a^a}{\Gamma(a)}\mu^{a-1}e^{-ar}dr$$

$$= \frac{1}{4}\left[\frac{1}{4} - \frac{1}{4}\left(\frac{a}{a+4t}\right)^a\right].$$

330 CHAPTER 11. MOLECULAR PHYLOGENETICS

For a site i at which the sequences are the same, we have

$$
\begin{aligned}
\int_0^\infty \pi_{x_i} P(y_i|x_i, tr)\rho(r)dr &= .25 \int_0^\infty \left(\frac{1}{4} + \frac{3}{4}e^{-4rt}\right) \frac{a^a}{\Gamma(a)}\mu^{a-1}e^{-ar}dr \\
&= \frac{1}{4}\left[\frac{1}{4} + \frac{3}{4}\left(\frac{a}{a+4t}\right)^a\right].
\end{aligned}
$$

Therefore,

$$
L(t) = \left(\frac{1}{4}\right)^{100}\left[\frac{1}{4} - \frac{1}{4}\left(\frac{a}{a+4t}\right)^a\right]^{11}\left[\frac{1}{4} + \frac{3}{4}\left(\frac{a}{a+4t}\right)^a\right]^{89}.
$$

Once we find the maximum likelihood value $\hat{t}$, our estimate of the genetic distance is

$$
\hat{D} = \bar{r}\hat{t} = \hat{t}.
$$

It is left as an exercise to investigate for a given value of a whether we obtain the same maximum likelihood value as the one obtained using the result in Theorem 10.8.

Ninio et al. [2006] discuss methods for estimating the parameter a in the gamma distribution of r.

EXERCISES

Section 11.2

Exercise 11.1 Suppose we have four OTUs A, B, C, and D and the distances in the following table:

OTU	A	B	C
B	6		
C	5	8	
D	9	4	7

1. Use the UPGMA method to determine a rooted tree.

2. Use the neighbors-relation method to determine an unrooted tree.

Section 11.3

Exercise 11.2 Suppose $v_W = .4$ in the tree in Figure 11.8. Assuming the Jukes-Cantor Model, determine the conditional probability distribution of W given $Y = G$.

Exercise 11.3 Prove Theorem 11.1.

Exercise 11.4 Suppose we have the tree $\mathbb{T}$ in Figure 11.8, and

$$data_i = \{b_i, d_i, e_i, f_i, h_i\} = \{G, A, A, T, C\}.$$

Suppose further we have the following set of values of the parameters P:

$$\{v_B = .3, v_D = .1, v_E = .4, v_F = .2, v_H = .6, v_Z = .7, v_W = .8, v_Y = .5, v_X = .9\}.$$

Following the procedure outlined in Example 11.4, compute the likelihood of the data.

Exercise 11.5 Prove Lemma 11.1.

Exercise 11.6 Suppose we have 30 nodes $S_1, S_2, \ldots$ and S_{30}. Compute the number of different bifurcating unrooted trees that have these nodes as leaves.

Section 11.4

Exercise 11.7 Assume the Jukes-Cantor Model. Suppose two homologous sequences have length 200 and they differ on 30 sites. Using the formula in Theorem 10.4, obtain an analytical estimate of D. As in Example 11.6, compute the likelihood of the analytical estimate and the likelihood of values close to the analytical estimate.

Exercise 11.8 At the end of Example 11.8, it was left as an exercise to investigate for a given value of a whether we obtain the same maximum likelihood value as the one obtained using the result in Theorem 10.8. Investigate this.

Additional Exercises

Exercise 11.9 As mentioned at the end of Section 11.3.7, the Bayesian network approach to learning phylogenetic trees has been implemented as freeware in PHYLIP Version 3.6. You can link to PHYLIP and to other packages for learning phylogenetic trees at

http://evolution.genetics.washington.edu/phylip/software.html.

1. If you have not already done so, complete Exercise 10.16, which concerns the PAX-6 gene. Choose sequences from several species such as the human, the house mouse, the fly *Drosophila*, and so on.

2. Refine the alignment of your sequences using the multiple alignment package ClustalX, which is available at

www.clustal.org/.

3. Using PHYLIP create phylogenetic trees for your sequences using the maximum likelihood methods and several of the other methods discussed in this chapter. Compare your trees.

Hall [2004] presents a simple introduction to creating phylogenetic trees using these packages.

Chapter 12

Analyzing Gene Expression Data

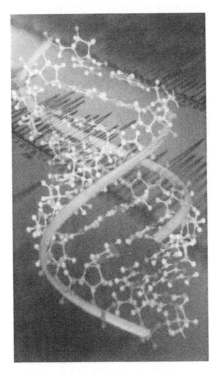

Recall from Section 4.2.2 that the protein transcription factor produced by one gene can have a causal effect on the level of mRNA (called the **gene expression level**) of another gene. In recent years, microarray technology has enabled researchers to measure the expression level of all genes in an organism, thereby providing us with the data to investigate the causal relationships among the genes. Early tools for analyzing microarray data used clustering algorithms (see, e.g., [Spellman et al, 1998]). These algorithms determine groups of genes

333

that have similar expression levels in a given experiment. Thus they determine correlation but tell us nothing of the causal pattern. By modeling gene interaction using a Bayesian network, we can learn something about the causal pattern.

Typically, there are thousands of genes, but we have at most a few thousand data items. In cases like this, there are often many structures that are equally likely. So, choosing one particular structure is somewhat arbitrary. However, as discussed in Section 8.5, when analyzing gene expression data, we are interested in the dependence and causal relationships between the expression levels of certain genes (see [Lander and Shenoy, 1999]), so model averaging seems appropriate. That is, by averaging over the highly probable structures, we may reliably learn there is a dependence between the expression levels of certain genes. We discuss the use of model averaging in Section 12.3. Before that, in Section 12.1 we briefly review how microarray technology enables us to simultaneously measure the expression levels of different genes, and in Section 12.2 we discuss a bootstrap approach to analyzing gene expression data. Section 12.4 presents a module network method for analyzing gene expression data.

12.1   DNA Microarrays

We start with some definitions. An **oligonucleotide** is a short segment of DNA or RNA that is synthesized using nucleotides. Each oligonucleotide represents a gene or family of genes that is capable of binding to the oligonucleotide. A **DNA microarray** is a two-dimensional array in which each array element is a **probe** that is capable of binding to a gene or family of genes. In current DNA microarrays, the probes are ordinarily oligonucleotides. A given DNA microarray can contain thousands of probes. Ordinarily, the set of probes corresponds to a portion of the genome for a given species, or if the genome is small, it could correspond to the entire genome. See www.bio.davidson.edu/Courses/genomics/chip/chip.html to view a microarray corresponding to an entire yeast genome that contains 6116 genes. That site also provides a very nice animation illustrating the use of DNA microarrays. Figure 12.1 (a) depicts a DNA microarray containing 16 probes. An actual DNA microarray would contain many more probes.

Next we explain how we use DNA microarrays to measure gene expression levels. First, we need another definition. **Complementary DNA (cDNA)** is DNA that is synthesized from mRNA using an enzyme called **reverse transcriptase**. cDNA is like the DNA that originally produced the mRNA except that it contains no introns. We first create a DNA microarray containing probes for the genes in which we are interested. We then extract mRNA from the cells we want to investigate. For example, if we are investigating the unicellular yeast *s. cerevisiae*, we would extract mRNA from a collection of these yeast cells. Next we synthesize cDNA from the mRNA and color the cDNA red or green using a fluorescent dye. The dyed cDNA is then applied to every probe on the microarray. A given cDNA sequence corresponding to gene X will bind to the probe for gene X. The more cDNA sequences corresponding to gene X, the

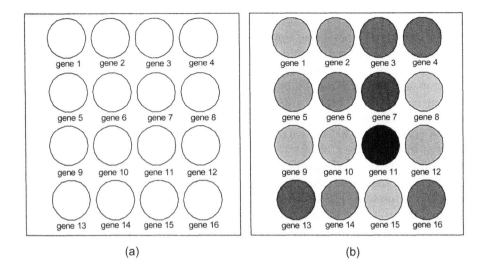

Figure 12.1: A DNA microarray is depicted in (a); the same array after cDNA is applied is depicted in (b).

greater will be the intensity of the fluorophore at the probe for gene X. Thus the intensity at the probe for gene X can be used to measure the expression level of gene X. Figure 12.1 (b) depicts this. For the sake of illustration, let's say that darker spots represent higher intensity. Then gene 11 is the most highly expressed gene. Gene expression level is often set as the ratio of measured expression to a control level. So, values greater than 1 would indicate a relatively high expression level, whereas values less than 1 would indicate a relatively low expression level.

The procedure just described would yield one data item comparing the expression levels of the various genes. We would repeat the procedure many times to obtain a set of data items, which we would then use to learn something about the causal relationships among the genes' expression levels.

We have described single-channel DNA microarray analysis, which is designed to give estimations of the absolute levels of gene expression (our purpose in this chapter). Two-channel DNA microarray analysis is used when we want to compare the expression levels of genes under two different conditions. For example, we might want to compare diseased tissue to healthy tissue. In this case, we would use two dyes, one for the sample obtained from the healthy tissue and one for the sample obtained from the diseased tissue.

12.2   A Bootstrap Approach

Making the causal faithfulness assumption, Friedman et al. [2000] investigated the presence of two types of features in the causal network containing the expression levels of the genes for a given species. See Section 8.5 for a discussion of features. The first type of feature, called a **Markov feature**, asks whether Y is in the Markov boundary (see Section 6.4) of X. Clearly, this relationship is symmetric. This relationship holds if two genes are related in a biological interaction. The second type of feature, called a **path feature**, asks whether X is an ancestor of Y in the causal network containing all variables. If this feature is present, X has a causal influence on Y. Friedman et al. [2000] note that the faithfulness assumption is not necessarily justified in this domain due to the possibility of hidden variables. So, for both the Markov and causal relations, they take their results to be indicative, rather than evidence, that the relationship holds for the genes.

As an alternative to using model averaging to determine the probability that a feature is present, Friedman et al. [2000] used the non-Bayesian bootstrap method to determine the confidence that a feature is present. A detailed discussion of this method appears in [Friedman et al., 1999]. Briefly, the method proceeds like this: We generate perturbed versions of the actual dataset D, learn Bayesian networks from these perturbed versions, and finally average over the learned Bayesian networks. The algorithm is as follows:

for $(i = 1; i <= m; i + +)$ {
    sample with replacement n data items from D;
    let D_i be the resultant dataset;
    learn DAG $\mathbb{G}_i$ from D_i using model selection;
}
for (each feature F)
    $confidence(F) = \frac{1}{m} \sum\limits_{i=1}^{m} P(F = present \mid \mathbb{G}_i);$

In this algorithm, n is the number of data items in D, and m can be assigned any value. In their study, Friedman et al. [2000] set $m = 200$. Furthermore, the conditional probability that a feature F is present is given by

$$P(F = present | \mathbb{G}_i) = \left\{ \begin{array}{ll} 1 & \text{if the feature is present in } \mathbb{G}_i \\ 0 & \text{if the feature is not present in } \mathbb{G}_i. \end{array} \right.$$

Friedman et al. [2000] applied this method to the dataset provided in [Spellman et al, 1998], which contains data on gene expression levels of *s. cerevisiae*. For each case (data item) in the dataset, the variables measured are the expression levels of 800 genes along with the current cell cycle phase. There are 76 cases in the dataset. The cell cycle phase was forced to be a root in all the networks, allowing the modeling of the dependency of expression levels on cell cycle phase.

They performed their analysis by (1) discretizing the data and using Equality 8.5 to compute the probability of the data given candidate DAGs, and by (2) assuming continuously distributed variables and using the Bayesian score for scoring Gaussian Bayesian networks, which appears in [Neapolitan, 2004], to compute the probability of the data given candidate DAGs. They discretized the data into the three categories, *under-expressed*, *normal*, and *over-expressed*, depending on whether the expression rate is respectively significantly lower than, similar to, or greater than control. The results of their analysis contained sensible relations between genes of known function. We show the results of the path feature analysis and Markov feature analysis in turn.

12.2.1 Analysis of Path Features

For a given variable X, they determined a dominance score for X based on the confidence X is an ancestor of Y summed over all other variables Y. That is,

$$d\_score(X) = \sum_{Y:C(X,Y)>t} (C(X,Y))^k,$$

where $C(X,Y)$ is the confidence X is an ancestor of Y, k is a constant rewarding high confidence terms, and t is a threshold discarding low confidence terms. They found that the dominant genes are not sensitive to the values of t and k. The highest-scoring genes appear in Table 12.1. This table shows some interesting results. First the set of high-scoring genes includes genes involved in initiation of the cell cycle and its control. They are CLN1, CLN2, CDC5, and RAD43. The functional relationship of these genes has been established [Cvrckova and Nasmyth, 1993]. Furthermore, the genes MCD1, RFA2, CDC45, RAD53, CDC5, and POL30 have been found to be essential in cell functions [Guacci et al., 1997]. In particular, the genes CDC5 and POL30 are components of prereplication complexes, and the genes RFA2, POL30, and MSH6 are involved in DNA repair. DNA repair is known to be associated with transcription initiation, and DNA areas that are more active in transcription are repaired more frequently [McGregor, 1999].

12.2.2 Analysis of Markov Features

The top-scoring Markov relations in the discrete analysis are shown in Table 12.2. In that table, most pairings involving the genes make sense biologically. Searches using Psi-Blast [Altschul et al., 1997] have substantiated some of this. Several of the pairs are physically close on the chromosome and therefore perhaps regulated by the same mechanism. Overall, there are 19 biologically sensible pairs out of the 20 top-scoring relations.

Comparison to Clustering Friedman et al. [2000] determined conditional independencies that are beyond the capabilities of the clustering method. For example, CLN2, RNR3, SVS1, SRO4, and RAD51 all appear in the same cluster, according to the analysis done by Spellman et al. [1998]. From this, we can conclude only that they are correlated. Friedman et al. [2000] found with high confidence that CLN2 is a parent of the other four and that there are no other causal paths between them. This means that each of the other four is conditionally independent of the remaining three given CLN2. This agrees with biological knowledge because it is known that CLN2 has a central role in cell cycle control, and there is no known biological relationship among the other four.

Gene	Cont. *d_score*	Discrete *d_score*	Comment
MCD1	525	550	Mitotic chromosome determinant
MSH6	508	292	Required for mismatch repair in mitosis
CS12	497	444	Cell wall maintenance, chitin synthesis
CLN2	454	497	Role in cell cycle start
YLR183C	448	551	Contains fork-headed associated domain
RFA2	423	456	Involved in nucleotide excision repair
RSR1	395	352	Involved in bud site selection
CDC45	394	–	Role in chromosome replication initiation
RAD43	383	60	Cell cycle control, checkpoint function
CDC5	353	209	Cell cycle control, needed for mitosis exit
POL30	321	376	Needed for DNA replication and repair
YOX1	291	400	Homeodomain protein
SRO4	239	463	Role in cellular polarization
CLN1	–	324	Role in cell cycle start
YBR089W	–	298	

Table 12.1: Dominant genes in the path feature analysis

Conf.	Gene-1	Gene-2	Comment
1.0	YKL163W-PIR3	YKL164C-PIR1	Close locality on chromosome
.985	PRY2	YKR012C	Close locality on chromosome
.985	MCD1	MSH6	Both bind to DNA during mitosis
.98	PHO11	PHO12	Nearly identical acid phosphatases
.975	HHT1	HTB1	Both are histones
.97	HTB2	HTA1	Both are histones
.94	YNL057W	YNL058C	Close locality on chromosome
.94	YHR143W	CTS1	Both involved in cytokinesis
.92	YOR263C	YOR264W	Close locality on chromosome
.91	YGR086	SIC1	Both involved in nuclear function
.9	FAR1	ASH1	Both part of a mating type switch
.89	CLN2	SVS1	Function of SVS1 unknown
.88	YDR033W	NCE2	Both involved in protein secretion
.86	STE2	MFA2	A mating factor and receptor
.85	HHF1	HHF2	Both are histones
.85	MET10	ECM17	Both are sulfite reductases
.85	CDC9	RAD27	Both involved in fragment processing

Table 12.2: The highest-ranking Markov relations in the discrete analysis

12.3 Model Averaging Approaches

An obvious way to determine our confidence in a feature is to actually compute its probability using model averaging, as discussed in Section 8.5. That is, we compute

$$P(F = present|\mathrm{D}) = \sum_{\mathbb{G}} P(F = present|\mathbb{G})P(\mathbb{G}|\mathrm{D}),$$

where

$$P(F = present|gp) = \begin{cases} 1 & \text{if the feature is present in } gp \\ 0 & \text{if the feature is not present in } gp. \end{cases}$$

We could average over DAGs as shown here or over DAG patterns as shown in Section 8.5. Arguments for choosing one or the other appear in Section 8.6.2. However, since in general the number of DAGs or DAG patterns is prohibitively large, we could not average over all of them. Alternatively, we could approximate the average using MCMC as shown in Section 8.6.2. There are two problems with using this straightforward application of MCMC: (1) The search space is enormous if the number of nodes is not small; and (2) the posterior probability distribution of the structures is often quite peaked, with structures close to a peak having much lower scores than the scores at the peaks. Thus it is hard to escape from a local maximum, which means that the MCMC is slow to mix. Ellis and Wong [2008] illustrate this concept with some simulations.

Friedman and Koller [2003] therefore took a different approach. Namely, they used MCMC to approximate averaging over orderings of the nodes instead of over all DAGS. There are only $n!$ orderings, which is much less than the number of DAGs. The following example illustrates this idea. We denote an ordering by $\prec$, and if X_i precedes X_j in the ordering, we write

$$X_i \prec X_j.$$

Example 12.1 *Suppose we have three nodes X_1, X_2, and X_3. Then there are 25 possible DAGs, but the only orderings are as follows:*

$$[X_1, X_2, X_3]$$

$$[X_1, X_3, X_2]$$

$$[X_2, X_1, X_3]$$

$$[X_2, X_3, X_1]$$

$$[X_3, X_1, X_2]$$

$$[X_3, X_2, X_1].$$

In the ordering $[X_2, X_3, X_1]$, for example,

$$X_2 \prec X_3 \prec X_1.$$

Definition 12.1 *A DAG $\mathbb{G}$ is consistent with an ordering $\prec$ if for each node X, all parents of X come before X in the ordering.*

Example 12.2 *Consider the ordering $[X_1, X_2, X_3]$. The DAGs*

$$X_1 \to X_2 \to X_3 \text{ and } X_2 \leftarrow X_1 \to X_3$$

are consistent with the ordering. The DAG

$$X_1 \leftarrow X_2 \to X_3$$

is not consistent with it because X_1's parent X_2 comes after X_1 in the ordering.

Next we present the approach that averages over orderings, and we show experimental results comparing its performance to the bootstrap method discussed in the previous section.

12.3.1   The Probability of a Feature Given an Ordering

If we compute the probability that a feature is present by averaging over all orderings of the nodes, we have

$$P(F = present|\mathsf{D}) = \sum_{\prec} P(F = present|\prec, \mathsf{D})P(\prec|\mathsf{D}).$$

To estimate this sum, we need to compute $P(F = present|\prec, \mathsf{D})$. To that end, we have that

$$P(F = present|\prec, \mathsf{D}) = \frac{P(F = present, \mathsf{D}|\prec)}{P(\mathsf{D}|\prec)}.$$

First we discuss computing the denominator.

Computing $P(\mathsf{D}|\prec)$

We have that

$$P(\mathsf{D}|\prec) = \sum_{\mathbb{G} \in \mathsf{G}_{\prec}} P(\mathsf{D}|\mathbb{G})P(\mathbb{G}|\prec),$$

where $\mathsf{G}_{\prec}$ is the set of all DAGs consistent with ordering $\prec$. We will average over orderings, which means that our prior probability distribution will be based on orderings rather than DAGs. Since all orderings have the same number of consistent DAGs, we can therefore define

$$c = P(\mathbb{G}|\prec),$$

where $\mathbb{G}$ is a DAG consistent with $\prec$, to be a constant which does not depend on $\mathbb{G}$ or $\prec$. We then have that

$$P(\mathsf{D}|\prec) = c \sum_{\mathbb{G} \in \mathsf{G}_{\prec}} P(\mathsf{D}|\mathbb{G}). \tag{12.1}$$

Recall from Section 8.6.1 that the Bayesian score, in the case of multinomial variables, which is

$$P(\mathsf{D}|\mathbb{G}) = \prod_{i=1}^{n} \prod_{j=1}^{q_i^{\mathbb{G}}} \frac{\Gamma(N_{ij}^{\mathbb{G}})}{\Gamma(N_{ij}^{\mathbb{G}} + M_{ij}^{\mathbb{G}})} \prod_{k=1}^{r_i} \frac{\Gamma(a_{ijk}^{\mathbb{G}} + s_{ijk}^{\mathbb{G}})}{\Gamma(a_{ijk}^{\mathbb{G}})},$$

is the product of the following local scores:

$$score(X_i, \mathsf{PA}_i^{\mathbb{G}} : \mathsf{D}) = \prod_{j=1}^{q_i^{\mathbb{G}}} \frac{\Gamma(N_{ij}^{\mathbb{G}})}{\Gamma(N_{ij}^{\mathbb{G}} + M_{ij}^{\mathbb{G}})} \prod_{k=1}^{r_i} \frac{\Gamma(a_{ijk}^{\mathbb{G}} + s_{ijk}^{\mathbb{G}})}{\Gamma(a_{ijk}^{\mathbb{G}})}.$$

Next let the possible parent sets of a node X_i given an ordering $\prec$ be denoted as follows:

$$\mathsf{U}_{i,\prec} = \{\mathsf{U} \text{ such that every node in } \mathsf{U} \text{ comes before } X_i \text{ in } \prec\}.$$

Example 12.3 *Suppose* $\prec = [X_2, X_3, X_1, X_4]$. *Then*

$$\mathsf{U}_{1,\prec} = \{\{X_2\}, \{X_3\}, \{X_2, X_3\}\}.$$

The following theorem obtains the probability of data given an ordering.

Theorem 12.1 *Suppose we have an ordering $\prec$ of a set of random variables* V *and data* D*, where* D *is a set of data items such that each data item is a vector of values of all variables in* V*. Then if we assume the Bayesian score for multinomial variables, and we assume for every DAG $\mathbb{G} = (\mathsf{V}, \mathsf{E}_{\mathbb{G}})$ consistent with $\prec$ that*

$$c = P(\mathbb{G}| \prec),$$

we have that

$$P(\mathsf{D}|\prec) = c \prod_{i=1}^{n} \sum_{\mathsf{U} \in \mathsf{U}_{i,\prec}} score(X_i, \mathsf{U}|\mathsf{D}).$$

Proof. *Owing to Equality 12.1, we have that*

$$
\begin{aligned}
P(\mathsf{D}|\prec) &= c \sum_{\mathbb{G} \in \mathbf{G}_{\prec}} P(\mathsf{D}|\mathbb{G}) \\
&= c \sum_{\mathbb{G} \in \mathbf{G}_{\prec}} \prod_{i=1}^{n} score(X_i, \mathsf{PA}_i^{\mathbb{G}} : \mathsf{D}) \\
&= c \prod_{i=1}^{n} \sum_{\mathsf{U} \in \mathsf{U}_{i,\prec}} score(X_i, \mathsf{U} : \mathsf{D}).
\end{aligned}
$$

This completes the proof. ∎

The transformation from summing over DAGs to summing over parent sets enables us to compute $P(\mathsf{D}|\prec)$ efficiently. Ordinarily, we put a bound b on the number of parents a node could have. We do this because Bayesian networks are usually reasonably sparse. Assuming this bound, the number of possible parent sets that a node could have is

$$\binom{n}{b} \leq n^b.$$

Therefore, the total number of scores we must compute to calculate $P(\mathsf{D}|\prec)$ using the result in Theorem 12.1 is bounded above by $n \times n^b = n^{b+1}$.

We obtained Theorem 12.1 using the Bayesian score for multinomial variables. It holds for any scoring criterion that can be decomposed to local scores involving each node and its parents. For example, it also holds for the Bayesian score for scoring Gaussian Bayesian networks, which can be found in [Neapolitan, 2004].

Computing Probabilities of Features

Next we show how to compute the conditional probabilities of several important features. Recall that

$$P(F = present|\prec, \mathsf{D}) = \frac{P(F = present, \mathsf{D}|\prec)}{P(\mathsf{D}|\prec)}.$$

So, we need to compute the numerator in the expression on the right for each of the features.

1. Compute the probability that X_j has a specific set of parents U. It is left as an exercise to show that Theorem 12.1 implies that

$$P(\mathsf{PA}_j^{\mathbb{G}}) = \mathsf{U}, \mathsf{D}|\prec) = score(X_j, \mathsf{U}|\mathsf{D}) \times c \prod_{i \neq j} \sum_{\mathsf{U} \in \mathsf{U}_{i,\prec}} score(X_i, \mathsf{U} : \mathsf{D}).$$

Therefore,

$$
\begin{aligned}
P(\mathsf{PA}_j^{\mathbb{G}}) = \mathsf{U}|\prec) &= \frac{P(\mathsf{PA}_j^{\mathbb{G}}) = \mathsf{U}, \mathsf{D}|\prec)}{P(\mathsf{D}|\prec)} \\
&= \frac{score(X_j, \mathsf{U}|\mathsf{D}) \times c \prod_{i \neq j} \sum_{\mathsf{U} \in \mathsf{U}_{j,\prec}} score(X_j, \mathsf{U} : \mathsf{D})}{c \prod_{i=1}^{n} \sum_{\mathsf{U} \in \mathsf{U}_{i,\prec}} score(X_i, \mathsf{U} : \mathsf{D})} \\
&= \frac{score(X_j, \mathsf{U}|\mathsf{D})}{\sum_{\mathsf{U} \in \mathsf{U}_{j,\prec}} score(X_j, \mathsf{U} : \mathsf{D})}.
\end{aligned}
$$

2. Compute the probability that a specific edge $X_k \rightarrow X_j$ is present. It is left as an exercise to show that

$$P(X_k \rightarrow X_j, \mathsf{D}|\prec)$$

$$= \sum_{\mathsf{U} \in \mathsf{U}_{j,\prec} \text{ and } X_k \in \mathsf{U}} score(X_j, \mathsf{U}:\mathsf{D}) \times c \prod_{i \neq j} \sum_{\mathsf{U} \in \mathsf{U}_{j,\prec}} score(X_j, \mathsf{U}:\mathsf{D}).$$

Therefore, in the same way as shown in (1),

$$P(X_k \rightarrow X_j|\prec, \mathsf{D}) = \frac{P(X_k \rightarrow X_j, \mathsf{D}|\prec)}{P(\mathsf{D}|\prec)}$$

$$= \frac{\displaystyle\sum_{\mathsf{U} \in \mathsf{U}_{j,\prec} \,:\, X_k \in \mathsf{U}} score(X_j, \mathsf{U}:\mathsf{D})}{\displaystyle\sum_{\mathsf{U} \in \mathsf{U}_{j,\prec}} score(X_j, \mathsf{U}:\mathsf{D})}.$$

3. Compute the probability that X_k is in the Markov boundary of X_j. Recall that the Markov boundary includes all parents of X_j, all children of X_j, and all other parents of children of X_j. Since the Markov boundary relationship is symmetric, without loss of generalization we can assume that $X_k \prec X_j$. Then we need concern ourselves only with whether X_k is a parent of X_j and whether X_k is a parent of a child of X_j. Now the probability that X_k is not a parent of X_j is equal to

$$1 - P(X_k \rightarrow X_j|\prec, \mathsf{D}),$$

which we can compute using (2). Furthermore, the probability that X_k and X_j are not both parents of X_m is equal to

$$1 - \sum_{\mathsf{U}:X_k,X_j \in \mathsf{U}} P(\mathsf{PA}_m^{\mathbb{G}} = \mathsf{U}|\prec, \mathsf{D}).$$

The probabilities in this expression can be computed using (1). Therefore, the probability that X_k is not in the Markov boundary of X_j is given by

$$(1 - P(X_k \rightarrow X_j|\prec, \mathsf{D})) \prod_{X_j \prec X_m} \left(1 - \sum_{\mathsf{U}:X_k,X_j \in \mathsf{U}} P(\mathsf{PA}_m^{\mathbb{G}} = \mathsf{U}|\prec, \mathsf{D}) \right).$$

We can multiply probabilities as we just did because the event that a given node has a particular parent set is independent of a different node having a particular parent set. We therefore finally have that the probability that X_k is in the Markov boundary of X_j is given by

$$1 - (1 - P(X_k \rightarrow X_j|\prec, \mathsf{D})) \prod_{X_j \prec X_m} \left(1 - \sum_{\mathsf{U}:X_k,X_j \in \mathsf{U}} P(\mathsf{PA}_m^{\mathbb{G}} = \mathsf{U}|\prec, \mathsf{D}) \right).$$

4. Compute the probability that there is a directed path from X_k to X_j. There is no formula for this probability. However, we can estimate the probability by sampling entire networks based on the posterior distribution of the networks given the data and the order. To obtain one sample item, we sample a parent set for each node using the posterior distribution in (1).

12.3.2 MCMC over Orderings

Next we discuss using MCMC to approximate averaging over orderings.

The Algorithm

We assume a uniform prior probability distribution $P(\prec)$ over all orderings $\prec$. Therefore, as discussed in Section 8.6.2, we can use $P(\mathsf{D}|\prec)$ as our stationary distribution instead of $P(\prec|\mathsf{D})$.

For each ordering $\prec_i$, we say $\prec_j$ is in the Nbhd($\prec_i$) if $\prec_j$ can be obtained from $\prec_i$ by one of the following operations:

1. Flip two nodes:

$$[X_1, \ldots, X_k, \ldots, X_m, \ldots, X_n] \Rightarrow [X_1, \ldots, X_m, \ldots, X_k, \ldots, X_n].$$

   Cut the deck:

$$[X_1, \ldots, X_k, X_{k+1}, \ldots, X_n] \Rightarrow [X_{k+1}, \ldots, X_n, X_1, \ldots X_k].$$

Either of these sets of operations is complete for the search space. That is, for any two orderings $\prec_i$ and $\prec_j$, there exists a sequence of operations that transforms $\prec_i$ to $\prec_j$. However, the flip operation is more efficient to compute (which we discuss in the next subsection), but it takes smaller steps in the search space. So, at each trial, we choose a flip operation with probability p and a cut operation with probability $1 - p$. We can heuristically search for a good value of p.

Let $N_1(\prec_i)$ be the number of orderings that differ from $\prec_i$ by a flip and $N_2(\prec_i)$ be the number of orderings that differ from $\prec_i$ by a cut. Similar to what we did in Section 8.6.2, we let our transition matrix $\mathbf{Q}$ be as follows: For each pair of states $\prec_i$ and $\prec_j$

$$q_{ij} = \begin{cases} \dfrac{p}{N_1(\prec_i)} & \text{if } \prec_j \text{ differs } \prec_i \text{ from by a flip} \\[2ex] \dfrac{1-p}{N_2(\prec_i)} & \text{if } \prec_j \text{ differs } \prec_i \text{ from by a cut} \\[2ex] 0 & \mathbb{G}_j \notin \mathsf{Nbhd}(\mathbb{G}_i) \end{cases}$$

The steps in our application of the MCMC method are then as follows:

1. If the ordering at the kth trial is $\prec_i$, choose a new ordering according to the probability distribution in the ith row of $\mathbf{Q}$. Suppose that ordering is $\prec_j$.

2. Choose the ordering for the $(k + 1)$st trial to be $\prec_j$ with probability

$$\alpha_{ij} = \min\left(1, \frac{P(\mathsf{D}|\prec_j)q_{ji}}{P(\mathsf{D}|\prec_i)q_{ij}}\right),$$

   and to be $\prec_i$ with probability $1 - \alpha_{ij}$.

Computing Likelihoods of Candidate Orderings

When determining the probability of whether to choose a candidate ordering $\prec_j$, we need to compute $P(\mathsf{D}|\prec_j)$ using Theorem 12.1. We show the result in that theorem again:

$$P(\mathsf{D}|\prec) \approx \prod_{i=1}^{n} \sum_{\mathsf{U} \in \mathsf{U}_{i,\prec}} score(X_i, \mathsf{U} : \mathsf{D}).$$

When $\prec_j$ differs from our previous ordering $\prec_i$ by a flip there is no need to compute all terms in this formula since many of them have the same value for $\prec_i$ as they do for $\prec_i$. We show this next. Consider the following flip:

$$\prec_i = [X_1, \ldots, X_k, \ldots, X_m, \ldots, X_n] \Rightarrow \prec_j = [X_1, \ldots, X_m, \ldots, X_k, \ldots, X_n].$$

For all nodes X_r such that $X_r \prec_i X_k$ the set $\mathsf{U}_{r,\prec}$ is unchanged when we go from $\prec_i$ to $\prec_j$. The same is true for all nodes X_r such that $X_m \prec_i X_r$. So, the total score for these nodes remain unchanged. For any node X_r such that $X_k \prec_i X_r \prec_i X_m$, we have additional parent sets that include X_m, but we lose parent sets that include X_k. So, we must add the following term to the total score of each such node X_r:

$$\sum_{\mathsf{U} \in \mathsf{U}_{r,\prec_j} \,:\, X_m \in \mathsf{U}} score(X_r, \mathsf{U}|\mathsf{D}) - \sum_{\mathsf{U} \in \mathsf{U}_{r,\prec_i} \,:\, X_k \in \mathsf{U}} score(X_r, \mathsf{U} : \mathsf{D}).$$

Similarly, for node X_k we must add this term to its total score:

$$\sum_{\mathsf{U} \in \mathsf{U}_{k,\prec_j} \,:\, \text{at least one of } X_{k+1}, X_{k+2}, \ldots, X_m \in \mathsf{U}} score(X_k, \mathsf{U} : \mathsf{D}),$$

and for node X_m we must subtract this term from its total score:

$$\sum_{\mathsf{U} \in \mathsf{U}_{m,\prec_i} \,:\, \text{at least one of } X_k, X_{k+1}, \ldots, X_{m-1} \in \mathsf{U}} score(X_m, \mathsf{U} : \mathsf{D}).$$

On the other hand, for an ordering $\prec_j$ that differs from $\prec_i$ by a cut, we must recompute the entire score.

12.3.3 Experimental Results

Friedman and Koller [2003] compared their order-MCMC method for analyzing gene expression data to the bootstrap method [Friedman et al., 2000] discussed in Section 12.2. They used the same dataset concerning the gene expression levels of *s. cerevisiae* as was used in the studies in [Friedman et al., 2000] and was discussed in Section 12.2. They discretized the data into the three categories *under-expressed*, *normal*, and *over-expressed*, depending on whether the expression rate is respectively significantly lower than, similar to, or greater than control. They used the Bayesian score for multinomial variables to score the orderings in the case of their order-MCMC method and to score the DAGs in the case of the bootstrap method.

Their comparison proceeded as follows: Given a threshold $t \in [0.1]$, we say a feature F is present if $P(F = present|\text{d}) > t$ and otherwise we say it is absent. If a method says a feature is present when it absent, we call that a **false positive error**, whereas if a method says a feature is absent when it is present, we call that a **false negative error**. Clearly, as t increases, the number of false negative errors increases, whereas the number of false positive errors decreases. So, there is a trade-off between the two types of errors.

Friedman and Koller [2003] first used Bayesian model selection to learn a DAG $\mathbb{G}$ from the dataset provided in [Spellman et al, 1998]. They then used the order-MCMC method and the bootstrap method to learn the Markov features from $\mathbb{G}$. Using the presence of a feature in $\mathbb{G}$ as the gold standard, they determined the false positive and false negative rates for both methods for various values of t. Finally, for both methods they plotted the false negative rates versus the false positive rates. For each method, each value of t determined a point on its graph.[1] They used the same procedure to learn the path features from $\mathbb{G}$. The results appear in Figure 12.2. We see that in both cases of Markov features and path features, the graph for the order-MCMC method was significantly below the graph of the bootstrap method, indicating that the order-MCMC method makes fewer errors.

Friedman and Koller [2003] caution that their learned DAG is probably much simpler than the DAG in the underlying structure because it was learned from a small dataset relative to the number of genes. Nevertheless, their results are indicative of the fact that the order-MCMC method is more reliable in this domain.

12.3.4   Using a Different Prior Probability Distribution

Friedman and Koller [2003] place a uniform prior probability distribution on the orderings, which of course does not place a uniform prior probability distribution on the DAGs. As discussed in Section 8.6.2, there are reasonable arguments for assigning equal prior probabilities to all DAGs, and there are reasonable arguments for assigning equal probabilities to all DAG patterns. Similarly, we could argue that it is reasonable to assign equal probabilities to all orderings, because often a DAG that is consistent with more orderings entails fewer assumptions about time ordering and therefore should be considered more probable *a priori*. However, this argument does not seem as cogent. Consider the following two DAGs:

$$X_1 \leftarrow X_2 \leftarrow \cdots \leftarrow X_n \leftarrow Z \rightarrow Y_1 \rightarrow Y_2 \rightarrow \cdots \rightarrow Y_n$$

$$X_1 \rightarrow X_2 \rightarrow \cdots \rightarrow X_n \rightarrow Z \rightarrow Y_1 \rightarrow Y_2 \rightarrow \cdots Y_n.$$

The first DAG is consistent with an exponential number of orderings; the second is consistent with only one ordering. Yet the two DAGs are Markov equivalent

---
[1] You might notice that this graph is similar to a ROC curve [Fawcett, 2006].

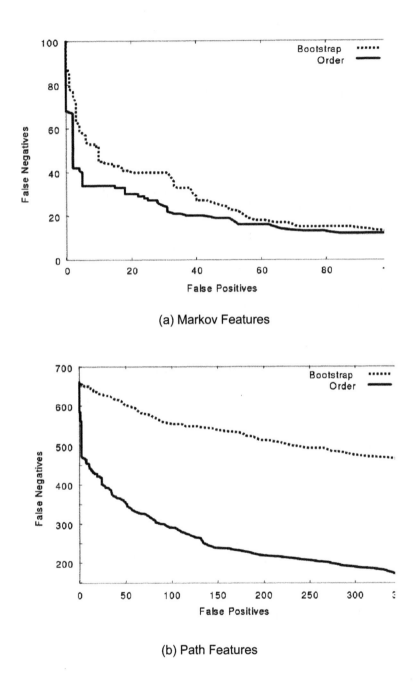

(a) Markov Features

(b) Path Features

Figure 12.2: Comparisons of the performance of the bootstrap method to the order-MCMC method.

and entail the same number of causal relationships.[2]

To address this difficulty, Ellis and Wong [2008] develop a computational adjustment that enables us to use the order-MCMC method, but with any prior distribution on the DAGs that satisfies structural modularity. A prior probability distribution on the DAGs containing n nodes $X_i, X_2, \ldots X_n$ satisfies **structural modularity** if for each DAG $\mathbb{G}$ we can represent $P(\mathbb{G})$ as the product of terms such that each term is a function of a node and the node's parents; that is,

$$P(\mathbb{G}) = \prod_{i-1}^{n} f(X_i, \mathsf{PA}_i^{\mathbb{G}}).$$

Clearly, the uniform probability distribution on the DAGs satisfies structural modularity. That is, for every node X_i and every set of nodes U not including X_i, we simply set

$$f(X_i, \mathsf{U}) = \frac{1}{N^{\frac{1}{n}}},$$

where N is the number of DAGs containing n nodes. Using this computational adjustment, we can take advantage of the speed of the order-MCMC method while choosing a prior probability distribution that might better represent our beliefs than a uniform distribution of orderings.

12.4   Module Network Approach

Many genes are organized into modules such that genes in the same module participate in the same biological process and are coregulated by the same set of other genes. If we develop a module network, such that each module contains a set of variables and variables in the same module have the same probabilistic relationships to their parents, we can represent the causal relationships among the genes more succinctly than we can with an ordinary Bayesian network. Since our models are more succinct, our learning algorithm will need to search a smaller space. Segal et al. [2005] took this approach.

First we define module networks, discuss their properties, and show how they can be used to analyze gene expression data. Then we show results of experiments in which module networks were used to analyze gene expression data.

12.4.1   Module Networks

Figure 12.3 depicts a module network consisting of modules whose random variables represent gene expression levels. The variables $gene_1$ and $gene_2$ have the same prior probability distribution. The variables $gene_3$ and $gene_4$ have the same parents, namely $gene_1$ and $gene_2$, and have the same conditional probability distribution given each instantiation of those parents. The same is true for the variables $gene_5$, $gene_6$, and $gene_7$ relative to their parent $gene_2$.

---

[2]I thank Daphne Koller for this example.

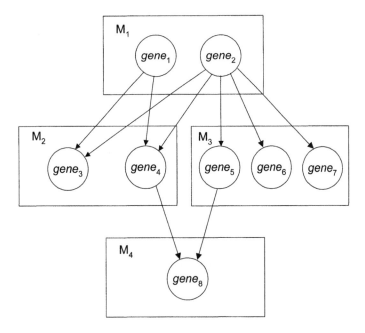

Figure 12.3: The structure of a module network.

Finally, the variable *gene*$_8$ is in a module by itself, and its parents are *gene*$_4$ and *gene*$_5$.

A module network is simply a Bayesian network with some special properties, as we see in the following definition.

Definition 12.2 *Suppose we have a Bayesian network* $(\mathbb{G}, P)$ *where* $\mathbb{G} = (\mathsf{V}, \mathsf{E})$, *and* $\mathsf{V} = \{X_1, X_2, \ldots X_n\}$ *with the following properties:*

1. *There is a set of modules* $\mathsf{C} = \{\mathsf{M}_1, \mathsf{M}_2, \ldots \mathsf{M}_m\}$, *such that each* $\mathsf{M}_i \subset \mathsf{V}$, *which partitions* V *into mutually exclusive and exhaustive subsets. That is,*

$$\bigcup_{i=1}^{m} \mathsf{M}_i = \mathsf{V}$$
$$\mathsf{M}_i \bigcap \mathsf{M}_j = \varnothing \qquad for\ i \neq j.$$

2. *If* X_h *and* X_l *are both in* M_j, *then*

   (a) X_h *and* X_l *have the same space. That is, they have the same set of possible values. Recall that the space of a random variables is its set of possible values.*

   (b) X_h *and* X_l *have the same set of parents* PA_j *in* $\mathbb{G}$ *and the same conditional probability distributions given each set of values of these* *parents. That is, for all values of x and* pa,

$$P(X_h = x | \mathsf{PA}_j = \mathsf{pa}) = P(X_l = x | \mathsf{PA}_j = \mathsf{pa}).$$

Such a Bayesian network is called a **module network***.*

Notice that a module network determines a graph $\mathbb{M} = (\mathsf{C}, \mathsf{F})$, where C is the set of nodes in the graph and F is the set of edges, such that the edge $\mathsf{M}_i \rightarrow \mathsf{M}_j$ is in F if and only if there is a variable $X_h \in \mathsf{M}_i$ and $X_h \in \mathsf{PA}_j$.

Recall from Section 5.2.2 that ordinarily a Bayesian network is constructed by specifying the random variables, DAGs, and conditional probability distributions. The same is true for a module network. We start with a set V of random variables. Then we specify a set of modules C such that each module $\mathsf{M}_j \in \mathsf{C}$ has a space similar to that of a random variable. Note that at this point we have not yet assigned variables to modules. So, these specifications are really module templates. Next, we specify for each module M_j a parent set PA_j that contains the parents of variables in M_j. Let

$$\mathsf{PA} = \{\mathsf{PA}_1, \mathsf{PA}_2, \ldots \mathsf{PA}_m\}.$$

Next we let θ_j be a set of conditional probability distribution templates needed to specify the conditional probability distributions of the variables in M_j, and we set

$$\boldsymbol{\theta} = \{\theta_1, \theta_2, \ldots \theta_m\}.$$

We call $\boldsymbol{\theta}$ the parameter set. Finally, we define a function A that assigns each variable in V to a module such that X_h is assigned to module M_j only if X_h and M_j have the same space. A module network is then fully specified, and we denote it as $(\mathsf{V}, \mathsf{C}, \mathsf{PA}, \boldsymbol{\theta}, A)$.

Example 12.4 *Suppose we have a set of random variables*

$$\mathsf{V} = \{X_1, X_2, X_3 X_4, X_5\},$$

and

$$space(X_1) = space(X_2) = space(X_3) = space(X_4) = \{1, 2\}$$

$$space(X_5) = \{1, 2, 3\}.$$

Next we specify a module network for these random variables.

   1. Specify C. The modules are as follows:

$$\mathsf{M}_1 \text{ has space } \{1, 2\}$$
$$\mathsf{M}_2 \text{ has space } \{1, 2\}$$
$$\mathsf{M}_3 \text{ has space } \{1, 2, 3\}.$$

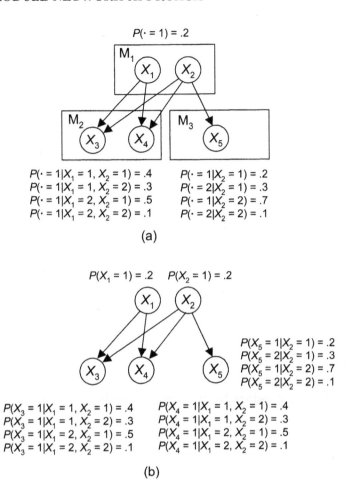

$P(\cdot = 1) = .2$

M_1

X_1 X_2

M_2

X_3 X_4

M_3

X_5

$P(\cdot = 1 | X_1 = 1, X_2 = 1) = .4$ $P(\cdot = 1 | X_2 = 1) = .2$
$P(\cdot = 1 | X_1 = 1, X_2 = 2) = .3$ $P(\cdot = 2 | X_2 = 1) = .3$
$P(\cdot = 1 | X_1 = 2, X_2 = 1) = .5$ $P(\cdot = 1 | X_2 = 2) = .7$
$P(\cdot = 1 | X_1 = 2, X_2 = 2) = .1$ $P(\cdot = 2 | X_2 = 2) = .1$

(a)

$P(X_1 = 1) = .2$ $P(X_2 = 1) = .2$

X_1 X_2

$P(X_5 = 1 | X_2 = 1) = .2$
$P(X_5 = 2 | X_2 = 1) = .3$
$P(X_5 = 1 | X_2 = 2) = .7$
$P(X_5 = 2 | X_2 = 2) = .1$

X_3 X_4 X_5

$P(X_3 = 1 | X_1 = 1, X_2 = 1) = .4$ $P(X_4 = 1 | X_1 = 1, X_2 = 1) = .4$
$P(X_3 = 1 | X_1 = 1, X_2 = 2) = .3$ $P(X_4 = 1 | X_1 = 1, X_2 = 2) = .3$
$P(X_3 = 1 | X_1 = 2, X_2 = 1) = .5$ $P(X_4 = 1 | X_1 = 2, X_2 = 1) = .5$
$P(X_3 = 1 | X_1 = 2, X_2 = 2) = .1$ $P(X_4 = 1 | X_1 = 2, X_2 = 2) = .1$

(b)

Figure 12.4: The specifications for a module network are in (a); the resultant Bayesian network is in (b).

2. *Specify* PA. *Our parent sets are as follows:*

$$
\begin{aligned}
PA_1 &= \varnothing \\
PA_2 &= \{X_1 X_2\} \\
PA_3 &= \{X_2\}.
\end{aligned}
$$

3. *Specify* $\boldsymbol{\theta}$. *Our parameters are as follows:*

$$\theta_1 = \{P(\cdot = 1) = .2\}$$

$$\theta_2 \;=\; \{P(\cdot = 1|X_1 = 1, X_2 = 1) = .4$$
$$P(\cdot = 1|X_1 = 1, X_2 = 2) = .3$$
$$P(\cdot = 1|X_1 = 2, X_2 = 1) = .5$$
$$P(\cdot = 1|X_1 = 2, X_2 = 2) = .1\}$$

$$\theta_3 \;=\; \{P(\cdot = 1|X_2 = 1) = .2$$
$$P(\cdot = 2|X_2 = 1) = .3$$
$$P(\cdot = 1|X_2 = 2) = .7$$
$$P(\cdot = 2|X_3 = 2) = .1\}.$$

The notation "·" stands for a variable assigned to the particular module.

4. *Specify A.*

$$A(X_1) \;=\; A(X_2) = \; \mathsf{M}_1$$
$$A(X_3) \;=\; A(X_4) = \; \mathsf{M}_2$$
$$A(X_5) = \; \mathsf{M}_3.$$

The specifications for the module network appear in Figure 12.4 (a), whereas the resultant Bayesian network appears in Figure 12.4 in (b).

12.4.2   The Bayesian Score for Module Networks

Our goal is to learn from data the structure of a module network. Since a module network is a Bayesian network, you might suspect that the Bayesian score of the structure of a module network is simply the Bayesian score of the DAG that the structure represents. However, this is not the case. The reason is that the parameters for variables in the same module are not independent. That is, we specify our prior belief concerning the conditional probability distributions of all variables in a given module using the same parameters, and we update these parameters using the data on all variables in the module. Given this, the following theorem obtains the score of the data, given a module network structure.

Theorem 12.2 *Suppose we have a module network structure* $\mathbb{S} = (\mathsf{V}, \mathsf{C}, \mathsf{PA}, A)$, *where the variables in* V *are multinomial, we assume exchangeability, and we use a Dirichlet distribution to represent our prior belief for each conditional probability distribution in the network. Suppose further that we have data* D *consisting of a set of data items such that each data item is a vector of values of all the variables in* V. *Then*

$$score(\mathbb{S} : \mathsf{D}) \equiv P(\mathsf{D}|\mathbb{S}) = \prod_{i=1}^{m}\prod_{j=1}^{q_i^{\mathbb{S}}} \frac{\Gamma(N_{ij}^{\mathbb{S}})}{\Gamma(N_{ij}^{\mathbb{S}} + M_{ij}^{\mathbb{S}})} \prod_{k=1}^{r_i} \frac{\Gamma(a_{ijk}^{\mathbb{S}} + s_{ijk}^{\mathbb{S}})}{\Gamma(a_{ijk}^{\mathbb{S}})},$$

where

1. *m is the number of modules.*

2. q_i^S *is the number of different instantiations of the parents of variables in module* M_i.

3. a_{ijk}^S *is our ascertained prior belief concerning the number of times a variable in module* M_i *took its kth value when the parents of that variable had their jth instantiation.*

4. s_{ijk}^S *is the sum, over all variables* X_h *in module* M_i, *of the number of times* X_h *took its kth value when the parents of* X_h *had their jth instantiation.*

5.

$$N_{ij} = \sum_k a_{ijk}$$

$$M_{ij} = \sum_k s_{ijk}.$$

Proof. *The proof can be found in [Segal et al., 2005].* ∎

Like the Bayesian score for multinomial variables, the score developed in Theorem 12.2 is the product of local scores. The local score for module M_i is

$$score(M_i, PA_i^S : D) = \prod_{j=1}^{q_i^S} \frac{\Gamma(N_{ij}^S)}{\Gamma(N_{ij}^S + M_{ij}^S)} \prod_{k=1}^{r_i} \frac{\Gamma(a_{ijk}^S + s_{ijk}^S)}{\Gamma(a_{ijk}^S)}. \tag{12.2}$$

Theorem 12.2 also holds for the Bayesian score for scoring Gaussian Bayesian networks, which is discussed in [Neapolitan, 2004].

Example 12.5 *Suppose we have the module network structure and prior beliefs in Figure 12.5, and we obtain the data D in the following table:*

Case	X_1	X_2	X_3	X_4	X_5
1	1	1	1	1	1
2	1	1	1	2	1
3	1	1	2	1	2
4	1	1	2	1	3
5	1	1	2	2	3
6	1	2	1	1	1
7	1	2	1	2	2
8	1	2	2	1	3
9	2	1	1	1	1
10	2	1	1	2	1
11	2	1	2	1	1
12	2	1	2	2	3
13	2	2	2	1	2
14	2	2	2	1	3
15	2	2	2	2	3

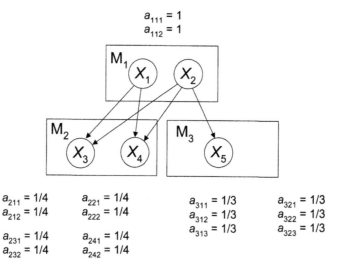

Figure 12.5: A module network structure with prior beliefs about the conditional probability distributions.

Then, owing to Theorem 12.2,

$$
\begin{aligned}
P(\mathsf{D}|\mathbb{S}) &= \prod_{i=1}^{3}\prod_{j=1}^{q_i^{\mathbb{S}}} \frac{\Gamma(N_{ij}^{\mathbb{S}})}{\Gamma(N_{ij}^{\mathbb{S}}+M_{ij}^{\mathbb{S}})} \prod_{k=1}^{r_i} \frac{\Gamma(a_{ijk}^{\mathbb{S}}+s_{ijk}^{\mathbb{S}})}{\Gamma(a_{ijk}^{\mathbb{S}})} \\[2mm]
&= \frac{\Gamma(2)}{\Gamma(2+30)} \frac{\Gamma(1+17)}{\Gamma(1)} \frac{\Gamma(1+13)}{\Gamma(1)} \times \\[2mm]
&\quad \frac{\Gamma(1)}{\Gamma(1+10)} \frac{\Gamma(1/4+5)}{\Gamma(1/4)} \frac{\Gamma(1/4+5)}{\Gamma(1/4)} \times \frac{\Gamma(1)}{\Gamma(1+6)} \frac{\Gamma(1/4+4)}{\Gamma(1/4)} \frac{\Gamma(1/4+2)}{\Gamma(1/4)} \times \\[2mm]
&\quad \frac{\Gamma(1)}{\Gamma(1+8)} \frac{\Gamma(1/4+4)}{\Gamma(1/4)} \frac{\Gamma(1/4+4)}{\Gamma(1/4)} \times \frac{\Gamma(1)}{\Gamma(1+6)} \frac{\Gamma(1/4+2)}{\Gamma(1/4)} \frac{\Gamma(1/4+4)}{\Gamma(1/4)} \times \\[2mm]
&\quad \frac{\Gamma(1)}{\Gamma(1+9)} \frac{\Gamma(1/3+5)}{\Gamma(1/3)} \frac{\Gamma(1/3+1)}{\Gamma(1/3)} \frac{\Gamma(1/3+3)}{\Gamma(1/3)} \times \\[2mm]
&\quad \frac{\Gamma(1)}{\Gamma(1+6)} \frac{\Gamma(1/3+1)}{\Gamma(1/3)} \frac{\Gamma(1/3+2)}{\Gamma(1/3)} \frac{\Gamma(1/3+3)}{\Gamma(1/3)} \\[2mm]
&= 2.7159 \times 10^{-33}.
\end{aligned}
$$

12.4.3   A Search Algorithm for Module Networks

We need a heuristic search algorithm for module network structure selection similar to the heuristic search algorithms for DAG selection discussed in Section 8.6.1. We show the algorithm developed in [Segal et al., 2005]. This algorithm fixes the set of modules C ahead of time. In general, we could also search over the module sets.

We assume that we have n random variables V and a set of modules C. The search space is then the set of all module network structures $\mathbb{S} = (V, C, PA, A)$ consisting of the modules in C and containing the variables in V. There are two components to the search:

1. We must determine a set of parent sets PA.

2. We must determine the assignment A of variables to the modules.

We develop an algorithm for each of these tasks. Then we repeatedly call the two algorithms in sequence until convergence.

The algorithm for Component 1 is very similar to Algorithm 8.2 in Section 8.6. Our set of operations is as follows:

1. If X_h is not in PA_i, add X_h to PA_i.

2. If X_h is in PA_i, remove X_h from PA_i.

The first operation is subject to the constraint that the resultant DAG in the Bayesian network does not contain a cycle. The set of all structures that can be obtained from $\mathbb{S}$ by applying one of the operations is called Nbhd($\mathbb{S}$). If $\mathbb{S}' \in$ Nbhd($\mathbb{S}$), we say $\mathbb{S}'$ is in the **neighborhood** of $\mathbb{S}$. Clearly, this set of operations is complete for the search space. That is, for any two structures $\mathbb{S}$ and $\mathbb{S}'$ containing the same assignments A there exists a sequence of operations that transforms $\mathbb{S}$ to $\mathbb{S}'$. The algorithm that searches using this set of operations is called PA_*search* in Algorithm 12.1 (which follows).

The algorithm for Component 2 above looks at each of the n variables in sequence. When investigating variable X_h, it checks each of the k modules M_i to see if reassigning X_h to module M_i increases the score. It then chooses the legal reassignment which increases the score the most. This procedure is repeated until no variable is reassigned. This algorithm is called A_*search* in Algorithm 12.1.

Our top-level algorithm repeatedly calls PA_*search* and A_*search* in sequence until the improvement in the scores is less than some preassigned threshold. The top-level algorithm is called *mod_net_search* in Algorithm 12.1. We need an initial assignment of variables to modules. This could be done at random, by expert judgement, or by some other technique. Segal et al. [2005] show a cluster technique, which they use for this initial assignment.

Algorithm 12.1 follows.

Algorithm 12.1 Module Network Search

Problem: Find a module network structure $\mathbb{S}$ that approximates maximizing *score* $(\mathbb{S} : D)$.

Inputs: A set V of n random variables; a set C of m modules; data D.

Outputs: A module network structure $\mathbb{S} = (V, C, PA, A)$ that approximates maximizing *score* $(\mathbb{S} : D)$.

```
void PA_search(module_net_struct& S = (V,C, PA, A))
{
  E = ∅; G = (V, E);
  do
    if (any structure in the neighborhood of our current structure
    increases score(S : D))
        modify PA according to the one that increases score(S : D) most;
    while (some operation increases score(S : D);
}

void A_search(module_net_struct& S = (V,C, PA, A))
{
  do
    for (h = 1; h <= n; h + +)
      for (i = 1; i <= m; i + +) {          // m is the number of modules.
        A' = A except A(X_h) = M_i;
        if (space(X_h) = space(M_i) and the Bayesian network is acyclic)
          if (score((V, C, PA, A') : D)) > (score((V, C, PA, A) : D))
            A = A';
      }
    while (some variable X_h is reassigned);
}

void mod_net_search(module_net_struct& S = (V,C, PA, A))
{
  Assign each variable in V to a module in C;
  for (i = 1; i <= m; j + +)      // m is the number of modules.
    PA_i = ∅;
  do
    PA_search(S);
    A_search( S);
  while (score improvement ≥ threshold);
}
```

It is left as an exercise to show that algorithms PA_search and A_search both have local scoring updating.

12.4.4 Using Class Probability Trees in Module Networks

If our variables were all discrete, we could used discrete probability distributions in our module network, represent our prior belief concerning these distributions using Dirichlet distribution, and then use the score developed in Theorem 12.2. However, gene expression levels are continuous, not discrete. We could discretize them as we did in Section 12.3.3. However, Segal et al. [2005] chose instead to use normal (Gaussian) distributions of the variables. Furthermore, they used a conditional probability model called a **class probability tree** to represent the family of distributions for each variable given values of its parents. So, first we provide a review of class probability trees, which is taken from [Neapolitan and Jiang, 2007].

Class Probability Trees

Suppose that we are interested in whether an individual might buy some particular product. Suppose further that income, sex, and whether the individual is mailed a flyer all have an influence on whether the individual buys, and we articulate three ranges of income: low, medium, and high. Then our variables and their values are as follows:

Target Variable	Values
Buy	$\{no, yes\}$

Predictor Variables	Values
Income	$\{low, medium, high\}$
Sex	$\{male, female\}$
Mailed	$\{no, yes\}$

There are $3 \times 2 \times 2 = 12$ combinations of values of the predictor variables. We are interested in the conditional probability of *Buy* given each of these combinations of values. For example, we are interested in

$$P(Buy = yes | Income = low, Sex = female, Mailed = yes).$$

We can store these 12 conditional probabilities using a class probability tree. A **complete class probability tree** has stored at its root one of the predictors (say *Income*). There is one branch from the root for each value of the variable stored at the root. The nodes at level 1 in the tree each store the same second predictor (say *Sex*). There is one branch from each of those nodes for each value of the variable stored at the node.

We continue down the tree until all predictors are stored. The leaves of the tree store the target variable along with the conditional probability of the target variable given the values of the predictors in the path leading to the leaf. In our current example there are 12 leaves, one for each combination of values of the predictors. If there are many predictors, each with quite a few values, a complete class probability tree can become quite large. Aside from the storage problem, it might be hard to learn a large tree from data. However, some conditional probabilities may be the same for two or more combinations of values of the predictors. For example, the following four conditional probabilities may all be the same:

$$P(Buy = yes | Income = low, Sex = female, Mailed = yes)$$

$$P(Buy = yes | Income = low, Sex = male, Mailed = yes)$$

$$P(Buy = yes | Income = high, Sex = female, Mailed = yes)$$

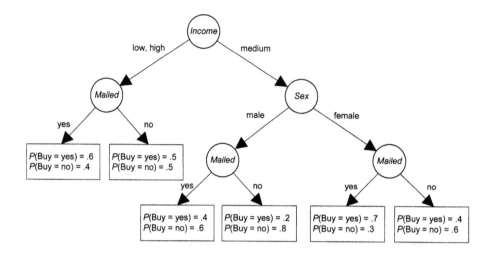

Figure 12.6: A class probability tree.

$$P(Buy = yes | Income = high, Sex = male, Mailed = yes).$$

In this case we can represent the conditional probabilities more succinctly using the class probability tree in Figure 12.6. There is just one branch from the *Income* node for the two values *low* and *high* because the conditional probability is the same for these two values. Furthermore, this branch leads directly to a *Mailed* node because the value of *Sex* does not matter when *Income* is *low* or *high*.

We call the set of edges emanating from an internal node in a class probability tree a **split**, and we say there is a split on the variable stored at the node. For example, the root in the tree in Figure 12.6 is a split on the variable *Income*.

To retrieve a conditional probability from a class probability tree, we start at the root and proceed to a leaf, following the branches that have the values on which we are conditioning. For example, to retrieve $P(Buy = yes | Income = medium, Sex = female, Mailed = yes)$ from the tree in Figure 12.6, we proceed to the right from the *Income* node, then to right from the *Sex* node, and, finally, to the left from the *Mailed* node. We then retrieve the conditional probability value .7. In this way, we obtain

$$P(Buy = yes | Income = medium, Sex = female, Mailed = yes) = .7.$$

The purpose of a class probability tree learning algorithm is to learn from data a tree that best represents the conditional probabilities of interest. This process is called **growing the tree**. Trees are grown using a greedy one-ply lookahead search strategy and a scoring criterion to evaluate how good the tree appears based on the data. The classical text in this area is [Breiman et al., 1984]. Buntine [1993] presents a tree learning algorithm that uses a Bayesian scoring

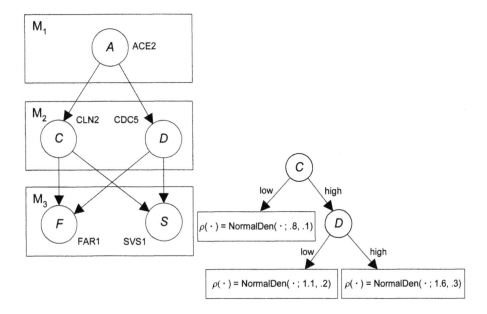

Figure 12.7: A module network whose variables are Gaussian and whose conditional probability distributions are determined by class probability trees. Only the distributions for module M_3 is shown.

criterion. This algorithm is used in the IND Tree Package [Buntine, 2002], which comes with source code and a manual.

An Application to Analyzing Gene Expression Level

Gene expression level is often set as the ratio of measured expression to a control level. So, values greater than 1 would indicate a relatively high expression level, whereas values less than 1 would indicate a relatively low expression level. We assume that each variable is normally distributed given values of its parents. However, we assume that each parent has only two values, namely *high* and *low*, which determine the conditional distribution of the child. The value *high* represents all expression levels greater than 1; the value *low* represents all expression levels less than or equal to 1. We represent the conditional distribution for each module in our module network using a class probability tree. Such a module network appears in Figure 12.7.

Scoring a Module Network Containing Class Probability Trees

A structure $\mathbb{S}'$ for a module network whose conditional probability distributions are modeled using class probability trees includes the structure of the class probability tree for each module. Segal et al. [2005] developed a Bayesian score $score(\mathbb{S}' : D)$ for these structures, similar to the score in Theorem 12.2, which is the product of local scores for each class probability tree in the module. Recall that for the score in Theorem 12.2, the total score is the product of the scores of the modules. Furthermore, each module's score is the product of scores, each of which is a score for one instantiation of the parents of a module (see Equality 12.2). These two results hold for $score(\mathbb{S}' : \mathsf{D})$. That is, the total score is the product of the scores of the modules. When the conditional probability distributions are represented using a class probability tree, each instantiation of the parents leads to one leaf in the class probability tree. Each module's score is a product of scores, each of which is a score of one leaf in the module. Next we show a leaf's score.

Let

L_{ji} be the ith leaf in Module M_j.

$L_j[r]$ be the leaf reached by the instantiations in PA_j for the rth data item.

$x_i[r]$ be the value of X_i for the rth data item.

Next set

$$S_{ji}^0 = \sum_{r\,:\,L_j[r]=L_{ji}} \sum_{X_i \in \mathsf{M}_j} 1$$

$$S_{ji}^1 = \sum_{r\,:\,L_j[r]=L_{ji}} \sum_{X_i \in \mathsf{M}_j} x_i[r]$$

$$S_{ji}^2 = \sum_{r\,:\,L_j[r]=L_{ji}} \sum_{X_i \in \mathsf{M}_j} x_i^2[r].$$

Note that the term $\sum_{X_i \in \mathsf{M}_j} 1$ in the first sum yields the number of variables in M_j.

Example 12.6 *Suppose $X_1, X_4,$ and X_5 are the variables in module M_3, and the sixth and eighth data items are the only data items that have values of the variables in PA_3 that lead to leaf L_{31} in the class probability tree in module M_3. Suppose further that*

$$
\begin{aligned}
x_1[6] &= 21 \\
x_1[8] &= 14 \\
x_4[6] &= 11 \\
x_4[8] &= 17 \\
x_5[6] &= 24 \\
x_5[8] &= 19
\end{aligned}
$$

Then

$$
\begin{aligned}
S_{31}^0 &= \sum_{r\,:\,L_3[r]=L_{31}} \sum_{X_i \in \mathsf{M}_3} 1 \\
&= (1+1+1) + (1+1+1) = 6
\end{aligned}
$$

$$S_{31}^1 = \sum_{r\,:\,L_3[r]=L_{31}} \sum_{X_i \in \mathsf{M}_3} x_i[r]$$

$$= (21 + 11 + 24) + (14 + 17 + 29) = 116$$

$$S_{31}^2 = \sum_{r\,:\,L_3[r]=L_{31}} \sum_{X_i \in \mathsf{M}_3} x_i^2[r]$$

$$= (21^2 + 11^2 + 24^2) + (14^2 + 17^2 + 29^2) = 2464.$$

The log of the Bayesian score of leaf L_{ji} is given by

$$\log\left(score(L_{ji} : \mathsf{D})\right) = -\frac{1}{2}S_{ji}^0 \log \pi + \frac{1}{2}\log\left(\frac{\lambda}{\lambda + S_{ji}^0}\right) + \log\left(\Gamma\left(\alpha + \frac{1}{2}S_{ji}^0\right)\right)$$

$$- \log\left(\Gamma\left(\alpha\right)\right) + \alpha \log \beta - \left(\alpha + \frac{1}{2}S_{ji}^0\right)\log \beta' \qquad (12.3)$$

where

$$\beta' = \beta + \frac{1}{2}\left(S_{ji}^2 - \frac{\left(S_{ji}^1\right)^2}{S_{ji}^0}\right) + \frac{S_{ji}^0 \lambda \left(\frac{S_{ji}^1}{S_{ji}^0} - \mu\right)^2}{2\left(\lambda + S_{ji}^0\right)}.$$

To obtain Equality 12.3, we assume that our uncertainty in the conditional probability distribution at leaf L_{ji} is modeled using Normal-Gamma priors, and the constants α, β, λ, and μ are the parameters in these priors. It is beyond the scope of this text to discuss this model or to develop the score in Equality 12.3. See [Neapolitan, 2004] for this material. Briefly, we only note that minimal prior belief can be modeled using small positive values of α and β, $\lambda = \alpha + 1$ and by setting $\mu = 0$.

Modification to the Algorithm

When we use class probability trees to model our conditional distributions in the module networks, we must modify Algorithm 12.1 to search over such trees when we evaluate a particular module network structure. We mentioned a reference that discusses learning such trees at the end of section 12.4.4. Segal et al. [2005] used the learning algorithm developed in [Chickering et al., 1997].

12.4.5 Experimental Results

Segal et al. [2005] evaluated this method using the dataset on gene expression level provided in [Gasch et al., 2000]. That dataset contains data items on gene expression levels in yeast when the yeast are subjected to various stress conditions. There are 6,157 genes and 173 data items. Based on prior knowledge of which genes are likely to play a regulatory role, they restricted the possible parents to 466 genes. They then investigated 2,355 genes whose expression levels varied significantly in the dataset. Since all variables are continuous, the spaces for all modules were the same, which means that every gene could be

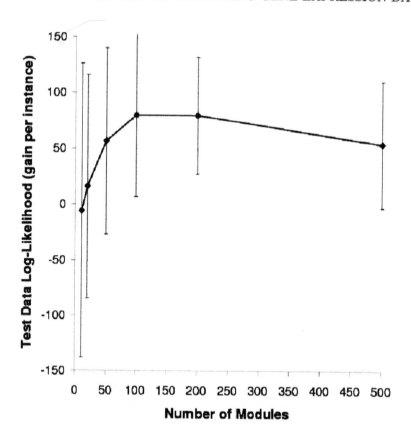

Figure 12.8: The average gain in the log of the score of the learned module network structure relative to the log of the score of the learned Bayesian network structure, taken over 10 trials. The error bars show the standard deviations.

assigned to every module. They initialized the modules using the clustering technique mentioned just before Algorithm 12.1.

To test whether the module network approach performs better than the approach of simply learning a Bayesian network structure, they used each approach to learn a structure from a portion of the dataset. For each approach, they then computed the likelihood of the remaining data given the structure learned. If $\mathbb{S}$ is the module network learned and $\mathbb{G}$ is the Bayesian network structure (DAG) learned, we set

$$gain = \log\left(P(\mathsf{D}|\mathbb{S})\right) - \log\left(P(\mathsf{D}|\mathbb{G})\right).$$

They did experiments using values of $m = 10, 25, 50, 100, 200,$ and 500 for the number of modules, and for each value of m they performed 10 trials. Figure 12.8 plots the value of *gain* for each value of m. We see that in all cases the module network approach outperforms the Bayesian network approach.

To evaluate whether the module network approach could learn biologically plausible modules, Segal et al. [2005] learned a module network in which the value of the number of modules m was equal to 50. They chose 50 modules because there were about 2,500 genes and in this domain modules ordinarily have 40 to 50 genes per module. They examined the extent to which genes in the same module share the same functional characteristics. They used the annotations of the genes' biological functions in the Saccharomyces Genome Database [Cherry et al., 1998]. By *annotation* we mean a function assigned to the gene such as "protein-folding." For each module they determined the p-value of each shared characteristic of the genes in the module. The following example illustrates the technique.

Example 12.7 *One module contained 10 genes, and 7 of these genes were annotated as protein-folding genes. Of the 2,345 genes investigated, only 26 were annotated as protein-folding genes. To obtain the p-value, we compute the probability of obtaining 7 protein-folding genes if 10 genes are sampled at random from the 2,345 genes. It is left as an exercise to show that this value is $p = 10^{-12}$.*

Segal et al. [2005] corrected the p-value using the Boneferroni correction for independent multiple hypothesis [Savin, 1980]. Their final results showed that of the 50 modules, 42 modules had at least one annotation that was significant at the $p = .005$ level. Furthermore, the annotations reflected the key biological processes expected in the dataset. They used these annotations to label each module with a meaningful biological name. They then compared their results to results obtained using the clustering technique in AutoClass [Cheeseman and Stutz, 1996]. They found that the module network approach discovers annotation clusters at a higher significance level than AutoClass much more often than it discovers them at a low significance level than AutoClass.

Segal et al. [2005] also found that the relationships between the modules were biologically plausible. For example, in the module network learned, the genes in the *cell-cycle* module were parents of the *histone* module. The **cell cycle** is the process by which a cell replicates its DNA and then divides. The cell cycle is known to regulate histones, which are proteins responsible for maintaining and controlling DNA structure. As another example, they found that the genes in the *nitrogen catabolite repression (NCR)* module were parents of the *amino acid, purine metabolism,* and *protein synthesis* modules. NCR is a cellular response that is activated when nitrogen sources are scarce. The three modules that are children of the genes in the *NCR* module all concern nitrogen-requiring processes and are therefore likely to be regulated by the *NCR* module.

EXERCISES

Section 12.3

Exercise 12.1 Consider the ordering $[X_2, X_1, X_3]$. Determine all DAGs consistent with this ordering.

Exercise 12.2 Consider the ordering $[X_1, X_2, X_3, X_4]$. Determine all DAGs consistent with this ordering.

Exercise 12.3 Suppose that

$$\prec = [X_5, X_3, X_1, X_2, X_1].$$

1. Determine $U_{1,\prec}$.

2. Determine $U_{2,\prec}$.

3. Determine $U_{3,\prec}$.

4. Determine $U_{4,\prec}$.

5. Determine $U_{5,\prec}$.

Section 12.4

Exercise 12.4 Suppose that we have the set of random variables

$$V = \{X_1, X_2, X_3 X_4, X_5\},$$

and

$$space(X_1) \;=\; space(X_2) = \{1, 2, 3\}$$

$$space(X_3) \;=\; space(X_4) = space(X_5) = \{1, 2\}.$$

Specify a module network for these random variables.

Exercise 12.5 Suppose that we have the module network structure $\mathbb{S}$ and prior beliefs in Figure 12.5, and we obtain the data D in the following table:

Case	X_1	X_2	X_3	X_4	X_5
1	1	2	2	2	1
2	2	1	1	1	3
3	1	1	2	1	2
4	2	2	1	2	1
5	1	1	2	1	1
6	2	2	1	1	1
7	1	1	2	2	3
8	1	2	2	2	2
9	2	1	1	1	1
10	2	2	1	1	2
11	2	1	2	2	1
12	1	1	1	2	3
13	2	2	1	1	2
14	1	1	2	2	1
15	2	2	2	2	3

Compute $P(\mathsf{D}|\mathbb{S})$.

Exercise 12.6 Show that algorithms PA_*search* and *A_search* both have local scoring updating.

Exercise 12.7 Suppose that $\alpha = \beta = 1$, $\lambda = 2$, and $\mu = 0$. Determine the $\log\left(score(L_{31} : \mathsf{D})\right)$ in Example 12.6.

Exercise 12.8 Suppose that X_1 and X_5 are the variables in module M_2, and the sixth, seventh, and eighth data items are the only data items that have values of the variables in PA_2 that lead to leaf L_{23} in the class probability tree in module M_2. Suppose further that

$$
\begin{aligned}
x_1[6] &= 5 \\
x_1[7] &= 4 \\
x_1[8] &= 9 \\
x_5[6] &= 3 \\
x_5[7] &= 11 \\
x_5[8] &= 12
\end{aligned}
$$

Assuming $\alpha = \beta = 1$, $\lambda = 2$, and $\mu = 0$, determine the $\log\left(score(L_{23} : \mathsf{D})\right)$.

Exercise 12.9 In Example 12.7 it was left as an exercise to show that $p = 10^{-12}$. Show this.

Chapter 13

Genetic Linkage Analysis

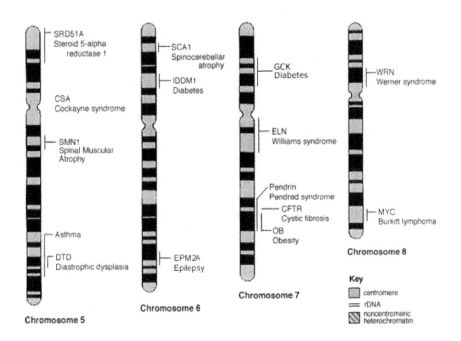

Recall from Section 4.2 that a gene is a locus (section of a chromosome) that is functional in the sense that it is responsible for some biological expression in the organism. For example, in Example 4.1 we discussed the *bey*2 gene, which is located on chromosome 15 in humans and is responsible for eye color. We are interested in knowing not only that a gene is on a particular chromosome but also its location on the chromosome. A **genetic linkage map** shows the order of genes on a chromosome and the relative distances between those genes. A genetic linkage map is not to be confused with a genome's DNA sequence.

DNA sequencing consists of determining the sequence of nucleotide bases in the chromosome. You might have heard that the human genome has been fully sequenced. However, knowing the sequence of nucleotides does not reveal to us the function of the nucleotides at a particular locus. This is the purpose of a genetic linkage map.

We create a genetic linkage map using a technique called **genetic linkage analysis**. An important application of genetic linkage analysis in humans is the determination of the locus of genes associated with hereditary diseases. In Section 13.1 we introduce genetic linkage analysis by analyzing the fruit fly *Drosophila*. Section 13.2 concerns genetic linkage analysis in humans. Finally, Section 13.3 discusses an application of Bayesian networks to genetic linkage analysis.

13.1   Introduction to Genetic Linkage Analysis

Although it is not immediately apparent, we can perform genetic linkage analysis by analyzing the results of crossing-over. Recall from Section 4.2 that crossing-over occurs during meiosis, when the chromatids from the two homologs that are next to each other exchange corresponding segments of genetic material. Genes that are close together on the same chromosome stay together for most occurrences of crossing-over, whereas genes that are far apart separate for most occurrences of crossing-over. The degree to which the genes stay together is called their **linkage**. So, by creating a large number of offspring and investigating what fraction of those offspring has genes that separated, we can obtain a measure of how close the genes are on the chromosome.

Next we show an example illustrating this technique.

13.1.1   Analyzing the Fruit Fly *Drosophila*

Like mammals, the fruit fly *Drosophila* has a pair of chromosomes, labeled X and Y, which share a small region of homology. This chromosome determines the sex of the individual. Females have two X chromosomes; males have an X chromosome and a Y chromosome. A characteristic determined by a gene that is located on the X chromosome is called a **sex-linked characteristic**. It is particularly simple to investigate linkage for a sex-linked characteristic because the phenotype of the male is determined entirely by the allele on the X chromosome.

The following table shows three sex-linked characteristics in *Drosophila*:

Allele	Characteristic
y^+ (dominant)	Gray body
y	Yellow body
w^+ (dominant)	Red eyes
w	White eyes
m^+ (dominant)	Normal wings
m	Miniature wings

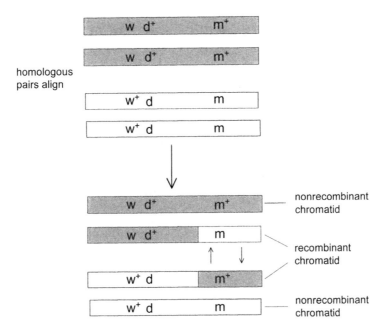

Figure 13.1: In this instance, the m and m$^+$ alleles cross over, whereas the y and y$^+$ and w and w$^+$ do not cross over.

Figure 13.1 shows the approximate locations of these alleles on the X chromosome. That figure also depicts an instance of crossing-over. Notice that the m and m$^+$ alleles cross over, but the y and y$^+$ and w and w$^+$ do not cross over. We say that we have **recombination** for the m and m$^+$ alleles relative to the w and w$^+$ alleles. Similarly, we have recombination for the m and m$^+$ alleles relative to the y and y$^+$ alleles. We do not have recombination for the y and y$^+$ alleles relative to the w and w$^+$ alleles. The m and m$^+$ alleles will recombine relative to the w and w$^+$ alleles only for crossovers that occur at points between the loci of these alleles. So, the number of recombinations is proportional to the distance between the loci. Similarly, the y and y$^+$ alleles will recombine relative to the w and w$^+$ alleles only for crossover that occurs at points between the loci of these alleles. Therefore, by measuring the **recombination frequency** between allele pairs in multiple offspring, we can measure the distance between the loci of the alleles. The following examples illustrates the technique.

Example 13.1 *An experiment, taken from [Hartl and Jones, 2006], estimating the recombination frequency of the m and m$^+$ alleles relative to the w and w$^+$ alleles, appears in Figure 13.2. That experiment is described next. First, we obtain a large number of female Drosophila that are homozygous for w (white eyes) and m$^+$(normal wings). This could be done by selective breeding until we find in repeated generations that all flies have white eyes and normal wings. That is, once we see in many generations that there are no miniature-winged*

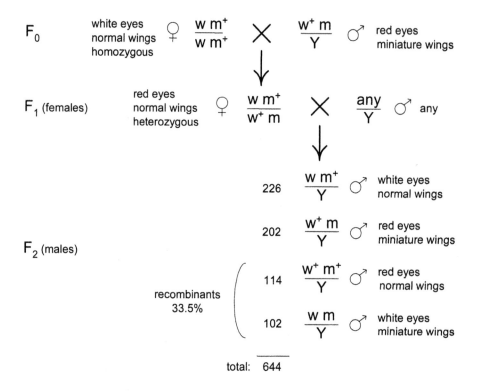

Figure 13.2: An experiment estimating the recombination percentage of the m and m^+ alleles relative to the w and w^+ alleles.

females, we can become confident that the m allele has been eliminated from our subpopulation.

Next we obtain a large number of male *Drosophila* that have w^+ (red eyes) and m (miniature wings) on the X chromosomes. These flies can be obtained by inspection. The two groups of flies together constitute our F_0 generation. The top of Figure 13.2 depicts the flies in this generation. The notation

$$\frac{wm^+}{wm^+}$$

depicts the allele configuration on homologous X chromosomes in the females. Henceforth, in the text we we will use the notation wm^+/wm^+ in-line. Similarly, the notation w^+m/Y depicts the alleles on the single X chromosome in the males and shows the corresponding Y chromosome.

We mate these flies, producing our F_1 generation. We select only the females from this generation. It is straightforward that every female has allele configuration wm^+/w^+m. Our purpose in this first mating was to produce this set of females. We mate these females with a group of arbitrary males, producing our F_2 generation, and we select only the males from this generation. The

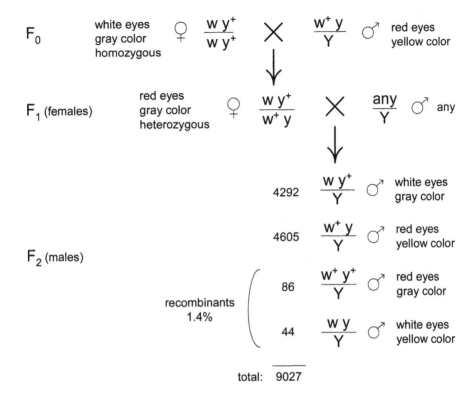

Figure 13.3: An experiment estimating the recombination percentage of the y and y^+ alleles relative to the w and w^+ alleles.

characteristics (relative to the alleles we are investigating) of these males are determined only by their mothers. So, for example, if a male has white eyes and normal wings, he must have received wm^+ from his mother. This means that there was no recombination of the alleles in this male, because w and m^+ occur on the same homolog in the mother. Similarly, if the male has red eyes and miniature wings, the male must have received w^+m from the mother and again there was no recombination. On the other hand, if the male has red eyes and normal wings, he must have received w^+m^+ from his mother. This means that there must have been recombination of the alleles in this male, because w^+ and m^+ occur on the different homologs in the mother. The same is true if the male has white eyes and miniature wings. The bottom of Figure 13.2 shows the percentage of recombinants. This recombination percentage of 33.5% reflects the distance between the loci of the alleles.

Example 13.2 An experiment, taken from [Hartl and Jones, 2006], estimating the recombination percentage of the y and y^+ alleles relative to the w and w^+ alleles appears in Figure 13.3. The details of the experiment are identical

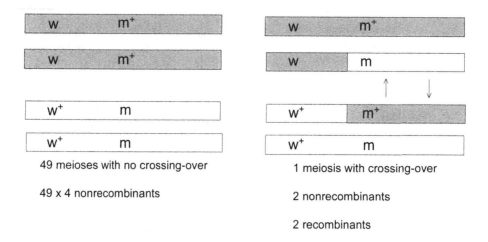

Figure 13.4: Only two of 200 gametes are recombinants, but one out of 50 meioses has a crossing-over.

to those discussed in Example 13.1, so we will not repeat them. The bottom of Figure 13.3 shows the percentage of recombinants. This recombination percentage of 1.4% reflects that the distance between the loci of these alleles is much smaller than the distance between the loci of the alleles for white eyes and miniature wings. This is in agreement with Figure 13.1.

13.1.2   Creating a Genetic Map

The genetic distance, called the **map distance**, between the loci of two genes is defined as one-half of the expected value of the number of crossovers that take place in the region between their loci per meiosis.

Example 13.3 *Suppose that we have two genes* $gene_1$ *and* $gene_2$ *and there are 640 crossovers in the region between their loci in 1000 meioses. Then we have that*

$$map\_distance(gene_1, gene_2) \approx \left(\frac{1}{2}\right)\left(\frac{640}{1000}\right) = .32.$$

The map distance is ordinarily shown as a whole number. So, we would say that the map distance is 32 in the previous example. One map unit is called a **centimorgan (cM)**. So, the map distance between $gene_1$ and $gene_2$ in the previous example is 32 cM.

You might wonder why we have the factor of one-half. We estimate the map distance by the recombination frequency that occurs in experiments like those discussed in Examples 13.1 and 13.2. In a given meiosis that has a crossover, two recombinants will be produced, but two nonrecombinants will also be produced. In a meiosis that does not have a crossover, four nonrecombinants will

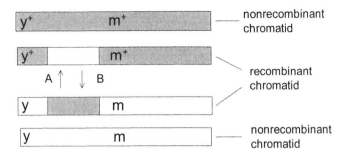

Figure 13.5: The two crossovers cancel each other out, resulting in no recombinants of alleles y and y^+ relative to alleles m and m^+.

be produced. Let's say, for example, that we have one crossover in 50 meioses, and every gamete results in an offspring. Then the fraction of crossovers per meiosis is

$$\frac{1}{50} = .02.$$

However, 50×4 offspring will be produced, and only two will be recombinants. The recombination frequency observed in the offspring will therefore be

$$\frac{2}{50 \times 4} = .01,$$

which is one-half of the fraction of crossovers. This situation is illustrated in Figure 13.4.

As just mentioned, we estimate the map distance using the recombination frequency in experiments. However, if two genes are far apart on the chromosome, there could be two crossovers between, thereby negating the recombination. This situation is depicted in Figure 13.5. At a computational level, the following is what happens in this situation: Each recombinant chromatid breaks at point A and recombines with the loose segment from the other recombinant chromatid. Then they break again at point B and recombine again with the new loose segment from the other recombinant chromatid. The result is that the alleles y and y^+ do not recombine relative to the alleles m and m^+. If two genes are far apart on the chromosome, this can occur quite often, which means that we cannot accurately estimate their map distance by studying their recombination frequency.

It is not hard to see that the map distance is additive. That is, if $gene_2$ is between $gene_1$ and $gene_3$, then

$map\_distance(gene_1, gene_3)$

$$= map\_distance(gene_1, gene_2) + map\_distance(gene_2, gene_3).$$

Therefore, owing to the problem concerning frequent multiple crossovers when genes are far apart, we ordinarily only estimate map distances from recombination frequencies for genes that are fairly close together (less than 10 cM).

Position	Gene
0	yellow body (y)
1.5	white eyes (w)
3,0	facet eyes (fa)
5.5	echinus eyes (ec)
7.5	ruby eyes (rb)
13.7	crossveinless wing (cv)
20	cut wings (ct)
21	single bristles (sn)
27.5	tan body (t)
33	vermilion eyes (v)
36.1	miniature wings (m)
43	sable body (s)
44	garnet eyes (g)
51.5	scalloped wings (sd)
56.7	forked bristles (f)
57	bar eyes (B)
59.5	fused veins (fu)
62.5	carnation eyes (ca)
66.0	bobbed bristles (b)

Table 13.1: Abbreviated genetic map of the X chromosome of Drosophila Melanogaster

We then create a genetic linkage map of the entire chromosome from these estimates.

An abbreviated genetic map of the X chromosome of *Drosophila* appears in Table 13.1.

13.2   Genetic Linkage Analysis in Humans

In the first step of Example 13.1 we did selective breeding using a large population of flies to obtain flies that are homozygous for w (white eyes) and m^+ (normal wings). In our final step we again needed to use a large population so that the observed recombination frequency was a good estimate of the map distance. We could not do either of these in human populations because (1) researchers do not control the breeding in humans, and (2) humans live long and have few offspring. Two techniques have been developed that handle each of these difficulties. First, we have developed ways to measure certain phenotypes from which we can infer the genotypes. That is, each phenotype has only one genotype associated with it. This solves the problem of knowing the genotypes of the individuals, which means that we do not need to do selective breeding in repeated generations. Second, we pool different pedigrees to obtain a reasonably large sample. Here we discuss each of these topics in turn.

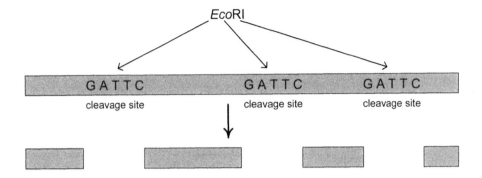

Figure 13.6: The restriction enzyme *Eco*RI cleaves DNA wherever the sequence GAATTC appears. In this example there are three such sites, resulting in four DNA fragments.

13.2.1 Polymorphic DNA Sequences as Genetic Markers

There are many minor differences in the DNA of different individuals. For example, if one individual has one allele for a gene and another individual has a different allele, there will be minor differences in the DNA at the locus of this gene. In general, any difference in a locus in a given genome that is relatively common in the population is called a **polymorphism**. Most polymorphisms exist in regions of DNA that do not code for any protein and that therefore have no effect on the individual's structure or processes. Regardless, we can use a locus that contains polymorphisms in genetic linkage analysis. If a gene of interest (e.g., one associated with an inherited disease) can be shown to be genetically linked to a given polymorphism, we can investigate the gene of interest by investigating the locus of the polymorphism. We say that the locus is used as a genetic marker. In general, a **genetic marker** is a locus such that the genotype can be inferred from the phenotype. Next we discuss one such class of polymorphisms.

Restriction fragment length polymorphisms (RFLPs) are recognized by a type of enzyme called **restriction endonuclease**, which cleaves DNA wherever a particular nucleotide sequence appears. For example, the restriction enzyme **EcoRI** cleaves DNA wherever the sequence GATTC appears. This is illustrated in Figure 13.6. So, when two different individuals have different numbers of a given sequence (e.g., GATTC), the restriction enzyme will produce a different number of fragments.

The various DNA fragments can be separated by an electric field in a supporting gel. Longer fragments are separated from shorter ones because they move through the gel more slowly. This idea is illustrated in Figure 13.7. By analyzing the gel, we can see how many fragments are produced and therefore learn the polymorphisms that are present in the individual. As shown in Figure 13.7, an individual with genotype *Aa* produces three bands in the gel. An individual with genotype *AA* would produce one band, and an individual with

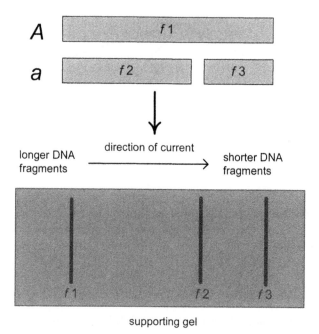

Figure 13.7: The two alleles A and a differ in the presence of one cleavage site. Allele A yields one long DNA fragment, whereas allele a yields two shorter DNA fragments. The fragments are labeled $f1$, $f2$, and $f3$. The longer fragments migrate through the gel more slowly.

genotype aa would produce two bands. *The phenotype of each individual is the pattern of bands in the gel,* and we can infer the genotype from the phenotype. So, instead of doing selective breeding (as done in Example 13.1), we can learn the genotype of an individual directly from the phenotype of the individual.

A second type of DNA polymorphism results when there are different numbers of copies of a short nucleotide sequence at a particular site. When a DNA molecule is cleaved at sites that enclose the copies, the size of the DNA molecule containing more copies will be longer and will therefore move more slowly through the gel. A genetic polymorphism of this kind is called a **simple sequence repeat (SSR)**.

Example 13.4 *Suppose that we have the following nucleotide sequences comprising two alleles of the same gene:*

$$A \quad A \quad T \quad G \quad \mathbf{C} \quad \mathbf{A} \quad \mathbf{C} \quad \mathbf{A} \quad \mathbf{C} \quad \mathbf{A} \quad \mathbf{C} \quad \mathbf{A} \quad T \quad C$$

$$A \quad A \quad T \quad G \quad \mathbf{C} \quad \mathbf{A} \quad \mathbf{C} \quad \mathbf{A} \quad T \quad C$$

The top allele has four copies of the sequence C A, whereas the bottom one has only two copies. So, the fragment from the top allele will move slowly through the gel.

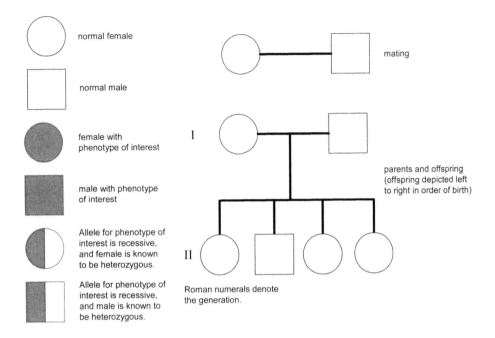

Figure 13.8: Conventionals symbols used in pedigrees.

Another type of polymorphism that occurs in the human genome and in other species is the **single-nucleotide polymorphism (SNP)**. In an SNP the nucleotide at a single site can differ between individuals. Rare mutants exist at almost every site, but these rare cases are not useful to our endeavors. So, for a site to qualify as an SNP, at least 5% of the population must have the less frequent nucleotide. According to this definition, the human genome has about one SNP for every 1300 base pairs.

13.2.2 Pedigrees

A **pedigree** is a family tree that shows the phenotype of each individual. The tree may include phenotypes corresponding to only one locus, or it may include phenotypes corresponding to more than one locus. Figure 13.8 shows conventional symbols used in a pedigree. The shaded and partially shaded symbols would only be used when the tree includes phenotypes corresponding to only one locus. Here we first present some examples of pedigrees, then we discuss how they can be used in genetic linkage analysis.

13.2.3 Examples of Pedigrees

Here we show several examples of pedigrees.

Example 13.5 *Figure 13.9 shows a pedigree in which the phenotype of inter-*

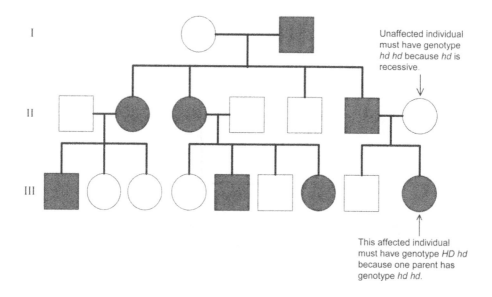

Figure 13.9: A pedigree in which the phenotype of interest is Huntington's disease.

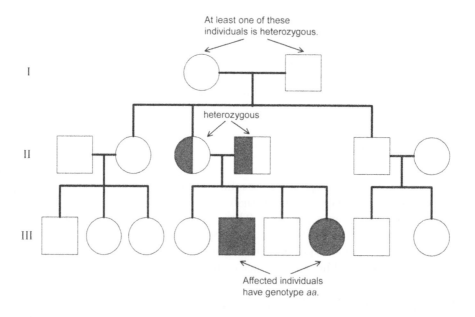

Figure 13.10: A pedigree in which the phenotype of interest is albinism.

ested is **Huntington's disease**, *which involves progressive nerve degeneration, usually starts in middle age, and results in severe physical and mental disability. The allele HD for Huntington's disease is dominant over the hd allele. So, for an individual to not have the disease, the individual would have to have genotype hd hd. Therefore, from the pedigree in Figure 13.9 we can infer the genotype of every individual except the male in generation I. Every individual who does not have the disease must have genotype hd hd, whereas every individual (other than the male in generation I) who has the disease must be HD hd because one of the individual's parents is hd hd.*

Example 13.6 *Figure 13.10 shows a pedigree in which the phenotype of interest is albinism, which is absence of pigment in the skin, hair, and iris of the eyes. The allele a for this condition is recessive to allele A. So, an individual would need to have the homozygous genotype aa to have albinism. Two individuals in generation III have albinism, so their genotype must be aa. Neither of their parents has albinism, but they both must carry it. Therefore, both their parents have genotype Aa. Since the female parent has genotype Aa, she must have obtained the a allele from one of her parents. Since neither of her parents has albinism, at least one of them must have genotype Aa. No other individuals have albinism. All we can conclude about them is that they do not have genotype aa.*

Next we show an example of a pedigree concerning an SSR.

Example 13.7 *Suppose some particular locus contains an SSR that, throughout a population, has six different numbers of copies of some DNA sequence. From each phenotype, we can determine the genotype (number of copies of the short DNA sequence in each allele). Figure 13.11 shows a pedigree that we might obtain when the phenotype of interest is the pattern in the gel for this SSR. In that figure the alleles are numbered 1 through 6. The genotype of each individual appears under the symbol for the individual. The phenotype is shown in the gel. Note that from each phenotype we can determine the genotype exactly.*

13.2.4 Pedigrees and Genetic Linkage Analysis

Assume that we have two loci on the same chromosome and we want to estimate their map distance as we did for the genes concerning eye color and wing size in Example 13.1. We could obtain a pedigree for each locus from the same family tree. For example, suppose the SSR discussed in Example 13.7 is on the same chromosome as the gene for albinism. We could obtain a pedigree concerning albinism for the individuals, whose pedigree concerning only the SSR is shown in Figure 13.11. We then analyze the recombination frequency of the alleles of the SSR and the alleles of the gene for albinism for the individuals in the family. Since the number of individuals in a human family is small, we would pool our results from many family trees.

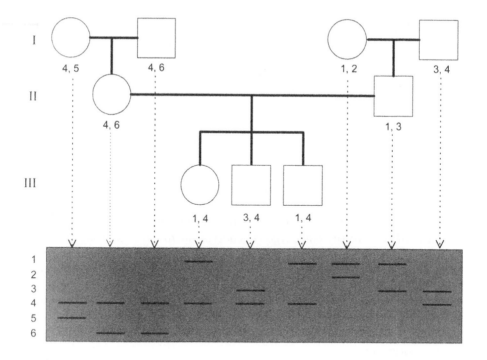

Figure 13.11: A pedigree in which the phenotype of interest is the band pattern in a gel resulting from the genotype of an SSR.

A problem with the method just discussed is that for many individuals we do not know the genotype, and so we do not make use of available information concerning these individuals. Another method is to model the problem probabilistically and use all available information. We now discuss this approach.

13.3   A Bayesian Network Model

Here we discuss a Bayesian network model for genetic linkage analysis. We assume a pedigree that represents phenotypes corresponding to one or more loci. For each individual i in the pedigree and each locus j, we have the following nodes:

$G_{i,j}^{(f)}$: This node represents the allele at the jth locus that individual i receives from its father.

$G_{i,j}^{(m)}$: This node represents the allele at the jth locus that individual i receives from its mother.

$PH_{i,j}$: This node represents the phenotype of individual i determined by the gene at the jth locus.

For each individual i who is not a root, we also have the nodes that follow.

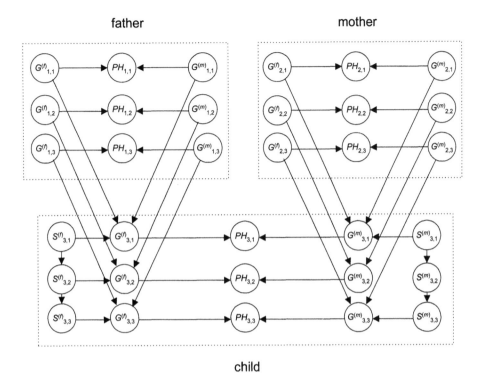

Figure 13.12: A fragment of a Bayesian network used to perform genetic linkage analysis. Three loci are modeled in this network.

$S_{i,j}^{(f)}$: This is a selector node, which tells us whether the allele that individual i received from the father came from the father or the mother of the father of individual i. That is, if individual k is the father of individual i, then

$$G_{i,j}^{(f)} = \begin{cases} G_{kj}^{(f)} & \text{if } S_{i,j}^{(f)} = 0 \\ G_{kj}^{(m)} & \text{if } S_{i,j}^{(f)} = 1 \end{cases}$$

$S_{i,j}^{(m)}$: This is a selector node, which tells us whether the allele that individual i received from the mother came from the father or the mother of the mother of individual i. That is, if individual r is the mother of individual i, then

$$G_{i,j}^{(m)} = \begin{cases} G_{rj}^{(f)} & \text{if } S_{r,j}^{(m)} = 0 \\ G_{rj}^{(m)} & \text{if } S_{r,j}^{(m)} = 1 \end{cases}$$

The Bayesian network fragment in Figure 13.12 shows the relationship between the nodes for an individual and the nodes for its parents. The entire Bayesian network consists of a set of nodes for each individual in the pedigree, where the relationship between its nodes and its parents' nodes are those shown in Figure 13.12.

Next we discuss the parameters in the Bayesian network.

- *Penetrance probabilities*:

$$P\left(PH_{i,j}|G_{i,j}^{(f)},G_{i,j}^{(m)}\right)$$

This is the probability of individual i having a given phenotype at the jth locus conditional on values of its genotypes at that locus. Ordinarily, these probabilities would be close to 1 or 0. For example, if an individual had alleles HD and hd, the individual would almost certainly have Huntington's disease. However, environmental factors could make them deviate from 0 and 1 in some cases. Ordinarily, the values of these parameters are obtained from previous analysis.

- *Transmission probabilities*:

$$P\left(G_{i,j}^{(f)}|G_{k,j}^{(f)},G_{k,j}^{(m)},S_{i,j}^{(f)}\right)$$

$$P\left(G_{i,j}^{(m)}|G_{r,j}^{(f)},G_{r,j}^{(m)},S_{i,j}^{(m)}\right)$$

In these probability expressions, individual k is the father of individual i and individual r is the mother of individual i. These probabilities are all close to 1 or 0. Mutations would make them deviate slightly from 1.

- *Recombination probabilities*:

$$P\left(S_{i,j}^{(f)}|S_{i,j-1}^{(f)},\theta_{j-1}\right)$$

$$P\left(S_{i,j}^{(m)}|S_{i,j-1}^{(m)},\theta_{j-1}\right)$$

In these probability expressions, θ_{j-1} is the recombination frequency between locus $j-1$ and locus j. These probabilities are unknown and depend on the unknown recombination frequencies θ_j.

Once the Bayesian network is constructed, our goal is to estimate the values of θ_j for all j from the data, where the data consist of the known phenotypes of the individuals in the pedigree plus any known genotypes. Before discussing this matter, we show an example that should clarify the recombination probabilities.

Example 13.8 *Suppose that the allele at the third locus, which individual i received from the father, came from the father of the father of individual i. Then if there was no crossing-over concerning the third and fourth loci in the chromosome individual i received from the father, the allele at the fourth locus, which individual i received from the father, would also come from the father of*

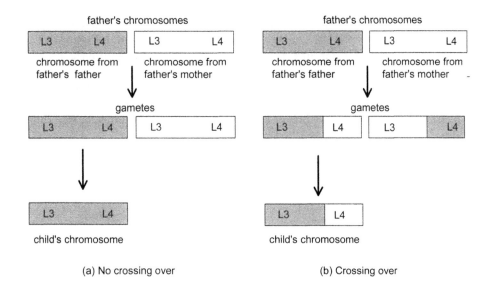

Figure 13.13: If there is no crossing over (a), the alleles at loci L3 and L4 both come from the father's father (or mother). If there is crossing over (b), one allele comes from the father's father; the other comes from the father's mother.

the father of individual i. This situation is illustrated in Figure 13.13 (a). If the recombination frequency were 0, this would always be the case. So

$$P\left(S_{i,j}^{(f)} = 1 | S_{i,j-1}^{(f)} = 0, \theta_{j-1} = 0\right) = 0.$$

However, if crossing-over for the two loci occurred, the allele at the fourth locus, which individual i received from the father, would come from the father's mother when the allele at the third locus came from the father's father. This situation is illustrated in Figure 13.13 (b). If the recombination frequency was .34, this event would occur 34% of the time. So,

$$P\left(S_{i,j}^{(f)} = 1 | S_{i,j-1}^{(f)} = 0, \theta_{j-1} = .34\right) = .34.$$

Recall that our goal is to estimate the values of θ_j for all j from the data. One way to do this would be to first represent our prior belief concerning the values of θ_j for all j using the method discussed in Section 7.1. If we have no prior belief, we could represent prior ignorance. We then perform inference in the Bayesian network to determine the maximum a posterior probability (MAP) estimate of θ_j for each j. Note that now we are doing inference to determine the posterior probability of parameters in the network. An approximate inference technique called **Gibb's sampling** can be used to perform this inference. Gibb's sampling and other methods for doing this inference are discussed in [Neapolitan, 2004].

Alternatively, we could determine the maximum likelihood estimate (MLE) of θ_j for each j. That is, we determine the values that maximize the probability of the data. This is the approach taken in [Fishelson and Geiger, 2002] and

[Fishelson and Geiger, 2004]. They develop an efficient exact inference algorithm specifically for the class of Bayesian networks represented in Figure 13.12. For specific values of θ_j for all j, they use this algorithm to compute the probability of the data. They then search for the MLE values of θ_j for all j. Furthermore, they show that their inference method is more efficient than other known methods. It has been implemented in the computer program SUPERLINK, which is available free to researchers at http://bioinfo.cs.technion.ac.il/superlink-online/. SUPERLINK is described in [Silberstein et al., 2006].

EXERCISES

Section 13.1

Exercise 13.1 Explain why, in Example 13.1, it was necessary to do selective breeding to obtain a large number of female *Drosophila* that are homozygous for w (white eyes) and m$^+$(normal wings), whereas we were able to obtain a large number of male *Drosophila* that have w$^+$ (red eyes) and m (miniature wings) on the X chromosomes by inspection.

Exercise 13.2 Suppose that we have two genes, and there are 320 crossovers in the region between their loci in 2000 meioses. What is their approximate map distance? Assuming no crossovers are cancelled out, what is the recombination frequency observed in the offspring?

Exercise 13.3 Suppose that

$$map\_distance(gene_1, gene_2) = 15$$

$$map\_distance(gene_2, gene_3) = 17.$$

Suppose further that $gene_2$ is between $gene_1$ and $gene_3$ on the chromosome. What is $map\_distance(gene_1, gene_3)$?

Section 13.2

Exercise 13.4 Suppose that we have the pedigree in Figure 13.14. A darkly shaded node indicates that the individual has Huntington's disease, a white node indicates that the individual does not have Huntington's disease, and a node containing a question mark indicates that we do not know the phenotype of the individual. Deduce all that is possible concerning the phenotypes and genotypes of all individuals in the pedigree.

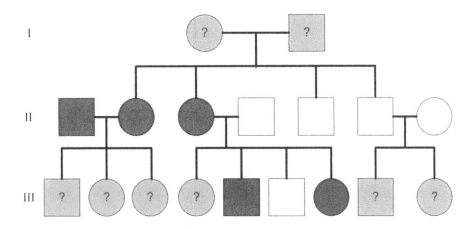

Figure 13.14: A pedigree in which the phenotype of interest is Huntington's disease.

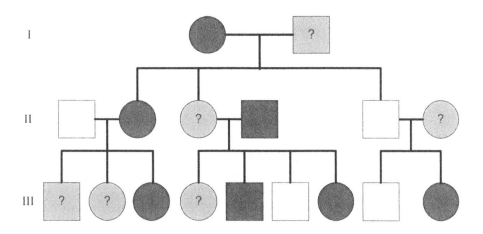

Figure 13.15: A pedigree in which the phenotype of interest is albinism.

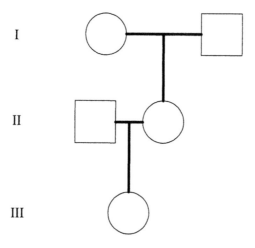

Figure 13.16: A pedigree.

Exercise 13.5 Suppose that we have the pedigree in Figure 13.15. A darkly shaded node indicates that the individual has albinism, a white node indicates that the individual does not have albinism, and a node containing a question mark indicates that we do not know the phenotype of the individual. Deduce all that is possible concerning the phenotypes and genotypes of all individuals in the pedigree.

Exercise 13.6 In Example 13.7, we discussed an SSR that, throughout a population, has six different numbers of copies of some DNA sequence.

1. How many phenotypes (band patterns in a gel) are there for this SSR?

2. How many genotypes are there for this SSR?

Section 13.3

Exercise 13.7 Consider the pedigree in Figure 13.16. Create a Bayesian network for genetic linkage analysis using fragments like the one shown in Figure 13.12. Assume that we are modeling two loci.

Exercise 13.8 As noted previously, SUPERLINK is available at

http://bioinfo.cs.technion.ac.il/superlink-online/.

1. Study several of the examples offered by SUPERLINK.

2. Either obtain a pedigree or develop a fictitious pedigree.

3. Use SUPERLINK to estimate map distances from your pedigree.

Bibliography

Altschul et al., 1997   Altschul, S., L. Thomas, A. Schaffer,
                         J. Zhang, W. Miller, and D. Lipman,
                         "Gapped Blast and Psi-blast: A New
                         Generation of Protein Database Search
                         Programs," *Nucleic Acids Research*, Vol.
                         25, 1997.

Anderson et al., 2007   Anderson, D., D. Sweeney, and T.
                         Williams, *Statistics for Business and
                         Economics*, South-Western, 2007.

Ash, 1970   Ash, R. B., *Basic Probability Theory*, Wi-
                         ley, 1970.

Bedell et al., 2003   Bedell, J., I. Korf, and M. Yandell,
                         BLAST, O'Reilly & Associates, Inc.,
                         2003.

Beinlich et al., 1989   Beinlich, I. A., H. J. Suermondt, R.
                         M. Chavez, and G. F. Cooper, "The
                         ALARM Monitoring System: A Case
                         Study with Two Probabilistic Inference
                         Techniques for Belief Networks," *Pro-
                         ceedings of the Second European Confer-
                         ence on Artificial Intelligence in Medi-
                         cine*, London, 1989.

Bentler, 1980   Bentler, P. N., "Multivariate Analysis
                         with Latent Variables," *Review of Psy-
                         chology*, Vol. 31, 1980.

Berry, 1996   Berry, D. A., *Statistics: A Bayesian Per-
                         spective*, Wadsworth, 1996.

Blaisdell, 1985   Blaisdell, V. E., "A Method of Esti-
                         mating from Two Aligned Present-Day
                         Sequences Their Ancestral Composition
                         and Subsequent Rates of Substitution,

	Possibly Different in Two Lineages, Corrected for Multiple and Parallel Substitutions at the Sites," *Journal of Molecular Evolution*, Vol. 22, 1985.
Breiman et al., 1984	Breiman, L., J. Friedman, R. Olshen, and C. Stone, *Classification and Regression Trees*, Wadsworth and Brooks, 1984.
Buneman, 1971	Buneman, P., "The Recovery of Trees From Measurements of Dissimilarity," in Hodson, E. R., D. G. Kendall, and P. Tautu (eds.): *Mathematics in the Archeological and Historical Sciences*, Edinburgh University Press, 1971.
Buntine, 2002	Buntine, W., "Tree Classification Software," *Proceedings of the Third National Technology Transfer Conference and Exposition*, Baltimore, 2002.
Cheeseman and Stutz, 1996	Cheeseman, P., and J. Stutz, "Bayesian Classification (Autoclass): Theory and Results," in Fayyad, D., G. Piatetsky-Shapiro, P. Smyth, and R. Uthurusamy (eds.): *Advances in Knowledge Discovery and Data Mining*, AAAI Press, 1996.
Cherry et al., 1998	Cherry, J. M., C. Ball, K. Dolinski, S. Dwight, M. Harris, J. C. Matese, F. Sherlock, G. Binkley, H. Jin, S. Weng, and D. Botstein, "Saccharomyces Genome Database," *Nucleic Acid Research*, Vol. 26, 1998.
Chickering et al., 1997	Chickering, D., D. Heckerman, and C. Meek, "A Bayesian Approach to Learning Bayesian Networks with Local Structure," in Geiger, D., and P. Shenoy (Eds.): *Uncertainty in Artificial Intelligence; Proceedings of the Thirteenth Conference*, Morgan Kaufmann, 1997.
Cooper, 1990	Cooper, G. F., "The Computational Complexity of Probabilistic Inference Using Bayesian Belief Networks," *Artificial Intelligence*, Vol. 33, 1990.
Cooper and Herskovits, 1992	Cooper, G. F., and E. Herskovits, "A Bayesian Method for the Induction of

	Probabilistic Networks from Data," *Machine Learning*, Vol. 9, 1992.
Cvrckova and Nasmyth, 1993	Cvrckova, F., and K. Nasmyth, "Yeast GI Cyclins CLN1 and CLN2 and a GAP-like Protein Have a Role in Bud Formation," *EMBO. J.*, Vol. 12, 1993.
Dagum and Chavez, 1993	Dagum, P., and R. M. Chavez, "Approximate Probabilistic Inference in Bayesian Belief Networks," *IEEE Transactions on Pattern Analysis and Machine Intelligence*, Vol. 15, No. 3, 1993.
Dagum and Luby, 1993	Dagum, P., and M. Luby, "Approximate Probabilistic Inference in Bayesian Belief Networks Is NP-hard," *Artificial Intelligence*, Vol. 60, No. 1, 1993.
de Finetti, 1937	de Finetti, B., "La prévision: See Lois Logiques, ses Sources Subjectives," *Annales de l'Institut Henri Poincaré*, Vol. 7, 1937.
Diez and Druzdzel, 2002	Diez, F.J., and M.J. Druzdzel, "Canonical Probabilistic Models for Knowledge Engineering," Technical Report IA-02-01, Dpto. Inteligencia Artificial, UNED, Madrid, 2002.
Eells, 1991	Eells, E., *Probabilistic Causality*, Cambridge University Press, 1991.
Ellis and Wong, 2008	Ellis, B. and W. H. Wong, "Learning Causal Bayesian Network Structures from Experimental Data," *Journal of the American Statistical Association*, Vol. 103, 2008.
Fawcett, 2006	Fawcett, T. "An Introduction to ROC Analysis," *Pattern Recognition Letters*, Vol. 27, 2006.
Feller, 1968	Feller, W., *An Introduction to Probability Theory and Its Applications*, Wiley, 1968.
Felsenstein, 2004	Felsenstein, J., *Inferring Phylogenies*, Sinauer Associates, 2004.

Felsenstein, 2007   Felsenstein, J., *Theoretical Evolution-ary Genetics*, online book available at http://evolution.genetics. washing-ton.edu/pgbook/pgbook.html, 2007.

Fishelson and Geiger, 2002   Fishelson, M., and D. Geiger, "Exact Ge-netic Linkage Computations for General Pedigrees," *Bioinformatics*, Vol. 18 (sup-plement 1), 2002.

Fishelson and Geiger, 2004   Fishelson, M., and D. Geiger, "Opti-mizing Exact Genetic Linkage Computa-tion," *Journal of Computational Biology*, Vol. 11, No. 2–3, 2004.

Friedman and Koller, 2003   Friedman, N., and K. Koller, "Being Bayesian about Network Structure: A Bayesian Approach to Structure Dis-covery in Bayesian Networks," *Machine Learning*, Vol. 20, 2003.

Friedman et al., 1999   Friedman, N., M. Goldszmidt, and A. Wyner, "Data Analysis with Bayesian Networks: A Bootstrap Approach," in Laskey, K. B., and H. Prade (eds.): *Un-certainty in Artificial Intelligence; Pro-ceedings of the Fifteenth Conference*, Morgan Kaufmann, 1999.

Friedman et al., 2000   Friedman, N., M. Linial, I. Nachman, and D. Pe'er, "Using Bayesian Networks to Analyze Expression Data," in *Proceed-ings of the Fourth Annual International Conference on Computational Molecular Biology*, 2000.

Friedman et al., 2002   Friedman, N., M. Ninio, I Pe'er, and T. Pupko, "A Structural EM Algorithm for Phylogenetic Inference," *Journal of Com-putational Biology*, Vol. 9, No. 2, 2002.

Fukami-Kobayashi and Tateno, 1991   Fukami-Kobayashi, K., and Y. Tateno, "Robustness of Maximum Likelihood Tree Estimation Against Different Pat-terns of Base Substitutions," *Journal of Molecular Evolution*, Vol. 32, No. 1, 1991.

Fung and Chang, 1990   Fung, R., and K. Chang, "Weighing and Integrating Evidence for Stochastic Simu-lation in Bayesian Networks," in Henrion,

M., R. D. Shachter, L. N. Kanal, and J. F. Lemmer (eds.): *Uncertainty in Artificial Intelligence 5*, North-Holland, 1990.

Gardner-Stephen and Knowles, 2004   Gardner-Stephen, P., and G. Knowles, "DASH: Localizing Dynamic Programming for Order of Magnitude Faster, Accurate Sequence Alignment," *Proceedings of IEEE Computational Systems Bioinformatics Conference*, 2004.

Gasch et al., 2000   Gasch, A. P., P. T. Spellman, C. M. Kao, O. Carmel-Harel, M. B. Eisen, G. Storz, D. Botstein, and P. O. Brown, "Genomic Expression Programs in the Response of Yeast Cells to Environmental Changes," *Molecular Biology of the Cell*, Vol. 11, No. 12, 2000.

Geman and Geman, 1984   Geman, S., and D. Geman, "Stochastic Relaxation, Gibb's Distributions and the Bayesian Restoration of Images," *IEEE Transactions on Pattern Analysis and Machine Intelligence*, Vol. 6, 1984.

Gilks et al, 1996   Gilks, W. R., S. Richardson, and D. J. Spiegelhalter (eds.): *Markov Chain Monte Carlo in Practice*, Chapman & Hall/CRC, 1996.

Griffiths et al., 2007   Griffiths, J. F., S. R. Wessler, R. C. Lewontin, and S. B. Carroll, *An Introduction to Genetic Analysis*, W. H. Freeman and Company, 2007.

Gu, Fu, and Li, 1995   Gu, X., Y.-X Fu, and W.-H Li, "Maximum Likelihood Estimation of the Heterogeneity of Substitution Rates Among Nucleotide Sites," *Molecular Biology and Evolution*, Vol. 12, 1995.

Guacci et al., 1997   Guacci, V., D. Koshland, and A. Strunnikov, "A Direct Link Between Sister Chromatid Cohesion and Chromosome Condensation Revealed Through the Analysis of MCDI in *s. cerevisiae*," *Cell*, Vol. 9, No. 1, 1997.

Hall, 2004   Hall, B. G., *Phylogenetic Trees Made Easy*, Sinauer Associates, 2004.

Hartl and Jones, 2006

Hartl, D. L., and E. W. Jones, *Essential Genetics*, Jones and Bartlett, 2006.

Hasegawa and Fujiwara, 1993

Hasegawa, M., and M. Fujiwara, "Relative Efficiencies of the Maximum Likelihood, Maximum Parsimony, and Neighbor-Joining Methods for Estimating Protein Phylogeny," *Molecular Phylogenetics and Evolution*, Vol. 2, No. 1, 1993.

Hastings, 1970

Hastings, W. K., "Monte Carlo Sampling Methods Using Markov Chains and Their Applications," *Biometrika*, Vol. 57, No. 1, 1970.

Heckerman, 1996

Heckerman, D., "A Tutorial on Learning with Bayesian Networks," Technical Report # MSR-TR-95-06, Microsoft Research, 1996.

Heckerman and Meek, 1997

Heckerman, D., and C. Meek, "Embedded Bayesian Network Classifiers," Technical Report MSR-TR-97-06, Microsoft Research, Redmond, Washington, 1997.

Herskovits and Cooper, 1990

Herskovits, E. H., and G. F. Cooper, "Kutató: An Entropy-Driven System for the Construction of Probabilistic Expert Systems from Databases," in Shachter, R. D., T. S. Levitt, L. N. Kanal, and J. F. Lemmer (eds.): *Uncertainty in Artificial Intelligence; Proceedings of the Sixth Conference*, North-Holland, 1990.

Hogg and Craig, 1972

Hogg, R. V., and A. T. Craig, *Introduction to Mathematical Statistics*, Macmillan, 1972.

Huelsenbeck, 1995

Huelsenbeck, J.P., "The Robustness of Two Phylogenetic Methods: Four-Taxon Simulations Reveal a Slight Superiority of Maximum Likelihood Over Neighbor-Joining," *Molecular Biology and Evolution*, Vol. 12, No. 5, 1995.

Hume, 1748

Hume, D., *An Inquiry Concerning Human Understanding*, Prometheus, 1988 (originally published in 1748).

Iversen et al., 1971   Iversen, G. R., W. H. Longcor, F. Mosteller, J. P. Gilbert, and C. Youtz, "Bias and Runs in Dice Throwing and Recording: A Few Million Throws," *Psychometrika*, Vol. 36, 1971.

Jin and Nei, 1990   Jin, L., and M. Nei, "Limitations of the Evolutionary Method of Phylogenetic Analysis," *Molecular Biology and Evolution*, Vol. 7, No 1, 1990.

Joereskog, 1982   Joereskog, K. G., *Systems Under Indirect Observation*, North Holland, 1982.

Jones et al., 1992   Jones, D. T., W. R. Taylor, and J. M. Thornton, "The Rapid Generation of Mutation Data Matrices from Protein Sequences," *Computer Applications in the Biosciences*, Vol. 8, 1992.

Keefer, 1983   Keefer, D.L., "3-Point Approximations for Continuous Random Variables," *Management Science*, Vol. 29, 1983.

Kenny, 1979   Kenny, D. A., *Correlation and Causality*, Wiley, 1979.

Kerrich, 1946   Kerrich, J. E., *An Experimental Introduction to the Theory of Probability*, Einer Munksgaard, 1946.

Kimura, 1980   Kimura, M., "A Simple Model for Estimating Evolutionary Rates of Base Substitutions Through Comparative Studies of Nucleotide Sequences," *Molecular Evolution*, Vol. 16, 1980.

Korf, 1993   Korf, R., "Linear-Space Best-First Search," *Artificial Intelligence*, Vol. 62, 1993.

Kuhner and Felsenstein, 1994   Kuhner, M. K., and J. Felsenstein, "A Simulation Comparison of Phylogeny Algorithms Under Equal and Unequal Evolution Rates," *Molecular Biology and Evolution*, Vol. 11, No. 3, 1994.

Lanave et al., 1984   Lanave, C., G. Preparata, C. Saccone, and G. Serio, "A New Method for Calculating Evolutionary Substitution Rates,"

                                        Journal of Molecular Evolution, Vol. 20,
                                        1984.

Lander and Shenoy, 1999   Lander, D. M., and P. Shenoy, "Mod-
                                        eling and Valuing Real Options Using
                                        Influence Diagrams," School of Business
                                        Working Paper No. 283, University of
                                        Kansas, 1999.

Lauritzen and Spiegelhalter, 1988   Lauritzen, S. L., and D. J. Spiegelhalter,
                                        "Local Computation with Probabilities in
                                        Graphical Structures and Their Applica-
                                        tions to Expert Systems," *Journal of the
                                        Royal Statistical Society B*, Vol. 50, No.
                                        2, 1988.

Li, 1997   Li, W., *Molecular Evolution*, Sinauer As-
                                        sociates, 1997.

Li and D'Ambrosio, 1994   Li, Z., and B. D'Ambrosio, "Efficient In-
                                        ference in Bayes' Networks as a Combina-
                                        torial Optimization Problem," *Interna-
                                        tional Journal of Approximate Inference*,
                                        Vol. 11, 1994.

Li and Nei, 1990   Li, J., and M. Nei, "Limitations of the
                                        Evolutionary Parsimony Method of Phy-
                                        logenetic Analysis," *Molecular Biology
                                        and Evolution*, Vol. 7, No. 1, 1990.

Lindley, 1985   Lindley, D. V., *Introduction to Probabil-
                                        ity and Statistics from a Bayesian View-
                                        point*, Cambridge University Press, 1985.

Lugg et al., 1995   Lugg, J. A., J. Raifer, and C. N. F.
                                        González, "Dehydrotestosterone Is the
                                        Active Androgen in the Maintenance of
                                        Nitric Oxide-Mediated Penile Erection in
                                        the Rat," *Endocrinology*, Vol. 136, No. 4,
                                        1995.

McClennan and Markham, 1999   McClennan, K. J., and A. Markham, "Fi-
                                        nasteride: A Review of Its Use in Male
                                        Pattern Baldness," *Drugs*, Vol. 57, No. 1,
                                        1999.

McGregor, 1999   McGregor, W. G., "DNA Repair, DNA
                                        Replication, and UV Mutagenesis," *J. In-
                                        vestig. Determotol. Symp. Proc.*, Vol. 4,
                                        1999.

McKee and Lensberg, 2002   McKee, T.E., and T. Lensberg, "Genetic Programming and Rough Sets: A Hybrid Approach to Bankruptcy Classification," *Journal of Operational Research*, Vol. 138, 2002.

McLachlan and Krishnan, 1997   McLachlan, G. J., and T. Krishnan, *The EM Algorithm and Extensions*, Wiley, 1997.

Meek, 1995   Meek, C., "Strong Completeness and Faithfulness in Bayesian Networks," in Besnard, P., and S. Hanks (eds.): *Uncertainty in Artificial Intelligence; Proceedings of the Eleventh Conference*, Morgan Kaufmann, 1995.

Metropolis et al., 1953   Metropolis, N., A. Rosenbluth, M. Rosenbluth, A. Teller, and E. Teller, "Equation of State Calculation by Fast Computing Machines," *Journal of Chemical Physics*, Vol. 21, 1953.

Neal, 1992   Neal, R., "Connectionist Learning of Belief Networks," *Artificial Intelligence*, Vol. 56, 1992.

Neapolitan, 1990   Neapolitan, R. E., *Probabilistic Reasoning in Expert Systems*, Wiley, 1990.

Neapolitan, 1992   Neapolitan, R. E., "A Survey of Uncertain and Approximate Inference," in Zadeh, L., and J. Kacprzyk (eds.): *Fuzzy Logic for the Management of Uncertainty*, Wiley, 1992.

Neapolitan, 1996   Neapolitan, R. E., "Is Higher-Order Uncertainty Needed?" in *IEEE Transactions on Systems, Man, and Cybernetics Part A: Systems and Humans*, Vol. 26, No. 3, 1996.

Neapolitan, 2004   Neapolitan, R. E., *Learning Bayesian Networks*, Prentice Hall, 2004.

Neapolitan and Naimipour, 2004   Neapolitan, R. E., and K. Naimipour, *Foundations of Algorithms Using Java Pseudocode*, Jones and Bartlett, 2004.

Neapolitan and Jiang, 2007   Neapolitan, R. E., and X. Jiang, *Probabilistic Methods for Financial and Marketing Informatics*, Morgan Kaufmann, 2007.

Ninio et al., 2006   Ninio, M., E. Privman, T. Pupko, and N. Friedman, "Phylogenetic Reconstruction: Increasing the Accuracy of Pairwise Distance Estimation Using Bayesian Inference of Evolutionary Rates," *Bioinformatics*, Vol. 23, 2006.

Olesen et al., 1992   Olesen, K. G., S. L. Lauritzen, and F. V. Jensen, "aHUGIN: A System Creating Adaptive Causal Probabilistic Networks," in Dubois, D., M. P. Wellman, B. D'Ambrosio, and P. Smets (eds.): *Uncertainty in Artificial Intelligence; Proceedings of the Eighth Conference*, Morgan Kaufmann, 1992.

Pearl, 1986   Pearl, J., "Fusion, Propagation, and Structuring in Belief Networks," *Artificial Intelligence*, Vol. 29, 1986.

Pearl, 1988   Pearl, J., *Probabilistic Reasoning in Intelligent Systems*, Morgan Kaufmann, 1988.

Pearl, 2000   Pearl, J., *Causality: Models, Reasoning, and Inference*, Cambridge University Press, 2000.

Pearl et al., 1989   Pearl, J., D. Geiger, and T. S. Verma, "The Logic of Influence Diagrams," in R. M. Oliver and J. Q. Smith (eds.): *Influence Diagrams, Belief Networks and Decision Analysis*, Wiley, 1990. (a shorter version originally appeared in *Kybernetica*, Vol. 25, No. 2, 1989.)

Piaget, 1966   Piaget, J., *The Child's Conception of Physical Causality*, Routledge and Kegan Paul, 1966.

Pradhan and Dagum, 1996   Pradhan, M., and P. Dagum, "Optimal Monte Carlo Estimation of Belief Network Inference," in Horvitz, E., and F. Jensen (eds.): *Uncertainty in Artificial Intelligence; Proceedings of the Twelfth Conference*, Morgan Kaufmann, 1996.

Robinson, 1977 — Robinson, R. W., "Counting Unlabeled Acyclic Digraphs," in Little, C. H. C. (ed.): *Lecture Notes in Mathematics, 622: Combinatorial Mathematics V,* Springer-Verlag, 1977.

Salmon, 1997 — Salmon, W., *Causality and Explanation,* Oxford University Press, 1997.

Sattach and Tversky, 1977 — Sattach, S., and A. Tversky, "Additive Similarity Trees," *Psychometrika,* Vol. 42, 1977.

Savin, 1980 — Savin, N. E., "The Bonferroni and the Scheffe Multiple Comparison Procedures," *Review of Economic Studies,* Vol. 47, No. 1, 1980.

Scheines et al., 1994 — Scheines, R., P. Spirtes, C. Glymour, and C. Meek, *Tetrad II: User Manual,* Erlbaum, 1994.

Segal et al., 2005 — Segal, E., D. Pe'er, A. Regev, D. Koller, and N. Friedman, "Learning Module Networks," *Journal of Machine Learning Research,* Vol. 6, 2005.

Shachter and Peot, 1990 — Shachter, R. D., and M. Peot, "Simulation Approaches to General Probabilistic Inference in Bayesian Networks," in Henrion, M., R. D. Shachter, L. N. Kanal, and J. F. Lemmer (eds.): *Uncertainty in Artificial Intelligence 5,* North-Holland, 1990.

Shenoy, 2006 — Shenoy, P. P., "Inference in Hybrid Bayesian Networks Using Mixtures of Gaussians," in Dechter, R., and T. Richardson (eds.): *Uncertainty in Artificial Intelligence: Proceedings of the Twenty-Second Conference,* AUAI Press, 2006.

Siepel and Haussler, 2004 — Siepel, A., and D. Haussler, "Phylogenetic Estimation of Context-Dependent Substitution Rates by Maximum Likelihood," *Molecular Biology and Evolution,* Vol. 21, No. 3, 2004.

Silberstein et al., 2006   Silberstein M., A. Tzemach, N. Dov-
golevsky, M. Fishelson, A. Schuster, and
D. Geiger, "On-line System for Faster
Linkage Analysis via Parallel Execution
on Thousands of Personal Computers,"
*The American Journal of Human Genet-
ics*, Vol. 78, No. 6, 2006.

Spellman et al, 1998   Spellman, P., G. Sherlock, M. Zhang, V.
Iyer, K. Anders, M. Eisen, P. Brown,
D. Botstein, and B. Futcher, "Com-
prehensive Identification of Cell Cycle-
Regulated Genes of the Yeast *saccharo-
momyces cerevisiae* by Microarray Hy-
bridization," *Molecular Biology of the
Cell*, Vol. 9, 1998.

Spirtes et al., 1993; 2000   Spirtes, P., C. Glymour, and R.
Scheines, *Causation, Prediction, and
Search*, Springer-Verlag, New York, 1993;
second edition, MIT Press, 2000.

Srinivas, 1993   Srinivas, S., "A Generalization of the
Noisy OR Model," in Heckerman, D.,
and A. Mamdani (Eds.): *Uncertainty in
Artificial Intelligence; Proceedings of the
Ninth Conference*, Morgan Kaufmann,
1993.

Tateno et al., 1994   Tateno T. Y., N. Takezaki and M. Nei,
"Relative Efficiencies of the Maximum-
Likelihood, Neighbor-Joining, and Maximum Parsimony Methods When Substitution Rate Varies with Site," *Molecular Biology and Evolution*, Vol. 11, No.
2, 1994.

Tierney, 1996   Tierney, L., "Introduction to General
State-Space Markov Chain Theory," in
Gilks, W. R., S. Richardson, and D.
J. Spiegelhalter (eds.): *Markov Chain
Monte Carlo in Practice*, Chapman &
Hall/CRC, 1996.

van Lambalgen, 1987   van Lambalgen, M., "Random Sequences," Ph.D. thesis, University of Amsterdam, 1987.

von Mises, 1919   von Mises, R., "Grundlagen der Wahr-scheinlichkeitsrechnung," *Mathematische Zeitschrift*, Vol. 5, 1919.

Waterman, 1984   Waterman, M.S., "General Methods of Sequence Comparisons," *Mathematical Biology*, Vol. 46, 1984.

Wright, 1921   Wright, S., "Correlation and Causation," *Journal of Agricultural Research*, Vol. 20, 1921.

Xiang et al., 1996   Xiang, Y., S. K. M. Wong, and N. Cercone, "Critical Remarks on Single Link Search in Learning Belief Networks," in Horvitz, E., and F. Jensen (eds.): *Uncertainty in Artificial Intelligence; Proceedings of the Twelfth Conference*, Morgan Kaufmann, 1996.

Yang, 1994   Yang, Z., "Maximum Likelihood Phylogenetic Estimation from DNA Sequences with Variable Rates Over Sites: Approximate Methods," *Journal of Molecular Evolution*, Vol. 39, No. 3, 1994.

Yang, 1996   Yang, Z., "Among-Site Rate Variation and Its Impact on Phylogenetic Analysis," *Trends in Ecology and Evolution*, Vol. 11, 1996.

Zadeh, 1995   Zadeh, L., "Probability Theory and Fuzzy Logic Are Complementary Rather Than Competitive," *Technometrics*, Vol. 37, 1995.

Zharkikh, 1994   Zharkikh, A., "Estimation of Evolutionary Distances Between Nucleotide Sequences," *Journal of Molecular Evolution*, Vol. 39, 1994.

Index

Printed and bound by CPI Group (UK) Ltd, Croydon, CR0 4YY

03/10/2024

01040313-0011